American Independent Inventors
in an Era of Corporate R&D

American Independent Inventors
in an Era of Corporate R&D

Eric S. Hintz

The MIT Press
Cambridge, Massachusetts
London, England

in association with
The Lemelson Center
Smithsonian Institution
Washington, DC

The MIT Press would like to thank the anonymous peer reviewers who provided comments on drafts of this book. The generous work of academic experts is essential for establishing the authority and quality of our publications. We acknowledge with gratitude the contributions of these otherwise uncredited readers.

This book was set in Stone Serif and Stone Sans by Westchester Publishing Services. Printed and bound in the United States of America.

Library of Congress Cataloging-in-Publication Data

Names: Hintz, Eric S., author.
Title: American independent inventors in an era of corporate R&D / Eric S. Hintz.
Description: Cambridge, Massachusetts : The MIT Press, [2021] |
    Series: Lemelson Center studies in invention and innovation series |
    Includes bibliographical references and index.
Identifiers: LCCN 2020040777 | ISBN 9780262542586 (paperback)
Subjects: LCSH: Inventors--United States. | Research, Industrial--United States. |
    Patents--United States.
Classification: LCC T39 .H55 2021 | DDC 609.73--dc23
LC record available at https://lccn.loc.gov/2020040777

10 9 8 7 6 5 4 3 2 1

For Emma, Patrick, and Gavin

Contents

Series Foreword

Think about all the stereotypes of inventors that have appeared in popular culture over the years. There is the heroic and stately inventor, typified by Thomas Edison, and the wild-haired eccentric, like Doc Brown in *Back to the Future*. Then there is the "accidental" inventor, such as Hedy Lamarr, who invented as a sideline to her Hollywood career. And most recently, we have seen the rise of a new stereotype in the cool and aloof inventor-entrepreneur of Silicon Valley, such as Steve Jobs.

What connects these disparate characterizations? It is the idea of the "lone" inventor, a "Eureka!"-shouting genius who single-handedly brings a new technology into the world. But "lone"—or better understood as independent—inventors are never really alone. To succeed, the independents have developed strategies that amplified their impact. They formed professional societies, lobbied for patent reforms, and offered their services to the government in times of war. Throughout, they found ways to bring their inventions to market, whether by manufacturing, licensing, or selling the rights to them entirely.

As Eric Hintz shows in this book, the independent inventor persisted outside the mainstream of corporate research and development laboratories. Once thought to have vanished in the first half of the twentieth century, the independents' contributions remained an important component of the invention landscape. By revealing the ways in which independents navigated the corporate, government, and social constraints of their times, Hintz moves us past romantic visions of inventors and enriches our appreciation of these diverse, talented, determined, indefatigable creators.

Invention and innovation have long been recognized as transformational forces in American history, not only in technological realms but also in politics, society, and culture. Since 1995, the Smithsonian's Lemelson Center has been investigating the history of invention and innovation from an interdisciplinary perspective. Books in the Lemelson Center Studies in Invention and Innovation extend this work to enhance public understanding of humanity's inventive impulse. Authors in the series raise new questions about the work of inventors and the technologies they create, while stimulating cross-disciplinary dialogue. By opening channels of communication between the various disciplines and sectors of society concerned with technological innovation, the Lemelson Center Studies advance scholarship in the history of technology, engineering, science, architecture, the arts, and related fields and disseminate it to a general-interest audience.

Joyce Bedi, Arthur Daemmrich, and Arthur P. Molella

Series editors, Lemelson Center Studies in Invention and Innovation

Acknowledgments

I have accrued many personal and professional debts during the fifteen years I have spent researching and writing this book. It is a pleasure to acknowledge and thank the many family members, colleagues, and friends who have sustained me along the journey.

I dedicate this book, with love, to my wife, Emma, and our sons, Patrick and Gavin. More than anyone, you have lived with the highs and lows associated with this project. Emma, you believed I could be a scholar before I believed it myself. Thank you for years of pep talks and for cheerfully making so many sacrifices to support our family. Patrick and Gavin, you are my beloved sons in whom I am well pleased (Matthew 3:17). I have been working on this book in one way or another for every moment of your lives; thank you for your patience and for being my biggest cheerleaders. Emma, Patrick, and Gavin—I am so proud to be your husband and father.

I hit the jackpot by being born into the Hintz family. My parents, John and Peggy Hintz, fueled my childhood curiosity with books and family field trips. More importantly, they encouraged me to pursue a career that did not feel like work. Thank you for your incredible generosity and for resisting the urge to ask about my progress on the book. For their ongoing love and support, I also thank the entire Hintz family, especially my brother, Mick; my beloved nieces, Giulia, Kathryn, and Caroline, and my nephew, David; my grandmother, Delores Slattery, and the whole Slattery clan; and my in-laws in the Sisson and Small families. Friends are the family you choose. From Notre Dame, to the Bay Area, to Manor Woods, I have too many comrades to thank individually, but please know that I appreciate your ongoing support and interest in the book.

This book began as a dissertation in the Department of the History and Sociology of Science at the University of Pennsylvania. I received outstanding historical training at UPenn and benefited from the guidance of several outstanding scholars. Ruth Schwartz Cowan was—and remains—a patient advisor and mentor. Thank you for dispensing practical advice and for modeling how to balance a distinguished scholarly career with devotion to your family. Susan Lindee continues to inspire me with her pragmatism, productivity, and commitment to her ideals. Walter Licht from the History Department introduced me to business and economic history, which has been a critical influence on my work. Walter, I hope you enjoy holding the book in your hands and "smelling the glue." Finally, thanks to my office mates Chris Jones and Matthew Hersch, and our fellow UPenn graduate students, for their warmth and continuing friendship.

The dissertation became a book thanks to my colleagues at the Smithsonian's Lemelson Center for the Study of Invention and Innovation. Art Molella and Joyce Bedi invited me to publish the book in MIT Press's Lemelson Center book series and, with Jeff Brodie, ensured that I had ample time for the project amid my other museum duties. Arthur Daemmrich first suggested this topic in 2004 while teaching a graduate seminar at UPenn. Since 2015, he has encouraged its completion as the Lemelson Center's second director and third book series editor. Joyce, in particular, has been a trusted colleague and guide at every stage of the project. She proposed new content, demanded clear writing, and read multiple drafts of every word. Joyce, thank you for challenging me to develop a more ambitious vision for the book—cheers mate!

I have benefited enormously from the wisdom and encouragement of several colleagues in the wider scholarly community. Several panelists and audience members commented on excerpts during presentations at the Hagley Museum and Library, the Institute of Electrical and Electronics Engineers, the Science History Institute, Rutgers University—Camden, Harvard Business School, the University of Pennsylvania, the University of Maryland, the Society for the History of Technology (SHOT), the Business History Conference (BHC), the National Museum of American History (NMAH), and the National Air & Space Museum. For offering suggestions and commenting on drafts, I thank my NMAH colleagues, including Ellen Feingold, Joan Fragaszy Troyano, Jon Grinspan, Peggy Kidwell, Peter Liebhold, Ryan

Lintelman, Margie Salazar-Porzio, Monica Smith, Carlene Stephens, and Deb Warner. Thanks especially to the community of scholars who think and write about invention, innovation, patents, and industrial research, including Chris Beauchamp, Bernie Carlson, Lisa Cook, Eric Dahlin, Gerardo Con Díaz, Ray Fouché, Robert Friedel, David Hounshell, Paul Israel, Richard John, Zorina Khan, Naomi Lamoreaux, Tom Lassman, Joris Mercelis, Tom Nicholas, Patricia Sluby, and Kathy Steen. I owe a special debt of gratitude to colleagues who read and commented on the penultimate drafts of various chapters, including Jonathan Coopersmith, Kathy Franz, Meg Graham, Kara Swanson, and members of Matt Wisnioski's graduate seminar at Virginia Tech. Matt generously read multiple drafts of several chapters and became a great friend in the process. These colleagues have inspired much of what is good in the book. I take sole responsibility for any factual errors, questionable interpretations, or awkward writing.

It is a pleasure to thank the many librarians and archivists who helped me discover the stories in this book. I am grateful to Alison Oswald, Craig Orr, Kay Peterson, and several colleagues at the NMAH Archives Center; Trina Brown, Alexia MacClain, Jim Roan, and Lilla Vekerdy at Smithsonian Libraries; Janice Goldblum at the National Academy of Sciences; Ron Brashear and the team at the Science History Institute; Phil Scranton, Roger Horowitz, Carol Lockman, the late Lynn Catanese, and the entire team at the Hagley Museum and Library; John Alviti, Irene Coffey, Virginia Ward, and Susannah Carroll at the Franklin Institute; and the reference staffs at the American Philosophical Society, the University of Pennsylvania Library, the Library of Congress, the National Archives and Records Administration, the New York Public Library, and the New-York Historical Society.

I am deeply appreciative to several institutions for their financial and intellectual support. Through various research and teaching fellowships, the University of Pennsylvania provided funding for seven years of graduate school. As a student, I received generous research support from the Science History Institute; the Smithsonian's Lemelson Center for the Study of Invention and Innovation; the Hagley Museum and Library; the Consortium for the History of Science, Technology, and Medicine; the Gilder Lehrman Institute of American History; and the Business History Conference's John E. Rovensky Fellowship in Business and Economic History. Since I joined the Smithsonian Institution in 2010, the Lemelson Center has provided time and financial support for additional research travel, writing, and revisions. This book, and the other titles in MIT Press's Lemelson Center book series, are generously supported by Barbara Hiatt in honor of her mentor, Father John Scott, the former president of St. Martin's University, who had a deep love of American history. Finally, I would like to thank editor Katie Helke, the press's anonymous readers, and the entire team at MIT Press for believing in this book.

1 Introduction

There is a sharp and intriguing conflict of opinion between those who hold that the day of the individual inventor is done and those who consider that not merely is he very much with us but that, from present appearances, he will continue to play an active part in technical progress.
—John Jewkes, David Sawers, and Richard Stillerman,
The Sources of Invention, 1958

"The independent inventor today has little chance against the formidable research organizations of modern industry." That was the judgment of Maurice Holland, director of the Division of Engineering and Industrial Research at the National Research Council, an organization representing America's elite academic and industrial scientists. In a December 1927 speech titled "The Vanishing American Genius," Holland told the American Institute of New York that the days of the solitary "genius in the garret" had passed. Instead, Holland explained, highly trained scientists and engineers now worked together as "members of a team producing pieces of group research" in hundreds of industrial laboratories in operation across the United States. These teams of salaried researchers, Holland reported, earned hundreds of patents every year for corporations such as General Electric (GE), American Telephone & Telegraph (AT&T), Eastman Kodak, and DuPont.

The *New York Times* printed excerpts from Holland's speech, which initiated a month-long exchange of follow-up articles, editorials, and impassioned letters to the editor.[1] Assistant Patent Commissioner William A. Kinnan protested that the "day of the independent inventor has not passed. . . . Nothing could be further from the truth." This fallacy, said Kinnan, had emerged thanks to boosters like Holland, who tended to exaggerate the increasingly important role that corporate laboratories were playing in the introduction of new technologies. Milton Wright, an associate editor at *Scientific American*, likewise countered that "new ideas, big and little, are the work of individual genius," and "genius is something no corporation or group of corporations can monopolize." The independent inventor had not vanished from the scene, Wright argued; rather, "his opportunities today are greater than at any time in the history of the country."[2]

A few days later, the *Times* published its own op-ed. The editors reviewed Holland's statement and the commentary that had followed. Then, having carefully examined "the conditions under which new discoveries and machines are introduced," the *Times* argued that the lone "garret inventor" could no longer compete with the scale, resources, and accumulated scientific expertise of 30,000 corporate scientists, who collectively spent $200 million annually on research. According to the editors, the sophisticated technical problems of modern industry had become "so formidable" and demanded "experimenting on so vast a scale" that "the independent inventor is automatically excluded from attacking them."[3]

How could this be? Just a few decades earlier, Americans had witnessed the "Heroic Age of American Invention," when mythic nineteenth-century inventors such as Samuel Morse, Thomas Edison, and Alexander Graham Bell created entirely new industries while achieving widespread fame. However, many contemporary observers—and later many historians—joined Holland in the belief that corporate R&D labs had displaced independent inventors as the wellspring of innovation.[4] For example, historian Thomas P. Hughes has suggested that, after World War I, "the independents never again regained their status as the pre-eminent source of invention and development. . . . Industrial scientists, well publicized by the corporations that hired them, steadily displaced, in practice and in the public mind, the figure of the heroic inventor as the source of change in the material world."[5]

However, the historical patent data tell a different story.[6] Figure 1.1 shows patenting activity among US individuals and firms between 1901 and 2000. Patents issued to individual inventors outnumbered corporate patents until 1933 and closely mirrored corporate output through about 1950. In real terms, the 18,960 patents (46.5 percent) awarded to individual inventors in 1950 represented a substantial contribution of new inventions. Corporations eventually dominated patenting after 1950, but individual inventors continued to produce a stable output of 10,000 to 15,000 patents per year.

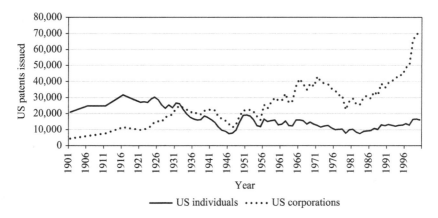

Figure 1.1

US patents issued to individual inventors outnumbered corporate patents until 1933, represented nearly 50 percent of patents through 1950, and held steady at 10,000 to 15,000 patents per year through 2000.

Source: Historic patent data from the United States Patent and Trademark Office (USPTO) have been conveniently compiled in Susan Carter et al., eds., *Historical Statistics of the United States: Earliest Times to the Present*, millennial ed. (New York: Cambridge University Press, 2006), 3:425–429, table Cg27–37.

In short, individual inventors exceeded or nearly equaled the output of corporate patents through midcentury, and they continued to contribute thousands of new inventions each year during the postwar boom in corporate patenting.[7]

In fact, the first half of the twentieth century was a long transitional period when lesser-known independents such as Chester Carlson (Xerox photocopier), Beulah Henry (bobbinless sewing machine), Samuel Ruben (Duracell batteries), and Earl Tupper (Tupperware) made important contributions to the overall context of American innovation. After 1950—when corporate R&D supposedly dominated the landscape of innovation—independent inventors developed Gore-Tex weatherproof fabrics (Robert W. Gore), intermittent windshield wipers (Robert W. Kearns), the self-wringing Miracle Mop (Joy Mangano), and Facebook (Mark Zuckerberg). The individual geniuses in the garrets, garages, and dorm rooms might have been less prominent in the popular imagination, but they never completely vanished.

Nevertheless, the emergence and growth of corporate research labs had far-reaching impacts for traditional, independent inventors. Maurice Holland

and his fellow industrial researchers eroded the independents' heroic reputation by characterizing them as obsolete, antiquated tinkerers. The enormous technical, legal, and financial resources of the big corporations made it difficult for the independents to compete in the marketplace, at the patent office, in the courts, and in the court of public opinion. Yet, many independents learned to survive by forging symbiotic partnerships with corporate R&D labs and offering their services to the federal government. In short, the first half of the twentieth century was a complex and difficult time for individual inventors. This is their story.

*   *   *

This book explores the changing fortunes of American independent inventors from approximately 1890 to 1950, a crucial transformational period marked by the expansion of corporate R&D, the Great Depression, and two world wars. Contrary to most interpretations of this period, I argue that individual "post-heroic" inventors were not supplanted by corporate R&D labs but instead persisted alongside them as an important, though less visible, source of inventions. The book documents how individual inventors navigated the turbulent early decades of the twentieth century as they competed (and sometimes partnered) with corporate rivals, fought for professional legitimacy, lobbied for political reforms, and mobilized for the national defense. More broadly, it explains how American independent inventors—once revered as heroes—gradually fell from public view as corporate brands increasingly became associated with high-tech innovation. It also traces how the postwar experiences of independent inventors mirrored the patterns of their predecessors. Overall, the book reframes the period from 1890 to 2015 as an era when both independent inventors and corporate R&D labs contributed substantially to technological innovation.

This interpretation challenges several long-held assumptions about the sources of invention during the twentieth century. Several studies have documented how General Electric, AT&T, DuPont, Kodak, Corning, Alcoa, and the Radio Corporation of America (RCA) established R&D laboratories after 1900 to bring the process of invention under corporate control. Echoing Maurice Holland, some of these studies suggest that the arrival of corporate R&D labs abruptly ended the era of independent inventors.[8] In contrast, this book views the emergence of corporate R&D as a gradual, intentional process with profound—though not existential—ramifications for independent inventors. Instead of focusing exclusively on innovation within corporate labs, the book documents how innovation occurred both inside and outside established firms and sometimes crossed their organizational boundaries through inventor-firm alliances.

The book builds on prior studies that have empirically confirmed the contributions of independent inventors. In *The Sources of Invention* (1958), economists John Jewkes, David Sawers, and Richard Stillerman conducted a survey of sixty-one important inventions that emerged during the first half of the twentieth century. They found that more than half the inventions in their study—including air conditioning, Kodachrome color film, and the jet engine—were originally developed by individuals, not corporations. They concluded that "the large industrial research organization cannot be considered, either actually or potentially, the sole and sufficient source of inventions."[9] Numerous quantitative studies have also established that independent inventors persisted alongside corporate R&D labs as an important source of inventions.[10] To flesh out this aggregate statistical picture, this book presents narrative episodes from the lives and careers of numerous individual inventors. Collectively, their stories provide important historical context and offer further confirmation of the independents' crucial role in the overall structure of technological development.

The book also extends our understanding of independent inventors beyond the heroic era. Scholars know comparatively little about twentieth-century inventors relative to their nineteenth-century predecessors. Hundreds of popular and scholarly biographies have chronicled heroic nineteenth-century inventors such as Samuel Colt, Thomas Edison, and Alexander Graham Bell,[11] yet only a handful of individual biographies consider twentieth-century independents and, until now, no synthetic study has interpreted their collective experiences.[12]

This book also complements traditional scholarship on the creative process of invention by taking a broader view of inventors in society.[13] While numerous studies have examined how inventors think, sketch, and work, this book sheds new light on their attitudes and activities by exploring a few key questions: How did individual inventors—once revered as geniuses—gradually lose their cultural authority? How did inventors survive economically in the face of competition from the corporate labs? How did otherwise independent inventors band together to advance their professional goals? How did inventors participate in civic debates and lobby for political reforms? How did independent inventors mobilize for the national defense during World War I and World War II? Overall, *what was it like* to be an independent inventor during the first half of the twentieth century?

To address these questions, the book draws on a wide range of sources to highlight common attitudes and experiences among multiple independent inventors. Unlike a biography of a single inventor—which could be discounted as an exceptional case—this book does not provide an in-depth portrait of any individual. Rather, it weaves together short vignettes and relevant episodes from the archival papers, diaries, and memoirs of dozens of inventors. Collectively, these stories document how independent inventors lived and worked during the twentieth century.

Corporate records—from big firms such as DuPont to small firms such as the Sheaffer Pen Company—describe how independent inventors partnered with a broad spectrum of companies and often figured prominently in their strategies for innovation. The institutional records of organizations such as the Franklin Institute and the Inventors' Guild illuminate how independents developed professional networks and lobbied collectively for political reforms. The records of scientific associations (e.g., National Research Council) and pro-research lobbying groups (e.g., National Association of Manufacturers) reveal the ideological tensions between independent inventors and the industrial scientists who sought to displace them. Government sources—including patent statistics, agency records, and congressional testimony—document the relative scale of independents' participation in the economy while shedding light on their political activities and wartime service. Finally, published primary sources, including daily newspapers and trade journals such as *Scientific American*, provide contemporary accounts of important events and suggest how the media and the public perceived independent inventors.

Unfortunately, the available sources provide few definitive answers regarding the population of independent inventors. Between 1870 and 1930, the US Census's occupational taxonomy tracked "designers, draftsmen, and inventors" under the general heading of "Architects." The 1930 census, for example, listed 102,730 designers, draftsmen, and inventors, but from the aggregate number it is impossible to know how many respondents were inventors or how many were corporate (vs. independent) inventors.[14] The number of patents issued to individual inventors (figure 1.1) fluctuated between a high of 31,742 in 1916 and a low of 7,444 in 1946.

These figures provide a rough but problematic estimate for the number of independents.[15] In any given year, most inventors in the population will not earn a patent; a few inventors may earn multiple patents; and, occasionally, multiple independents (e.g., Orville and Wilbur Wright) will be listed on a single patent. To account for these distortions, sociologist Barkev Sanders analyzed a sampling of patents issued in 1938, 1948, and 1952 and then extrapolated the total population of inventors. Sanders estimated approximately 225,000 "independent" and 235,000 "employee" patentees were at work in 1965.[16] The wide variance in these estimates, and the problems associated with each source, only underscore the difficulty of defining the population of independent inventors.

Perceptive readers may rightly ask: How did you select the inventors who appear in this book? Is this group representative of all independents from the first half of the twentieth century? These methodological concerns highlight the challenge of investigating a group of inventors once assumed to be extinct. There are precious few archival collections documenting the lives of independent inventors, so I selected the cases based on the availability of extant archival records.[17] I have also focused on independents who earned at least one US patent. One does not need a patent to self-identify as an inventor; conversely, some patentees may see themselves as scientists or engineers, not inventors. Nevertheless, I have abided by this standard of inclusion because earning a patent provided definitive evidence of invention. As a result, the book's case studies are somewhat biased toward successful inventors; that is, inventors with multiple patents, whose careers merited preservation in an archive or the publication of a memoir.

It is important to note that the book's case studies are dominated by American-born White men and include only a handful of women, African Americans, and immigrants. Although the sample is not diverse, it is representative of a time when women and African Americans were woefully underrepresented among inventors and earned no more than 1–2 percent of all patents.[18] Nevertheless, women inventors such as Martha Coston (signaling flares) and Marion O'Brien Donovan (leak-proof diaper cover) successfully developed industrial and household products.[19] Similarly, African American inventors such as George Washington Murray (agricultural equipment) and Garrett Morgan (gas mask) achieved some recognition and commercial success despite their otherwise marginal social status.[20] Immigrants, too, found plenty of opportunity in the United States as independent inventors and entrepreneurs. Belgian Leo H. Baekeland developed Bakelite, the first synthetic plastic, while Hungarian independent Charles Eisler successfully defended his patented bulb-making machinery against GE's powerful "lamp cartel."[21] During the early twentieth century, women, African Americans, and some immigrants were systematically excluded from membership in scientific and engineering societies and from employment in early industrial research labs.[22] Independent inventing was often the only option for these marginalized groups and was an important springboard for social mobility. In an era marked by limited suffrage, Jim Crow laws, and xenophobia, women, African Americans, and immigrants could nevertheless take advantage of America's patent system, which granted intellectual property rights to "the first and true inventor," regardless of sex, race, age, religion, or national origin.[23]

Some independents, such as Baekeland and radio pioneer Lee De Forest, held advanced scientific degrees, while others, such as battery inventor Samuel Ruben, claimed only a high school education. Most independents pursued invention as a full-time professional career, but some—such as realtor and flexible drinking straw inventor Joseph Friedman—moonlighted as inventors during evenings and weekends. Independents hailed from every geographic region in the country and worked in nearly every industrial sector, including agriculture, transportation, electricity, communications, military/defense, chemicals, and consumer products. Some of the independents, such as Earl Tupper of Tupperware fame, were moderately well known by their contemporaries and are still familiar today, but most of them, such as Charles Adler, Jr. (transportation safety devices) and Everett Bickley (optical sorting machines), remained obscure. It is therefore difficult to make too many generalizations about an average or typical independent inventor from this era.

Nevertheless, these inventors all shared one crucial thing in common: their independence. By choice or necessity, they continued to go it alone, even as industrial research labs increasingly influenced the course of innovation. Beyond their independence, these inventors also shared a common historical milieu. They all operated under the same patent system and economic conditions, and they all faced the defining challenges of their time, including the Great Depression and the two world wars. By examining the activities of several individuals, we can discern what it was like to be an independent inventor in an era of expanding corporate R&D.

From Heroic Individual Geniuses to Industrial R&D

From the perspective of the early twenty-first century, it might seem odd to devote a book-length study to the rise of corporate R&D and the difficulties of independent inventors. After all, many contemporary observers would probably recognize individual inventor-entrepreneurs and their start-ups—not big, conservative corporations—as the source of most cutting-edge innovations.[24] However, the early twentieth century was a somewhat anomalous period, when independent inventors struggled for recognition and respect, unlike their heroic predecessors and internet-era successors.

Consider this hypothetical experiment. Suppose we stopped an average person on the street and asked about a nineteenth-century invention; for example, who invented the telephone? Most respondents could probably identify Alexander Graham Bell. Similarly, we might ask about a twenty-first-century inventor-entrepreneur; for example, who is the founder of Facebook? Again, the average informed person could probably name Mark Zuckerberg, selected by *Time* magazine as its 2010 "Person of the Year."

Now, suppose we asked that same person about an important twentieth-century technology; for example, who invented the photocopier? Or, who invented the first synthetic plastic? Chances are that person would name the company that commercialized the copier (Xerox) but not its individual inventor (Chester F. Carlson). It is also doubtful that the average person could identify plastics inventor Leo H. Baekeland, even though he was also featured on the cover of *Time*, in September 1924. Instead, regarding plastics, we might expect our respondent to name a big multinational firm such as DuPont or Dow Chemical.

Why is that? Why did individual inventors recede from public view in the twentieth century? How did the public come to believe that they had disappeared? Furthermore, how did corporations and their sophisticated R&D labs increasingly become associated with high-tech innovation? Moreover, given the notoriety of a recent dorm room inventor like Zuckerberg, why do we still know so little about earlier independents like Carlson and Baekeland? Before exploring the changing fortunes of lesser-known twentieth-century inventors, it is important to review the history of their predecessors and contemporaries—the heroic individual inventors who loomed so large during the nineteenth century and the corporate R&D labs that emerged at the turn of the twentieth century.

Historian Robert Friedel has noted that during the nineteenth century the act of invention "was closely associated with the 'genius' of individuals."[25] During that century of rapid industrialization, dozens of gifted individual inventor-entrepreneurs built eponymous firms to commercialize their inventions and became famous as those new technologies were widely adopted. Indeed, the inventors' names—such as Samuel Colt (revolver), Isaac Singer (sewing machine), Samuel Morse (telegraph), Cyrus McCormick (mechanical reaper), Alexander Graham Bell (telephone), and Thomas Edison (incandescent lighting)—became synonymous with their inventions. In addition, thousands of lesser-known inventors sold or licensed their patents to the railroads or an established manufacturer instead of starting their own firms. Others took pride in simply earning their patents and never attempted to commercialize them. Regardless, nineteenth-century Americans admired inventors of all stripes for their "Yankee ingenuity." These homegrown American inventors were a source of pride for the fledgling republic, which was rapidly surpassing Great Britain and the rest of Europe as a hotbed of industrial activity.[26]

During the peak of the nineteenth-century Romantic movement, American authors and artists lionized inventors in a series of fawning biographies, children's books, and sentimental paintings (figure 1.2). These depictions portrayed inventors as heroic figures of high moral character, but they were often filled with apocryphal episodes and tended to obscure any details regarding the inventors' assistants, financial backers, and corporate ties.[27] Meanwhile, inventors such as Edison and Nikola Tesla contributed to their own myth making by conscientiously courting newspaper reporters and fashioning themselves as individual "wizards" of ineffable genius. The public came to revere individual inventors; for example, in 1922, a *New York Times* poll ranked Edison as the greatest living American.[28] The myths surrounding American inventors contained a few inaccuracies, but culturally they served as useful origin stories for America's technological ascendance. Consequently, nineteenth-century Americans understood invention as a process overseen by heroic individual geniuses.[29]

By the end of the nineteenth century, the locus of invention began to shift from individuals toward teams of scientists and engineers. In 1876, Edison opened a new laboratory in Menlo Park, New Jersey, and initiated a systematic, team-based approach to invention. Edison merged a traditional machine shop with sophisticated electrical and chemical laboratories. He employed a

Figure 1.2
In his 1862 painting *Men of Progress*, artist Christian Schussele depicted an imagined convocation of nineteen heroic American independent inventors, including Samuel Colt, Cyrus McCormick, Charles Goodyear, and Samuel Morse. National Portrait Gallery, Smithsonian Institution, Washington, DC.

dozen assistants to work in parallel on multiple carefully defined problems; however, Edison claimed any patents resulting from their work. At Menlo Park, Edison and his team could move inventions rapidly from experimentation to development and commercialization. Edison's "invention factory"— which produced his phonograph and incandescent lighting system—served as an important inspiration for the first industrial research laboratories.[30]

In 1892, Edison and inventor Elihu Thomson merged their companies to form General Electric, which gave the consolidated firm a full range of direct- and alternating-current products. With many of Edison and Thomson's original patents set to expire, GE needed to continue acquiring new electrical technologies, but the Sherman Antitrust Act of 1890 discouraged any further mergers that might create a monopoly in the electrical industry. Instead, GE turned inward; the firm decided to invent and commercialize its own new technologies to stay ahead of competitors. As a highly capitalized firm, GE was unwilling to bet its technological fortunes on the idiosyncrasies of individual inventors. Rather, it sought a more continuous and predictable form of innovation, one that could be managed and aligned with corporate prerogatives. In addition, the firm's leaders hoped to harness the latest advances in chemistry and physics to inform its process of invention. In 1900, GE's executives took inspiration from Edison's Menlo Park and established the General Electric Research Laboratory in Schenectady, New York, arguably the first corporate R&D lab in the United States. GE installed Willis R. Whitney, a PhD chemist from the Massachusetts Institute of Technology (MIT), as the lab's research director.[31]

Many large technical firms soon emulated GE and established their own industrial research laboratories. In these corporate labs, teams of salaried scientists and engineers pursued carefully defined problems chosen by a research director, with ownership of any resulting patents assigned to the firm. Rather than relying exclusively on an unpredictable supply of inventions purchased from outside independent inventors in the open market, these firms brought the inventive function inside the firm, where it could be carefully managed.[32] The corporatization of invention was part of a larger trend toward vertical integration and centralized managerial control that emerged among the nation's largest firms around the turn of the twentieth century. In the same way that US Steel vertically integrated by acquiring iron ore mines and Great Lakes shipping companies, many technical firms established in-house research laboratories to secure a steady supply of their own "raw materials"—the scientific discoveries and inventions that led to new products and streamlined operations.[33]

Several pioneers of industrial research, including AT&T, DuPont, and Kodak, held nearly monopolistic control over their respective industries. With their robust profits, these firms could easily afford the substantial investments in facilities and personnel to operate their laboratories, whose innovations, in turn, helped them maintain their dominant market positions. These laboratories operated at the cutting edge of science and technology while developing commercial products. For example, GE's Irving Langmuir studied the properties of inert gases and filaments to improve the performance of light bulbs. In 1932, he became the first corporate scientist to win the Nobel Prize, for his fundamental discoveries in surface chemistry. These discoveries, in turn, helped GE develop a new product line of X-ray and microwave tubes. Similarly, during the 1930s, DuPont researchers, led by Wallace Carothers, received carte blanche to study the chemistry of polymers. These pure scientific investigations eventually resulted in new commercial materials, such as neoprene and nylon.[34]

The emergence of industrial research coincided with the rise of commercial advertising and public relations (PR), and these corporate functions matured together. With extensive advertising and classic slogans—such as DuPont's "Better Things for Better Living Through Chemistry"—the big corporations introduced their R&D labs to the American public.[35] Although Langmuir and Carothers were well known within the scientific community, they were not household names on par with famous individual inventors like Morse or Edison. Rather than promote their researchers, the PR teams sold their brands as the abstract inventors of new products. DuPont, not Carothers, invented nylon; GE, not Langmuir, improved the light bulb.

Industrial research labs and their teams of salaried scientists represented a fundamental challenge to solitary individual geniuses, who had long defined what it meant to be an inventor. Not surprisingly, independent inventors resisted this encroachment on their professional turf. There were other sources of invention during the early twentieth century, including university and government laboratories (e.g., MIT Radiation Lab, National Bureau of Standards) and contract research firms (e.g., Arthur D. Little, Battelle). However, this book focuses on the professional tensions between the era's two main sources of invention: independent inventors and the corporate researchers who sought to displace them.

How would we characterize these competing modes of invention?[36] An "individual," "lone," or "independent" inventor was defined by his or her relative solitude and autonomy. Independents did not perform their work as salaried employees. Rather, they worked for themselves and made their own decisions about which problems to pursue. Most individual inventors usually worked alone, although some retained one or two assistants, often members of their family. Independents were responsible for setting up and equipping a suitable place to work, which often meant their kitchens, basements, or garages. Some independents invented full-time and some moonlighted as part-time inventors while working conventional jobs to support themselves and their families.

Independents tended to specialize in the act of *invention*, the process of investigation, experimentation, and creative mental activity resulting in new prototypes and processes. Some (but not all) independents also engaged in some form of entrepreneurship and *innovation*, the process of commercializing new inventions and bringing them into widespread use. To shepherd an invention from the lab bench to the store shelves, an inventor-entrepreneur had to assemble a coalition of business partners, including financiers, patent attorneys, designers, manufacturers, marketers, and salesmen. Alternatively, many independents preferred to avoid the perils of entrepreneurship and instead sold or licensed their patents to established manufacturing firms. Unless they chose to sell their patents outright, independents retained control over their intellectual property, although they might trade a percentage of their patent rights for access to start-up capital. Independents might keep all, or most, of any future profits from a blockbuster idea, but they also incurred all, or most, of the up-front financial risks involved in trying to bring it to market. The life of an independent inventor was often financially precarious, since he or she might spend years developing an idea before realizing any profits, if ever.

In contrast, an "industrial researcher," "corporate scientist," or "employee-inventor" performed his or her duties as a full-time, salaried employee. While the educational attainments of independents varied widely, an industrial researcher usually held an advanced degree in physics, chemistry, or engineering as a prerequisite for employment. Once hired, corporate scientists joined a team of fellow scientists and technicians who worked collaboratively in a fully equipped R&D laboratory (figure 1.3). Industrial researchers tackled specific problems assigned by their research director. Sometimes these were pure, open-ended scientific investigations, but corporate scientists more typically engaged in applied research to improve the firm's existing products and operations. As stipulated in their employment contracts, industrial researchers assigned any new inventions to their firms.

To commercialize an invention, a corporate scientist did not have to assemble a coalition of external partners. Rather, the researcher simply collaborated with the legal, marketing, manufacturing, and sales teams already operating within the vertically integrated firm. The corporation and its stockholders—not the individual researcher—incurred the risks and costs involved in commercialization. The life of an industrial researcher was usually financially stable, but there was a ceiling on his or her earning potential. Corporate scientists typically received a straight salary for their work, which was paid whether a particular line of research panned out or not. Industrial scientists might receive a bonus for earning a patent but none of the downstream

Figure 1.3
Analytical chemists at work in one of DuPont's research laboratories during the 1950s.
DuPont Company Product Information photographs (Accession no. 1972.341), Hagley
Museum and Library, Wilmington, DE.

revenues of a successful commercialization, which accrued to the firm's
equity partners and stockholders. Because the firm owned all intellectual
property rights, provided the up-front capital and material resources, and
assumed the risks of commercialization, it also kept all the profits.

In the early twentieth century, when industrial research was still maturing, independent inventors and industrial researchers often worked on
problems that were similar in scale and sophistication. Frank Sprague's
electric motors and Charles Eisler's automated bulb-making machinery,
for example, were comparable to equipment under development at GE or
Westinghouse. There was, however, truth in Maurice Holland's suggestion
that the sophisticated problems of modern industry had become too big
and expensive for an individual to solve. Alexander Graham Bell had transmitted his voice over a wire running between two rooms, but it required
enormous capital and decades of sustained innovation for AT&T to build
the nationwide long-distance telephone network.[37] As industrial research

matured, the independents had to cede large-scale projects in nuclear physics and process chemistry to the corporate labs. The R&D labs, in turn, left
toys and consumer novelties—such as Joseph Friedman's flexible straw—to
the independents.

As industrial research proved its worth, more and more firms established
new R&D laboratories. By the time Maurice Holland delivered his controversial speech in 1927, a registry compiled by the National Research Council
listed 999 corporate R&D labs in the United States, employing approximately 30,000 scientists and technicians.[38] With this growing momentum,
it was easy to understand why Holland and the *New York Times* argued that
corporate R&D labs had supplanted independent inventors as the wellspring of new inventions.

Independent Inventors in an Era of Corporate R&D

By the end of World War I, critics began claiming that independent inventors were on the brink of extinction. Chapter 2 examines the roots of this
misperception and suggests that independent inventors had not disappeared but were merely *less visible* than their heroic predecessors or corporate
rivals. Some women inventors and African American inventors intentionally concealed their own identities to circumvent discriminatory laws and
avoid prejudice in the marketplace. A few independents found their contributions obscured when Thomas Edison falsely claimed credit for their
inventions. Inventors who sold or licensed their patents to established firms
found their names erased from their own inventions in favor of corporate
brands. Meanwhile, industrial researchers marshaled persuasive rhetoric,
slick advertisements, and sophisticated public relations campaigns to celebrate science-based R&D, while characterizing the independents as obsolete.
In the process, industrial researchers replaced the mythology of the heroic
independent inventor with potent new myths about the corporate R&D lab.

Independent inventors learned to adapt and survive in this evolving
landscape. Chapter 3 describes how individual inventors evaluated and
deployed a variety of commercialization strategies to earn a living. Inventors could either "make, sell, or ally": that is, found a start-up to manufacture and market an invention; sell the patent to an established firm for a
onetime fee; or license the patent to a manufacturing partner to earn royalties and consulting fees. Despite their disparaging rhetoric, corporate R&D

labs continued to purchase and license patents from independent inventors to complement the inventions they developed in-house. Through these inventor-firm alliances, innovation occurred inside, outside, and across the firm's permeable organizational boundaries. Itinerant inventors migrated into and out of firms as they pursued multiple strategies simultaneously.

Individual inventors cherished their independence and autonomy yet willingly banded together to confront their professional challenges. Chapter 4 documents how independent inventors established a series of organizations to help them commercialize their inventions, lobby for political reforms, and garner respect from a skeptical public who assumed they were extinct. The Franklin Institute, for example, offered a range of services—including a technical library, exhibitions, and the evaluation of new ideas—to help independents develop and showcase their inventions. The American Association of Inventors and Manufacturers and the Inventors' Guild were political organizations that tried to remedy excessive Patent Office delays and curb corporate abuses of the system. Communities of socially marginalized inventors—including women, African Americans, and eccentric amateurs—banded together to obtain mutual support and public recognition. Unfortunately, inventors' eagerness to join these organizations made them vulnerable to predatory scams. Moreover, these fragile organizations usually disbanded within ten to twenty years, leaving inventors ill equipped to address their many challenges. In the end, inventors—unlike scientists and engineers—were unable to establish their occupation as a profession.

Without the benefit of a durable advocacy group to lobby for their concerns, independent inventors were unable to address inequities in the patent system that favored the corporations. Chapter 5 examines how patent reform became a key, if underappreciated, element of the New Deal antitrust movement. In 1938–1939, Congress's Temporary National Economic Committee (TNEC) convened two hearings to investigate how General Electric, General Motors, and other industrial giants had used their massive patent portfolios and aggressive litigation to crowd out independent inventors and exert monopolistic control over their industries. During the hearings, progressive New Dealers introduced a series of reform proposals to level the playing field. However, a combination of factors—including resistance from the National Association of Manufacturers, inventors' political disorganization, and the onset of World War II—resulted in little new legislation. Independent inventors achieved few of their desired reforms, resulting in a patent system that, even today, remains more agreeable to corporations.

The preparations for World War II may have derailed certain patent reforms, but the urgency of national defense energized independent inventors. Chapter 6 traces how the independents mobilized to serve their country during World War I and World War II. In 1915, an aging Thomas Edison marshaled "the natural inventive genius" of everyday Americans to develop new military technologies for the US Navy. In an early form of government crowdsourcing, Edison's Naval Consulting Board (NCB) evaluated 110,000 ideas submitted by America's grassroots inventors. Unfortunately, the navy implemented only one idea by the war's end, which saddled America's independent inventors with a bad reputation. Two decades later, the US Department of Commerce established the National Inventors Council (NIC) to mobilize the independents for World War II. This time, the independents rebounded to produce several key inventions, such as Samuel Ruben's improved walkie-talkie batteries and Charles Hedden's land mine detectors. In both conflicts, independent inventors—not scientists—initiated the first efforts at technical mobilization, and their patriotic contributions served as further proof that they had not been supplanted by their corporate rivals.

Those contributions did not cease with the conclusion of World War II. Chapter 7 explores how the postwar (1950–2015) experiences of independent inventors resembled those of their predecessors. For example, independents remained both energized and frustrated in their efforts to supply inventions for the US military and the federal government. Likewise, postwar independents continued to organize—and generally failed to sustain—professional organizations to advance their political agenda and assist with commercialization. In the policy arena, postwar independents benefited from a pro-patent judiciary and aggressive antitrust enforcement, but they were unable to block fundamental changes to the patent laws that favored the big corporations.

Nevertheless, independent inventors mounted a comeback at the turn of the twenty-first century and rehabilitated their public image. When corporate labs struggled with smaller research budgets and low productivity, they began to acquire inventions from independent inventors. These "open innovation" practices revived commercialization strategies from the early twentieth century and generated more opportunities for independents.[39]

Meanwhile, garage inventors such as Steve Jobs and Jeff Bezos emerged as a new generation of heroic inventor-entrepreneurs.

Despite these gains, independents continued to struggle with predatory scams, an inequitable patent system, and the persistent challenge of surviving as an individual inventor in an era of corporate R&D. Today's independents would easily recognize the hardships their brethren faced a century earlier.

2 The Invisible Inventor

> Most of us believe the independent inventor is dead and buried. Most of us believe, too, that invention today has become the exclusive stamping ground of the salaried Ph.D. working in the laboratories of large corporations surrounded by mysterious instrument panels, electronic brains, and other Ph.D.s. . . . It makes a difference whether the widely held view is true or false. If false, continued public belief in it will help make it true.
>
> —Economist Jacob Schmookler, "Inventors Past and Present," 1957

During the first half of the twentieth century, the American independent inventor was pronounced dead. For decades, observers had been predicting his extinction at the hands of sophisticated corporate research laboratories such as AT&T's Bell Labs, General Electric's Research Laboratories, and DuPont's Experimental Station. As early as 1930, *New York Times* science editor Waldemar Kaempffert suggested that these "splendidly equipped laboratories" could solve problems "that hopelessly baffle the lone, heroic inventor."[1] By 1951, Harvard president James B. Conant had observed that "the typical lone inventor of the eighteenth and nineteenth century has all but disappeared."[2] In a 1954 article, "The Inventor in Eclipse," *Fortune* reported that "the hired inventors who work in corporation laboratories" had already supplanted independent inventors in the chemical and electronics industries. The "lone, unaided inventor" could do little but "concentrate on the mechanical-type inventions, the better mousetraps" that "today are considered to be only gadgets."[3]

However, there was some disagreement on this question. In a 1957 study, economist Jacob Schmookler argued that the dominance of corporate R&D and death of independent inventors had both been overblown. "Most of us believe the independent inventor is dead and buried," Schmookler wrote, and that invention "has become the exclusive stamping ground of the salaried Ph.D. working in the laboratories of large corporations." However, in his empirical study, Schmookler found that individuals still earned 41 percent of US patents, compared with 57 percent by firms and 2 percent by government employees. "The contributions of independent inventors continue at an appreciable, though reduced, volume," he concluded, and the prevailing view of their demise was "a serious distortion of reality."[4]

Schmookler's study underscored a significant misperception. Statistically, it was clear that independent inventors remained an important source of new inventions. However, in the *New York Times*, in *Fortune* magazine, and generally among the populace, there was a widespread belief that industrial research labs had made independent inventors obsolete—and extinct. What caused this misperception? Schmookler attributed the public's false views to the "*well-advertised* increase in industrial research laboratories since World War I" and the fact that research directors had been "shouting their own progressiveness from the transmitting antenna-tops."[5]

As they tried to carve out a niche within industry, industrial researchers zealously tried to convince their executive patrons (and the public) that corporate R&D represented a more reliable and profitable mode of invention than the traditional reliance on individual geniuses. During the first half of the twentieth century, individual corporate researchers, their firms, and pro-research associations such as the National Research Council marshaled persuasive rhetoric, slick advertisements, and sophisticated public relations campaigns to characterize independent inventors as antiquated, unsophisticated "tinkerers," while celebrating their own team-based, scientific approach to invention.

Beyond this ideological rhetoric, independent inventors faced real social and economic challenges that obscured their contributions. Disenfranchised women and African American inventors were systematically underrepresented in the patent record, and many intentionally concealed their identities to avoid prejudice in the marketplace. Some independents struggled for recognition when more famous inventors falsely claimed credit for their inventions. As independent inventors strove to commercialize their patents, they increasingly sold them to the big firms, only to find their names erased from products and advertisements in favor of corporate brands.

Despite any rhetoric to the contrary, corporate R&D labs had not vanquished independent inventors during the early twentieth century. Rather, sophisticated advertising and public relations efforts made corporate researchers *more visible* to the public, while a combination of social and economic circumstances made independent inventors *less visible* than their heroic forefathers. The net effect was a growing misperception that (1) industrial research represented a superior form of invention and (2) that independent inventors were obsolete and extinct. Through their brilliant campaign to "sell the research idea," industrial researchers managed to replace the mythology of the heroic independent inventor with the new, equally potent myth of the corporate R&D lab.

Hidden by Prejudice

Independent inventors of all stripes struggled for recognition at the turn of the twentieth century. In particular, the contributions of women inventors and African American inventors were often obscured because of racial and gender discrimination. In an era when disenfranchised women and African Africans had limited civil rights, they could still take advantage of the US patent system, one of America's most democratic institutions. The original Patent Act of 1790 had no language—and thus no restrictions—limiting patentees based on gender, race, age, religion, nation of origin, or any other category.[6] Nevertheless, women inventors and African American inventors were systematically underrepresented in the patent record, and many intentionally concealed their identities to avoid prejudice in the marketplace.

The fear of social stigma and the practical effect of discriminatory laws deterred many women from pursuing patents in their own names. Writing in 1883, Matilda Joslyn Gage observed that many traditionalists considered it "unladylike" for a woman to perform technical labor. A woman inventor therefore might eschew patenting to avoid "the ridicule and contumely of her friends and a loss of position in society." Common law in many states also prohibited married women from owning or controlling any property, including patents. A married woman, Gage concluded, would have no legal standing to license her patent, sue an infringer, or otherwise administer the "work of her own brain."[7] These conditions deterred many women inventors from even pursuing patents; others elected to apply for patents in the names of their male relatives and attorneys. Writing in *The Woman Inventor* (1891), Charlotte Smith lamented that "thousands of such concealments exist in the list of patents granted," since women's inventive contributions were often "hidden under the names of fathers, husbands, brothers, and sons."[8]

Similarly, fear of discrimination among African Americans led to their systematic underrepresentation in the patent records. Henry E. Baker, an African American patent examiner, discovered this phenomenon when he began to assemble lists of Black patentees in the 1890s. Baker solicited information from thousands of registered patent agents and attorneys and eventually published a list of 370 patents issued to Black inventors for display at the "Negro Exhibit" of the 1900 Paris Exposition.[9] An expanded search in 1913, to mark the fiftieth anniversary of the Emancipation Proclamation, turned up 1,200 leads, but Baker could confirm only 800 patents as definitively earned by Black inventors. In his book *The Colored Inventor*, Baker explained that many African American inventors refused to acknowledge their heritage for fear that "publication of that fact might adversely affect the commercial value of their inventions." Given these uncertainties, Baker believed that perhaps half of all African American patents might remain hidden forever "in the unbreakable silence of official records."[10]

Although an issued patent provided no definitive indication of an inventor's gender or race, women and African Americans still found it necessary to conceal their identities to avoid discrimination in the marketplace. For example, in *The Woman Inventor*, Charlotte Smith described the struggles of Ellen Eglin, an African American housekeeper, government clerk, and inventor from Washington, DC. Eglin invented a mechanical clothes wringer but sold it to a patent agent in 1888 for only $18. Smith inquired why Eglin would sell the invention for such a small sum, especially since the wringer eventually became a great financial success for the buyer. "You know I am black," Eglin replied, "and if it was known that a negro woman patented the invention, white ladies would not buy the wringer."[11] Eglin chose to sell her patent and render herself invisible to avoid a prejudiced marketplace, and her story might have been lost forever if not for Smith's effort to publicize it.

Garrett A. Morgan, an African American inventor-entrepreneur from Cleveland, Ohio, went to even greater lengths to conceal his racial identity while commercializing his inventions. Between 1901 and 1960, Morgan invented an array of technologies, including a belt fastener for sewing machines; a comb and chemical solution for straightening curly hair; and one of the first traffic signals, which General Electric acquired for $40,000. However, Morgan is best known for inventing a gas mask. His fitted "smoke hood" used a long tube to draw fresh air from the ground and a second tube to vent exhaled air.[12]

Morgan patented the gas mask in 1914, but, unlike Eglin, he did not sell his invention under duress. Instead, Morgan established the National Safety Device Company and recruited prominent White businessmen from Cleveland to serve as its directors. Morgan carefully concealed—and purposefully obscured—his racial identity so he could sell the mask to the mostly White fire chiefs, engineers, and chemists that constituted his market. For example, Morgan featured a White man holding the gas mask in his firm's advertisements and trade literature. Morgan also hired a White associate to accompany him as he demonstrated the gas mask at inventors' fairs and trade shows. During these appearances, his White associate posed as the inventor "Garrett A. Morgan," while Morgan himself donned a Native American costume and adopted the persona of "Big Chief Mason." Inside a canvas tent, they built a noxious fire of tar, sulfur, formaldehyde, and manure; Morgan, as Big Chief Mason, would don his gas mask, enter the tent, and emerge unharmed twenty minutes later. Morgan's strategy was good for business, and sales flourished. Between 1914 and 1916, the National Safety Device Company's stock rose from $10 to $250 per share.

The Cleveland newspapers revealed Morgan's racial identity in July 1916 when he participated in a dangerous rescue mission. A natural gas explosion at the Cleveland waterworks trapped eleven laborers five miles from the shore of Lake Erie and two hundred feet below the surface. Morgan and his brother Frank entered the smoke-filled tunnel wearing and carrying extra safety masks; they brought two laborers out alive and recovered the remaining dead bodies (figure 2.1). Morgan was bestowed honorary awards by several Cleveland organizations and the International Association of Fire Engineers, and this led to additional gas mask orders. However, when dramatic newspaper photos from the rescue revealed that Morgan was an African American inventor, several fire departments canceled their orders. In a paradox driven by prejudice, Morgan was more commercially successful when he intentionally concealed his race but less successful when publicity favorable to his invention revealed his true identity.

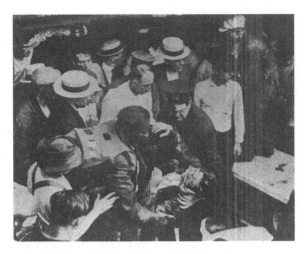

Figure 2.1
Garrett A. Morgan intentionally concealed his identity as an African American inventor to avoid prejudice. However, his identity was revealed during a daring, highly publicized rescue in 1916, which led racist customers to cancel their gas mask orders. The Western Reserve Historical Society, Cleveland, OH.

To be clear, not all women inventors and African American inventors chose to obscure their identities. This was especially true of inventors who hoped to target a specific gender or racial market. For example, Beulah Henry (1887–1973) earned forty-nine patents for improvements in clothing, kitchenware, and children's toys that appealed directly to other women. Widely known as the "Lady Edison," Henry founded two companies to exploit her inventions and served as a consultant for several other firms. Garrett Morgan used his own likeness in advertisements for inventions—such as his comb and chemical hair straightener—that were geared specifically toward African American consumers.[13] Overall, however, because of discriminatory laws and entrenched bigotry, women inventors and African American inventors were far less visible than their White male counterparts.

Overshadowed by Edison

Some inventors remained obscure because they were unable to retain proper recognition for their inventions. This was especially true of inventors in the electrical field, who had to contend with the outsized mythology surrounding Thomas Edison. Edison, of course, was the most heroic of American inventors, and, over time, the "Wizard of Menlo Park" became the very symbol of American ingenuity.[14] However, Edison (and his biographers) tended to exaggerate his accomplishments, and he contributed to the invisibility of his fellow inventors by falsely claiming credit for their inventions.

Edison's mythology obscured the legacy of inventor Frank J. Sprague (figure 2.2). Sprague (1857–1934) was born in Milford, Connecticut, and grew up in North Adams, Massachusetts. He studied engineering at the United States Naval Academy and then served as a technical officer. In 1882, he met Edison's associate Edward H. Johnson while reporting on London's Crystal Palace Electrical Exhibition. Johnson convinced Sprague to resign his commission and join Edison's construction department, where he installed municipal lighting systems. Sprague developed a mathematical

Figure 2.2
Frank J. Sprague was hailed as the "Father of Electric Traction" for developing electric motors, trolleys, subways, and elevators. He spent thirty years trying to regain credit for his accomplishments. Manuscripts and Archives Division, New York Public Library, New York, NY.

method for calculating the proper sizes of main and feeder lines in central power station installations and assigned ownership of the patent to the Edison Electric Light Company.[15] Sprague also conducted occasional experiments with an electric motor for powering industrial machinery. He believed these motors might be used for "traction" to propel locomotives, trolleys, and elevators.[16]

In 1880, Edison had erected an experimental electric railway at Menlo Park, and in 1883 he incorporated the Electric Railway Company of America. Edison admitted, however, that these developments were merely a "hobby" while he focused on electric lighting.[17] That changed in 1884, when Edison asked Sprague to concentrate more formally on his motor experiments. Sprague was reluctant and decided to break from Edison to retain his independence. In his resignation letter dated April 25, 1884, Sprague cited his desire to retain "individual title" to his experimental electric motor and eventually earn his own notoriety in the electrical field.[18] Initially, the split was amicable, and the newly independent Sprague spent the summer of 1884 perfecting his electric motor. In September 1884, Sprague impressed the electrical community with his demonstrations at the Franklin Institute's International Electrical Exhibition. Edison praised Sprague and told the *Philadelphia Press* that his former protégé had solved "the problem of transmission of electrical force."[19]

In November 1884, Sprague submitted the first of several electric motor and railway patent applications and established the Sprague Electric Railway and Motor Company with individual financial backing from Edward H. Johnson. Johnson, in turn, negotiated a contract with Edison's Machine Works to manufacture Sprague's patented motors and equipment. Sprague's big break came in 1887–1888 with his successful installation of the municipal trolley system in Richmond, Virginia. Business boomed as Sprague supplied approximately one hundred new municipal railways over the next two years. To raise $600,000 in working capital, Sprague and Johnson sold shares of the firm to the Edison Electric Light Company and other Edison associates. Sprague's manufacturing contract soon accounted for two-thirds of the motors being built by Edison's Schenectady works. Although Sprague was technically independent, his firm had become reliant on Edison's employees, capital, and manufacturing capabilities.[20]

In 1889, Edison consolidated his various ventures into Edison General Electric and installed Johnson as vice president. Johnson then moved to acquire Sprague's start-up for Edison. With the acquisition, Johnson recouped his personal investment in Sprague, secured one of Edison's best manufacturing customers as a subsidiary, and entered Edison General Electric into the booming electric railway business. Johnson held all the cards, so Sprague reluctantly sold out to Edison General Electric and agreed to stay on as a vice president.[21]

Sprague soon grew frustrated in his second turn as an Edison employee. Edison General Electric did not provide Sprague's railway department with its own experimental facilities but instead moved all railway work to Edison's West Orange laboratory. Edison directed those experiments away from Sprague's system of overhead wires and toward his low-voltage system that supplied power through the rails. Sprague could not abide watching Edison control the railway field he had pioneered, so he resigned on June 7, 1890. "I am too young in years," Sprague seethed, "too jealous of professional reputation, too well trained by education, experience and a natural taste, and too capable of earnest work to become a nonentity, and to stand still while others advance to the front as active workers in electrical science."[22]

Edison's officers convinced Sprague to remain affiliated as a consultant, but he fumed when the firm officially rebranded its acquired Sprague motors and railway products as the "Edison" system in August 1890. Sprague severed all ties in December 1890 and angrily accused Edison General Electric of purposefully doing "everything possible to wipe out the Sprague name and to give to Mr. Edison the reputation properly belonging to other men's work." Still frustrated a year later, Sprague used a pen to mark up the *Edison System of Electric Railways*, the firm's 1891 catalog. He crossed out the name "Edison," replaced it with a handwritten "Sprague," and added that Edison's renaming of the railway business marked "one of the most cold-blooded thefts of a man's reputation and work ever known."[23]

In the following decades, Sprague vigorously defended his record of achievement, often in direct, combative exchanges with Edison. For example, on August 10, 1919, the *New York Sun* published a syndicated interview in which Edison proudly claimed credit for "the perfection of the trolley." Sprague responded a few days later with a forceful letter to the editor. While granting that Edison had experimented with electric railways, Sprague asserted that his own installation of the Richmond railway in 1888 was widely considered "the pioneer of the modern trolley" and "the basis of modern practice." After Edison General Electric acquired his firm, he explained, "the Sprague name was arbitrarily wiped off from the list of 113 electric railways in the United States and abroad, the Edison name substituted therefore, and every detail of construction, equipment and control assigned to a new parentage."[24]

A new round of animus emerged in August 1928, as several laudatory articles appeared to mark the fiftieth anniversary of Edison's practical incandescent lamp. James Young of the *New York Times Magazine* wrote that "within one decade, Edison accomplished three things at Menlo Park—the phonograph, the incandescent light and the electric [trolley] car." In a rebuttal article, Sprague again asserted that any claims, by Edison or his biographers, "that he created the industrial motor and 'invented' or 'perfected' the electric railway and trolley system are quite unwarranted. . . . I must in common justice to myself," Sprague concluded, "protest the maintenance of an industriously cultivated fetish."[25]

In the years after her husband's death, Sprague's widow, Harriet, continued to witness the appropriation of her husband's accomplishments in fawning Edison biographies, official company publications such as the *Con-Edison News*, and even MGM's 1940 film *Edison, the Man*, starring Spencer Tracy. These distortions inspired Harriet to publish the 1947 pamphlet *Frank J. Sprague and the Edison Myth*. The pamphlet served as a brief biography of her husband and recounted a litany of examples in which Edison was wrongly credited for her husband's work. A promotional flyer for the book argued that ascribing Sprague's accomplishment to Edison gave false "credence to overzealous publicity and organized propaganda sources." Harriet hoped her pamphlet might "whittle down" the Edison record "to its proper proportions."[26]

Eventually, modern biographers, such as Paul Israel, began to interpret Edison's record more carefully. From Sprague (trolleys), to Edward Calahan (stock printer), to William K. L. Dickson and C. Francis Jenkins (motion pictures), Israel documented Edison's "tendency to aggrandize his role" in various technical developments with "claims that the historical record would find hard to justify."[27] Two recent biographies have validated Sprague's central role in the development of modern trolleys and subway systems, but these revisionist accounts were no help to him in his own lifetime.[28] As Frank Sprague himself could attest, the powerful mythology surrounding Edison helped assure his own relative obscurity.

Acquisition and Rebranding

Sprague's contributions were obscured when Edison was inaccurately credited for his inventions. However, there was an undeniable economic reality beneath these distortions: when Sprague reluctantly sold his motor and trolley developments to Edison, he relinquished control of the naming and marketing of his inventions. This occurred not once but twice in Sprague's career. In 1898, a cash-strapped Sprague sold the Sprague Electric Elevator Company to the Otis Elevator Company, which controlled 90 percent of the market. More than a century later, it is still common to ride in an Otis elevator, but as Harriet Sprague lamented, "the Sprague name was dropped, another substituted."[29] Even Edison was a victim of this process. In 1892, Edison, Elihu Thomson, and Edwin Houston saw their names erased when Edison General Electric merged with the Thomson-Houston Company to form General Electric.[30]

In the earlier heroic era, nineteenth-century inventor-entrepreneurs built eponymous firms and became well-known brand names as Colt revolvers, Singer sewing machines, and Bell telephones were widely adopted. However, if inventors lacked the financial means or entrepreneurial nerve to commercialize their own inventions, they sold or licensed their patents to larger firms (see chapter 3). By enlisting that help, however, inventors typically had to agree that the manufacturer's name (and not their own) would appear on the resulting products and advertisements. "To get your name on products across America takes big resources," historian George Wise once observed, "and an organization that has those resources generally doesn't want any name on the box except its own."[31]

This was the case with independent inventor Charles Adler, Jr. (figure 2.3). Born in Baltimore in 1899, Adler patented an electric automotive brake at age nineteen, studied engineering at Johns Hopkins University, and then served briefly in the US Army during World War I. In 1919, he went to work for the Maryland and Pennsylvania Railroad, where he developed crossing signals and other safety devices. Adler eventually left the firm to become an independent inventor, though he continued to serve as a consultant for several railroads. Adler received more than sixty patents, and his career generally mirrored the development of transportation technologies. He began making safety and signaling devices for railroads, then switched to traffic

Figure 2.3
Baltimore inventor Charles Adler, Jr., invented safety devices for railroads, automobiles, and airplanes, but he achieved little recognition after he sold or licensed his inventions to bigger firms. Charles Adler, Jr., Collection, Archives Center, National Museum of American History, Smithsonian Institution, Washington, DC.

lights and braking systems for automobiles, and ended his career developing beacons and anti-collision systems for airplanes. Although largely unknown outside his hometown, Adler was something of a minor celebrity in Baltimore. When he died in 1980, his obituary and picture ran on the front page of the *Baltimore Sun*.[32]

Unlike the more famous inventor-entrepreneurs of a previous generation, Adler was unable to attract the capital necessary to manufacture his own inventions. In an unpublished memoir, a frustrated Adler wrote that an independent inventor was "usually licked before he starts" unless he had "a manufacturing plant of his own."[33] Adler, however, collaborated with smaller manufacturing firms and initially managed to keep his name associated with his inventions. During the 1920s, Adler contracted first with the Railroad Accessories Corporation and later with the Signal Accessories

Corporation to manufacture his Adler Flashing Relay— a controller for the flashing red lamps on railroad crossing signals.[34]

Adler's name disappeared from his inventions when he began working with larger firms such as General Electric. In 1927–1928, Adler invented the first driver-actuated traffic signal, which allowed a driver waiting at a red light to change the signal to green by honking the horn. Adler's initial tests drew press coverage in the *Baltimore News*, *Time*, and *Popular Mechanics* and attracted the attention of GE executives, who invited him to perform additional demonstrations at their R&D headquarters in Schenectady, New York.[35] Although the exact terms are unknown, Adler struck a deal with GE to improve and commercialize the signal. GE's engineers added the familiar push-button actuator for pedestrians and eventually introduced the device as the Novalux Traffic Signal. The initial press coverage and GE's own internal publications clearly established Adler as the inventor and assigner of the horn actuator. However, Adler's name disappeared when he partnered with GE; only its Novalux brand appeared on the commercial product.[36]

This pattern of effacement by acquisition occurred again during the late 1930s after Adler patented a special double-filament traffic signal lamp for use on railroads and roadways. Under normal operating conditions, the signal's primary filament provided full illumination; when it burned out, a backup filament lit the lamp with a distinctive visual pattern that alerted safety officials that the lamp needed replacing.[37] Again, Adler was unable to manufacture the lamps himself, so he licensed his patent to both GE and Westinghouse. In 1896, GE and Westinghouse had entered into a patent cross-licensing agreement that enabled the firms to cross-promote similar products, such as tungsten filament lamps bearing the shared MAZDA trademark.[38] The two firms drew up separate but similar licensing agreements with Adler. The General Electric contract, for example, paid Adler a one-cent royalty for each of the first 25,000 lamps sold, with total royalties not to exceed $10,000 per year. The two agreements also contained nearly identical branding provisions. The Westinghouse agreement, for example, stipulated that Adler-licensed lamps might bear the MAZDA trademark at Westinghouse's discretion but would "at least bear the trademark of Westinghouse." Similarly, the General Electric contract included the MAZDA branding option but required that all Adler-licensed lamps "at least bear the trademark G.E."[39]

Adler's manufacturing contracts with GE and Westinghouse represented a significant trade-off. The contracts enabled Adler to earn royalties from his invention and see it widely disseminated. However, the contracts also required the removal of his name from the product in favor of corporate trademarks. This process of effacement was built right into the contractual agreements.

Royalties from Adler's airplane safety beacons and other inventions eventually made him a wealthy man by the end of his career, but he was only rich, not rich and famous. Reflecting on her father's career in 1987, Adler's daughter Amalie Ascher told the *Baltimore Evening Sun* that Adler's "whole life was a struggle for recognition." As an independent inventor of limited means, Adler "couldn't manufacture his own stuff," so he constantly battled with his licensees "to get his name on his things and even to get payment for some of his inventions." Adler, the *Sun* reporter concluded, "was forced to turn to major corporations for assistance" and, like many independent inventors, "his name became lost in the wide-scale promotion and production process."[40]

After an acquisition, large firms also removed the names of independent inventors from their advertising. For example, in 1914 New York independent inventor Henry J. Gaisman (1869–1974) invented the "autographic Kodak," a camera attachment that allowed a photographer to expose a portion of the negative and make notations with a stylus.[41] After securing his patent, Gaisman negotiated with Kodak president George Eastman and assigned the firm full rights to the invention for $300,000. On August 8, 1914, *Scientific American* published a feature on the deal titled "A Fortune for a Simple Invention." The article included pictures of Gaisman, Eastman, the autographic camera, and a facsimile of the $300,000 check, which the journal believed was one of the largest sums ever paid to acquire an invention.[42]

Although Gaisman earned a handsome sum for his invention, he did not receive any credit from Kodak in its subsequent advertisements. In September 1914, less than one month after reporting Gaisman's triumph, *Scientific American* ran its first advertisement for the autographic Kodak. Kodak's ad described the new invention as "the greatest photographic advance in twenty years," but Gaisman's name was entirely absent from the copy. Similar ads followed in nearly every issue of *Scientific American* over the next several months, but none of them mentioned Gaisman.[43] Although the August news article had clearly established Gaisman as the device's inventor, Kodak's subsequent advertisements effectively erased his contributions, rendering him invisible.

The cases of Charles Adler, Jr., and Henry J. Gaisman reveal a paradox concerning the relative invisibility of independent inventors. In order to become wealthy and famous, inventors needed their inventions to be widely adopted by the public. However, if inventors elected to sell or license their inventions to larger firms as the best path to widespread adoption, then the credit for the rebranded invention accrued to the manufacturing firm, not the original inventor. Of course, no one forced inventors to sell or license their ideas, although many probably felt compelled to do so given their financial limitations or the competitive circumstances of their particular markets. Nevertheless, independent inventors had to accept that their contributions would be obscured when they elected to sell or license their inventions to bigger firms. This process of acquisition and rebranding, in turn, further perpetuated the false impression that corporate R&D labs had become the sole source of all new inventions.

Advertising R&D

Meanwhile, extensive corporate advertising helped increase the visibility of industrial researchers. Corporate R&D labs and the professional public relations (PR) industry emerged simultaneously during the first few decades of the twentieth century, and the former eagerly employed the latter.[44] As firms such as DuPont and General Motors built the first corporate R&D labs, they also invested in national radio broadcasts, permanent and traveling exhibitions, and extensive print advertising to present their new research capabilities to a wide audience.

In 1920, Kodak's research director, C. E. Kenneth Mees, could already report that several American firms had conducted "publicity campaigns informing their customers of the existence of their research laboratories." Why advertise R&D? Mees, for one, believed that ads describing sophisticated corporate research efforts helped convince consumers of the quality of a firm's products. These PR efforts also helped firms recruit talented young scientists and engineers to join their corporate laboratories. For both audiences, R&D advertising helped individual firms draw on the progressive aura of science and establish what historian Roland Marchand has called a "corporate image of future-mindedness."[45]

Corporate executives also invested in R&D labs and associated advertising as a strategy for avoiding taxes and preempting antitrust prosecutions.

With ratification of the Sixteenth Amendment in 1913 and the first corporate income taxes, many executives preferred to reinvest in R&D—and purchase more advertising—rather than pay taxes on excess profits. Moreover, firms such as AT&T, DuPont, Kodak, and General Electric operated virtual monopolies in their industries and worried constantly about antitrust prosecutions. These huge firms strategically deployed advertising to "humanize" their research teams while demonstrating how R&D investments translated into better products and services for everyday consumers.[46]

These motivations led individual firms to advertise their R&D labs across several different formats, including the emerging medium of broadcast radio. Individual inventors—including Guglielmo Marconi, Lee De Forest, Reginald Fessenden, and Edwin Armstrong—had developed many of the key components of wireless communications. However, the largest R&D firms, including AT&T, General Electric, and Westinghouse, acquired these patents piecemeal and then—with the federal government's encouragement—pooled the patents under the Radio Corporation of America (RCA). Born a monopoly in 1919, RCA and its subsidiaries integrated the technologies of the individual pioneering inventors and commercialized the radio industry. As Steve Wurtzler has argued, these R&D firms then used the new medium to argue that radio itself and other high-tech advances were only possible because "the site of invention" had shifted decisively "from the workbench of the individual inventor to the well-capitalized research and development laboratory."[47]

For example, between 1942 and 1945, General Motors research director Charles Kettering delivered a series of weekly five-minute "Short Stories of Science and Invention" during the intermission of the GM-sponsored *Symphony of the Air* classical music program on the National Broadcasting Company (NBC), RCA's radio network. Belying his ample intellect and university training, Kettering presented himself as a "pliers and screwdriver man" and "professional amateur" whose folksy persona more closely resembled Thomas Edison or Henry Ford than the PhD-trained scientists working in GM's labs. On one broadcast, Kettering explained industrial research by comparing his team of GM researchers to maestro Arturo Toscanini's NBC Orchestra. Each violinist (and scientist) possessed unique individual talents, Kettering explained, but the orchestra (and the GM laboratory) could accomplish more when several scientists worked together. Kettering acknowledged that corporate researchers employed "expensive scientific apparatus" in "large, elaborately equipped buildings" but suggested that their inventive process was not especially mysterious. Rather, GM scientists tackled complex problems by "process of elimination," employing "the same procedure as in solving a crossword puzzle." In this plainspoken way, Kettering used radio to celebrate GM's labs while demystifying the concept of corporate R&D for millions of listeners.[48]

Research-oriented firms also created impressive exhibitions that celebrated their products and cutting-edge R&D capabilities. General Electric, DuPont, Ford, and General Motors each sponsored impressive pavilions along the Atlantic City boardwalk. Between 1916 and 1955, for example, DuPont's exhibition hall treated 1.5 million visitors annually to scientific demonstrations, lectures, and filmstrips celebrating "the Marvels of Chemistry."[49] These firms also advertised their research prowess at the Chicago (1933) and New York (1939) world's fairs with multiacre exhibit halls such as DuPont's Tower of Research and GM's Futurama. GM even took its show on the road, converting its Chicago exhibits into a traveling "circus for science" called the Parade of Progress (figure 2.4). Between 1936 and 1939, fifty men drove a caravan of thirty-three vehicles to 146 cities and amazed three million visitors with stroboscopic stop-motion cameras that froze fast-moving crankshafts and oscillographs that converted chassis squeaks into detectable light waves.[50] While lone inventors like Frank Sprague and Charles Adler, Jr., wrote letters to the editor and hoped for favorable press coverage, corporate PR departments could draw attention to their laboratories on a scale unattainable by any individual.

Beyond the radio programming, pavilions, and road shows, print advertising was a corporation's most potent medium for touting its research capabilities. In the decades after World War I, DuPont advertised extensively to repair its image after accusations of war profiteering.[51] In 1935, in the midst of a damaging Senate investigation, DuPont launched a massive advertising campaign featuring its new slogan: "Better Things for Better Living Through Chemistry." The print ads (and accompanying sponsorship of *The Cavalcade of America* national radio program) were designed to demonstrate "the importance of chemical research to the average person" while highlighting DuPont's diversification from munitions into a host of consumer products.[52] One ad from this series described how teenagers Amos Hunter and Susie Blossom prepared for their Saturday night date, unknowingly benefiting from DuPont's industrial research. Amos shined

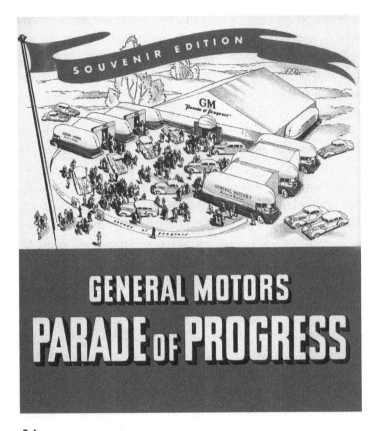

Figure 2.4
This 1936 brochure described the General Motors Parade of Progress, a traveling "circus for science" that promoted the firm's R&D prowess. General Motors Heritage Center, Sterling Heights, MI.

the family car with Duco polish while Susie sewed the last few "frills and furbelows on her new, peach-colored DuPont Rayon dress" before going to see a movie shot on DuPont celluloid film. The ad reminded readers that the "shiny car, the Rayon dress, and the movie film—all resulted from the work of chemists."[53]

The big firms sometimes used print ads to brag about the sheer size and scope of their expansive R&D operations. A 1925 AT&T ad boasted that Bell Laboratories had grown from "Bell and his assistant" to some 3,000 employees, whose daily experiments in chemistry, electricity, and magnetism advanced the telephonic arts. Similarly, a 1919 ad noted that

DuPont's 1,200 "graduate chemists" represented "about ten percent of the total number in the United States." The firm operated "four great research laboratories" plus another thirty analytical laboratories to test raw materials and ensure the quality of its finished products. With its $2 million annual investment in R&D, DuPont proudly claimed it was "the greatest chemical organization in America."[54]

Yet, in an interesting twist, the big firms appropriated the imagery of the independent inventor by featuring lone heroic figures—not huge teams—in their advertising. A series of ads in 1921 and 1922 lionized a lone DuPont chemical engineer as "today's Prometheus." The ads showed a solitary masculine figure, a bespectacled and aproned scientist with rolled-up sleeves, gazing heroically at a test tube in his raised hand. DuPont's chemical engineer combined "the man of science with the manufacturing expert." He and his colleagues, the ad claimed, had revolutionized industry by transforming "the laboratory's discoveries on an experimental scale . . . into production on the larger scale of commerce."[55] Similarly, a 1924 GM ad (figure 2.5) featured a solitary garage inventor—wearing overalls, holding a wrench, and standing over a disassembled engine—even as it proclaimed his obsolescence. The text suggested that automotive advances were "no longer dependent upon the work of isolated inventors." Instead, General Motors maintained "the largest research laboratory of its kind"—a six-acre plant in Dayton, Ohio—to ensure the continuous improvement of its vehicles.[56]

This advertising strategy served two purposes. First, by featuring solitary figures in their advertising, large corporations helped consumers understand a new and unfamiliar mode of invention—the industrial research laboratory—in terms of the well-established imagery of the independent inventor. While DuPont's ad featured a solitary chemical engineer, it was really a celebration of the firm's comprehensive approach to R&D: research in pure science, followed by practical development, commercialization, and scaled production, all within one vertically integrated organization. Yet, DuPont's ad had cleverly collapsed all the complexities and scale of its massive research operations into a more familiar image—the chemical engineer as heroic, individual inventor.

Second, executives believed that advertisements with solitary figures could "humanize" the big R&D firms and help them avoid antitrust prosecutions. Legendary advertising guru Bruce Barton recognized that the

Figure 2.5
This General Motors advertisement featured a solitary mechanic in his garage yet suggested that automotive developments were "no longer dependent upon the work of isolated inventors."
Source: *Scientific American* 131, no. 7 (July 1924): 65.

sheer size of modern corporations invited monopoly accusations and made them appear soulless, aloof, and impersonal to both consumers and federal antitrust officials alike. By featuring the folksy Kettering in his ads for GM, or the heroic Edison in his ads for GE, Barton helped make their corporate "bigness" seem more familiar, less threatening, even benevolent.[57] By appropriating the imagery of heroic, independent inventors, large technical firms and their R&D labs hoped to associate themselves with the best and most admirable figures of America's inventive past—the Yankee tinkerer, the jack-of-all-trades, the indefatigable Edison.

Overall, extensive advertising helped introduce the concept of industrial research to the public. Through nationally broadcast radio programs, permanent and traveling exhibitions, and especially print advertising, individual firms began to explain their new mode of invention to average Americans in a homespun and unthreatening manner. Although large firms sometimes co-opted the positive imagery of lone inventors, their ads suggested that organized industrial research had supplanted independent inventors as the new wellspring of innovation. Research-oriented corporations commanded huge advertising budgets, and as they saturated the airwaves and printed pages with their messages, corporate R&D became more visible and familiar.

Selling the Research Idea

Corporate efforts to legitimize R&D became even more potent when several firms banded together with the blessing and financial support of the federal government. In the decades following World War I, everyday taxpayers unwittingly promoted the expansion and increased visibility of industrial research through the federal government's National Academy of Sciences (NAS), its National Research Council (NRC), and the trade association it spawned, the Industrial Research Institute (IRI). Beginning in 1916, the NRC and its leaders embarked on a vigorous promotional campaign to introduce the concept of industrial research to businessmen and the public through popular lectures, magazine articles, books, and radio addresses. In the process, the public came to see corporate R&D as an alternative mode of invention that was supplanting the traditional individual inventor.[58]

World War I was a watershed event in the emergence and growth of industrial research, as both university-based and industrial scientists capitalized on the opportunity to gain enhanced funding, recognition, and status by serving the cause of military preparedness. In April 1916, the National Academy of Sciences, led by astronomer George Ellery Hale, drafted a proposal to create the National Research Council (NRC) within the NAS as a way of mobilizing American scientists to help solve the military's scientific and technical problems. The NRC appointed several industrial scientists to its ranks, including General Electric's research director, Willis R. Whitney, and AT&T's chief engineer, John J. Carty. The NRC's physicists and chemists thrived during the war effort, as they developed sonar for submarine detection and a host of synthetic chemical explosives (see chapter 6).[59]

In addition to committees on Physics, Chemistry, and Nitrate Supply, the NRC also established a Committee for the Promotion of Industrial Research, chaired by Carty, to encourage the "increased use of scientific research in the development of American industries." On May 11, 1918—six months before the armistice—President Woodrow Wilson signed an executive order establishing the NRC on a permanent, peacetime basis, and the council began to shift its emphasis from military to industrial service.[60] Just days after securing Wilson's executive order, Hale told a dinner audience that the peacetime council would work to "promote a wider appreciation of the national importance of scientific and industrial research."[61]

The NRC's leadership hoped that promoting industrial research would help secure employment and influence for American scientists in both universities and corporations. Its initial efforts were derailed by bureaucratic turmoil as the Committee for the Promotion of Industrial Research endured several reorganizations and name changes between 1916 and 1923.[62] The NRC's promotional work was revitalized in December 1923 when its executive board formed the consolidated Division of Engineering and Industrial Research under the leadership of Chairman Frank B. Jewett, vice president and chief of research at Western Electric and later Bell Labs. According to historian Rexmond Cochrane, the division thereafter embarked on "a massive speaking and publication effort" to extol the virtues of industrial research to industrial executives and the wider public. Because of its wide-ranging responsibilities, the newly consolidated division hired two full-time employees, Secretary William Spraragen and Executive Director Maurice Holland.[63]

Holland was the driving force behind the NRC's promotional efforts and the eventual establishment of the Industrial Research Institute (IRI).

Holland studied engineering at MIT's Lowell Institute, which provided public lectures and evening classes akin to today's university extension programs. During World War I, he flew bi-wing Curtis JN-4 "Jennies" and later served as chief of the Army Air Service's Industrial Engineering Branch. Holland then worked as director of the NRC's Division of Engineering and Industrial Research from 1924 to 1942 before finishing his career as an adviser and consultant to IRI and many of its corporate members.[64]

Following his June 1924 appointment, Holland spent the next year articulating the NRC's "Plan for the Promotion of Research," to be pursued along two fronts. First, the Division of Engineering and Industrial Research would pursue a "national campaign of selling the research idea" by popularizing the results of research "in terms which will reach the man in the street." This national campaign, in turn, would prepare the ground for targeted sales pitches to executives in industries that had "recognized the necessity for research." Holland believed the division's promotional campaign would resemble "missionary work," as it preached the gospel of industrial research to executives and the public at large.[65]

Holland's promotional efforts began close to home, at the division's general membership meetings. During a special metallurgical symposium in February 1925, division members from Alcoa and Union Carbide made presentations on the value of industrial research to potential converts, including the Bridgeport Brass Company and International Nickel. Holland also preached to the choir; at a November 1927 meeting, he taught division members how to "sell the research idea" to the public at large. Holland encouraged the assembled scientists to translate the esoteric results of their research into terms appropriate for public consumption, transforming them from "headaches to headlines" while avoiding "scientific jaw-breakers."[66]

The division's leaders also delivered popular public lectures on the value of industrial research. In March 1927, NRC chemist Harrison Howe drew considerable press attention with his lecture before a meeting of the American Institute of Electrical Engineers. Using familiar items such as artificial silk, dry ice, and synthetic leather as his demonstration props, Howe emphasized how industrial research had been responsible for much of the nation's material prosperity.[67] Later that year, Holland inspired some controversy by delivering a speech titled "The Vanishing American Genius" to New York's American Institute (see chapter 1). Holland suggested that the days of the individual "genius in the garret" had passed, since "the independent inventor of today has little chance against the 'formidable' research organizations of modern industry." Holland's speech initiated a month-long exchange of follow-up articles, editorials, and impassioned letters to the editor of the *New York Times*, all while generating greater public awareness of industrial research.[68]

The NRC also publicized industrial research over the emerging medium of broadcast radio. Between 1924 and 1926, the NRC's Committee on Radio Talks produced several ten-minute programs on industrial research that were broadcast locally in Washington, DC; the transcripts were subsequently reprinted in the *Scientific Monthly*. In the months prior to the 1933 Chicago world's fair, Holland delivered a syndicated radio talk on "Science in the Headlines," while William Spraragen contributed a segment on "Team Work in Research." Through radio's nationwide platform, Holland and the NRC sold the research idea to literally millions of Americans.[69]

Holland's division also distributed several brochures and pamphlets extolling the virtues of industrial research. In January 1930, the division published a sixteen-page pamphlet, *Research: A Paying Investment*. It summarized the results from the division's National Industrial Research Survey, which queried 800 firms and reported on the scope, costs, and benefits of their R&D operations. The division mailed the pamphlet to 1,000 executives and assured them that the NRC stood ready to help them initiate or expand their research capabilities.[70]

The division also pursued more substantial publication projects, including two full-length books. In 1928, Holland and Henry F. Pringle published *Industrial Explorers*, a series of short biographies of nineteen eminent research directors, including Willis R. Whitney of General Electric, C. E. Kenneth Mees of Eastman Kodak, and Charles Reese of DuPont. Following in the tradition of heroic inventor biographies, Holland painted romantic portraits of his subjects. In an extended metaphor, Holland compared his industrial explorers to geographic explorers like Lewis and Clark:

> No less important, no less thrilling, often no less dangerous, is the quest of men whose work is carried on behind the walls of quiet laboratories, in corners of great industrial plants, and in institutions dedicated to the pursuit of knowledge. These are the men reaching out to our industrial frontiers. . . . Impelled by the spirit of research to push beyond the borders of the unknown, these explorers open up new territories with the tools of science, stake out claims to their discoveries, and consolidate with practical application the new position in advance. They are, too, trail blazers on the path of progress.

Echoing themes from his American Institute speech, Holland concluded that these heroic research directors and their corporate laboratories had become the modern successors to the lone "Genius in the Garret," who no longer played "the leading role in the drama of modern industry."[71]

The division's next book reflected a subtle change in strategy. By September 1930, Spraragen could report that the division's promotional efforts had "produced admirable results in selling the research idea to the executives of the leading corporations." With the benefits of R&D now well established, Spraragen suggested that the division shift its message from "'why do research' to 'how [to] do research.'" In 1932, the division published *Profitable Practice in Industrial Research*, a volume that assembled the accumulated experience of research directors such as Frank Jewett (AT&T) and Charles Kettering (General Motors). In the volume, several executives described how they approached the administration of industrial research—from preparing budgets, to equipping the laboratory, to hiring scientific personnel. At the time, the book represented a remarkable accumulation of best practices in the nascent field of research management.[72]

The division also took its promotional efforts on the road in three highly publicized tours of prominent industrial laboratories. By inviting corporate executives to observe industrial research in action, the division hoped they would be inspired to establish laboratories of their own. The first week-long tour departed in October 1930 and took eighty-five executives to visit seven leading R&D labs, including Bell Laboratories in New York City and the General Motors labs in Detroit. The visitors were treated to fantastic demonstrations. For example, the staff at Bell Labs showed the executives a two-way television, akin to a modern videoconferencing system; a subsequent newspaper headline described the visit as a "Trip to Magic Land."[73] Immediately following the tour, the *New York Times* reported that some New England businessmen planned to create a research consortium "as a direct result of impressions" gained on the trip. The NRC followed the 1930 tour with two additional excursions, in 1931 and 1935.[74]

The tours were a public relations bonanza. The division hired a special publicist, Raymond C. Mayer, who in turn recruited reporters from the *New York Times*, the *Wall Street Journal*, and *Fortune* magazine to join the tours and arranged for local press coverage at each laboratory visit. Thanks to Mayer's efforts, the 1935 tour generated 400 local newspaper clippings plus coverage in national magazines such as *Time* and *Business Week*.[75] While the three tours had been a direct pitch to about 200 executives, the subsequent nationwide news coverage also familiarized the public with the benefits of industrial research.

The three tours also galvanized a core group of executives who formed the nucleus of Holland's plan for a National Research Laboratories Association (NRLA). In a 1935 prospectus, Holland acknowledged that rival firms would be unwilling to discuss proprietary research problems in a trade association, but he believed that most firms shared "kindred problems" in the *administration* of research, such as recruiting, salaries, and publication policies. The Division of Engineering and Industrial Research provided free advice on these practices through its books, pamphlets, inspection tours, and general membership meetings, but this model was becoming unsustainable as the NRC struggled financially during the Great Depression. However, Holland believed that firms would be willing to pay modest dues to an organization that could provide these essential services "in a more regular and systematic fashion."[76]

In fact, a similar group, the Directors of Industrial Research (DIR), had already emerged in 1923 from meetings of the NRC's Research Information Service. The DIR was an elite group—strictly capped at twenty-five members—that soon became independent of the NRC. It included the top R&D executives, such as Charles Reese (DuPont), Willis R. Whitney (General Electric), Frank Jewett (AT&T), and Charles Skinner (Westinghouse). Holland's efforts to form the NRLA—which were opposed by many DIR members as unnecessary and duplicative—were probably motivated by other research executives' inability to join the DIR's rarefied membership.[77]

On February 28, 1938, at the Engineers' Club in New York City, Holland and the new division chairman, MIT professor and Raytheon cofounder Vannevar Bush, presided over the kickoff meeting of what they now called the Industrial Research Institute (IRI). Bush explained that the new organization would initially operate under the division's aegis and operational support but would eventually become an independent trade association, "utterly autonomous and distinct" from the National Research Council. About forty-five interested researchers attended the IRI's inaugural meeting, and Holland signed on fourteen firms as charter members, including Colgate-Palmolive-Peet, the Hercules Powder Co., and Procter & Gamble. Henceforth, the IRI took on the former promotional and service functions of the division: hosting general membership meetings, conducting surveys, disseminating best practices, and arranging inspection tours of prominent labs.[78]

The IRI continued its affiliation with the NRC through the early 1940s while moving steadily toward independence. In 1941, the IRI ratified a formal constitution and by-laws; in 1942, the NRC abolished Holland's position as director of the Division of Engineering and Industrial Research and reassigned him to work on defense projects as the NRC returned to its original military support mission during World War II. Meanwhile, the IRI grew steadily to include seventy-three member firms by 1945. Having achieved a self-sustaining group of dues-paying members, the IRI formally separated from the NRC on July 1, 1945, and became an independent R&D trade association, which still exists today.[79]

What was the ultimate impact of the NRC's promotional efforts? In 1927, AT&T's Frank Jewett, chairman of the NRC's Division of Engineering and Industrial Research, acknowledged that it would be "difficult to measure how much more research is being done now than ten years ago" thanks to the division's promotional activities. Nevertheless, according to directories published regularly by the NRC, the number of industrial laboratories in the United States increased nearly eightfold from 297 labs in 1920 to 2,264 labs in 1940, with corresponding growth in R&D personnel from 9,300 to 70,000. More importantly, Jewett argued, the NRC had provoked a "gradual change in the minds of our captains of industry toward research," a change that reverberated through the public at large.[80] Overall, the NRC's zealous commitment to "sell the research idea" helped marginalize independents while making industrial researchers far more visible in the public consciousness.

A War of Words

The emergence of corporate R&D represented a fundamental challenge to independent inventors, who had been the sole arbiters of what it meant to pursue invention. Industrial research laboratories represented a new competing notion of invention that relied on organizational control and teamwork rather than the creative brilliance of individual wizards.[81] This new conception of invention did not appear overnight; it was carefully constructed by the boosters of industrial research in speeches, radio addresses, print ads, traveling road shows, and how-to books. As industrial researchers tried to establish themselves as credible inventors, they had to convince their executive patrons (and the public) that their team-based, scientific approach to invention was a more reliable, profitable, and superior mode of invention than the traditional reliance on individual "geniuses." Not surprisingly, independent inventors resisted this encroachment on their professional turf.

The result was a war of words in which industrial researchers argued for a broader, more inclusive notion of what it meant to be an inventor, while independents strove to maintain that exclusive title for themselves. In this process of professional "boundary work," both industrial researchers and independent inventors deployed argumentative rhetoric in their public speeches and writings as they renegotiated that definition for themselves and the broader public.[82]

Independents and industrial researchers waged their rhetorical battles along several fronts. First, industrial researchers underscored the value of teamwork over individual contributions in corporate invention. In 1919, NRC chairman James Angell complained that there had been "a fetish" among scientists who believed their community should rely on "individual inspiration and initiative" and resist "the corroding influence of adminis-trative organization." Rather, Angell heralded the scores of industrial scien-tists who cooperatively tackled problems so complex "that no one scientist" could solve them "single-handed."[83] Charles Kettering, research chief at General Motors and a former independent inventor, admitted in a radio address that "one individual cannot do much anymore." Early in his career, Kettering recalled, "I was everything from draftsman to superintendent in the factory," but now he and other research directors coordinated dozens of "physicists, chemists, engineers, and practical men" working together.[84]

In contrast, independents maintained that invention was, and always would be, an individual pursuit. Isolation, they argued, nurtured creativ-ity and freed inventors from the discouragement of skeptics. "My mind is sharper and keener in seclusion and uninterrupted solitude," Nikola Tesla told the *New York Times* in 1934. "No big laboratory is needed in which to think," he insisted, adding, "Be alone, that is the secret of invention; be alone, that is when ideas are born."[85] Similarly, in a 1952 interview, Charles Adler, Jr., worried that too many "scientifically minded young men are grabbed up by big laboratories and set to work on projects where they collaborate with scores of others." Adler believed it was becoming more difficult to be a "real inventor" like Thomas Edison. He could not imagine "the phonograph ever being created if one man had been assigned to work on the needle, another on the record, still another on the amplifier." Team-based inventing, Adler insisted, "robs a man of the creative urge."[86]

As Adler had intimated, independents considered themselves "real" inventors, unlike their corporate counterparts. They believed that corporate R&D labs were too conservative in their approach to innovation and generally unreceptive to new ideas. In a 1966 radio interview, Chester Carlson, the independent inventor of xerography, was asked, "Is there room for a loner today?" Carlson responded affirmatively, since "organization men tend to think in certain grooves, and there are plenty of areas that are just left unattended. It's usually one man that's behind the idea."[87] In a speech later that year, Carlson also credited his outsider status for inspiring his copier. "If I had been a photographic chemist working for Eastman Kodak Company," Carlson suggested, "it would never have occurred to me to try for such a far out, unconventional process as electrostatic copying."[88]

Independent inventors also portrayed industrial researchers as cowardly corporate pawns who traded away their autonomy and creativity for the security of a steady salary. In a 1966 speech, Duracell battery inventor Samuel Ruben observed that young scientists increasingly tried to "obtain the safest job with all insurances until death" and thus became "permanent organization men at an early age." Ruben believed that individual inventors would always be important because they worked "independently of popular trends with respect to a given project." Unlike an employee-inventor, an independent inventor never worried about "possibly jeopardizing his position if he is wrong." Ruben contrasted the conservative, corporate attitude with that of bold individuals like himself, who were "willing to take the risks to acquire the rewards of a successful invention."[89] For men like Carlson and Ruben, only independent mavericks—unconstrained by corporate inertia—could create truly groundbreaking discoveries.

In their frequent references to "organization men," Carlson and Ruben were no doubt inspired by William Whyte's 1956 book *The Organization Man*, a trenchant critique of corporate and social conformity in postwar America. Whyte believed that the goals of corporate managers and talented scientists were fundamentally at odds; while the former valued order, organization, and structure, the latter valued the freedom to pursue research wherever it might lead. Scientists, Whyte asserted, had comparatively little freedom within the corporate laboratory, where a research director dictated the agenda. To maintain harmony, Whyte argued that companies like DuPont, Monsanto, and Standard Oil preferred to hire mediocre team players rather than brilliant renegades. "Management has tried to adjust the scientist to the Organization rather than the Organization to the scientist," Whyte observed, adding, "It can do this with the mediocre and still have a harmonious group. It cannot do it with the brilliant; only freedom will make them harmonious." Whyte feared that this approach would have a stultifying effect on corporate innovation and undermine American economic progress.[90]

Industrial researchers did not deny their preference for cooperation and conformity. As Joseph Nocera has argued, corporate scientists tried to "defend that quality—to elevate it, to ennoble it, to make a virtue of it."[91] As early as 1919, the NRC's Angell had argued that in corporate laboratories "obedience, rather than initiative, is the first military virtue." New ideas were welcome, said Angell, but "each individual workman must play his previously assigned part, play it promptly and without debate, become in short a cog in the great machine."[92] Writing in 1941, GE engineer Konstantin K. Paluev likewise suggested that "the co-operation of all co-workers is essential" in a corporate lab since "no accomplishments of much importance are achieved single-handed." Then, anticipating Whyte's *Organization Man*, Paluev argued that "men less able in other respects often attain greater personal success because of their outstanding talent for co-operation."[93] Essentially, both Angell and Paluev believed that the sublimation of an individual's initiative was more critical for the success of an industrial laboratory than his ideas or natural ability.

In stark contrast, independents sought to preserve the notion that invention was the product of individual genius, achievable only by persons of superior intellect. By taking this rhetorical stance, independent inventors hoped to maintain the same high regard and cultural status enjoyed by their heroic forefathers. In a 1908 speech, explosives inventor Hudson Maxim argued that "there is no other intellectual activity so tangled up with the gray matter of the brain as . . . the creative act of invention." According to Maxim, inventors—more so than "litterateurs, artists and scientific research workers"—exercised "faculties of the highest order." Maxim confidently asserted that "the achievements of the inventor transcend those of any other class of men."[94]

Later, in a 1911 article for *The Youth's Companion*, Maxim pondered the question, can invention be taught? "Assuredly it can," Maxim wrote, "provided the person taught is possessed of certain requisite natural endowments." However, Maxim believed that most people "fall so far short of inventive mindedness that it would be impossible, with any amount of instruction and opportunity, to make them invent anything."[95] Maxim was a member of a famously inventive family; although he answered the opening question affirmatively, he clearly believed that inventive geniuses were born, not made.[96]

In contrast, industrial researchers sought to undermine the notion that the act of invention required special, individual genius. As long as the concept of "genius" served to define who could be an inventor, invention would remain a small, exclusive profession and unable to accommodate the thousands of scientists working in corporate labs. In his 1920 book *The Organization of Industrial Scientific Research*, Kodak research director C. E. Kenneth Mees rejected the necessity for genius in the context of the industrial research laboratory:

> Since a great many of the more important scientific discoveries have been made by men of transcendent genius, there is general belief that research work must depend upon such men for its execution and that the prosecution of such research depends upon the development of men of exceptional caliber. While it is true that men differ very greatly in ability and that their value for research work varies, the fact remains that much scientific research depends upon the accumulation of facts and measurements, an accumulation requiring many years of patient labor by numbers of investigators, but not demanding any special originality on the part of the individual worker. Providing that an investigator is well trained, is interested in his work, and has a reasonable proportion of new ideas, he can make valuable contributions to scientific research *even though he be entirely untouched by anything that might be considered as the fire of genius.* And in considering the planning of research laboratories we have no right to assume that we can obtain men who are geniuses; all we have a right to assume is that we can obtain at a fair rate of recompense, well trained, average men having a taste for research and a certain ability for investigation.[97]

Again and again, corporate researchers downplayed the necessity of genius and celebrated their preference for obedient "average men." For example, a Monsanto recruiting film from the 1950s showed three industrial scientists in white laboratory coats. "No geniuses here," the narrator proclaimed, "just a bunch of average Americans working together."[98] Whereas Maxim had argued that inventive geniuses were born, not made, industrial scientists argued that geniuses were not even necessary—or desirable—in the corporate lab.

Independent inventors also liked to characterize their discoveries as discrete "Eureka!" moments or "flashes of genius," a tendency noted by Joseph Rossman in his 1931 study *The Psychology of the Inventor*. Howard Parker, an inventor of machinery for making pulp and paper, noted that, "My best inventions always come to me in a flash. . . . I may be at a concert, church, maybe reading or talking when suddenly a new idea flashes through my mind." Likewise, M. M. Goldenstein, a control systems inventor, reported that "an idea may come to me from within, all of a sudden, like from the clear sky. . . . I would suddenly have before me . . . every minute detail of a complicated wiring diagram." For Rossman, attributing invention to these "sudden flashes and inspirations" was unhelpful for understanding the mental processes of invention, as it merely gave "a name to a thing which puzzles and mystifies us."[99] However, by describing the process in this way, independent inventors portrayed the act of invention as ineffable and incomprehensible to the average person, thereby securing their status as geniuses.

This rhetorical stance did not sit well with corporate executives, who could not afford to bet their highly capitalized technology firms on the unpredictable "Eureka!" moments of idiosyncratic inventors. Invention via a flash of genius was fundamentally discontinuous and unpredictable— who could say when the muses would speak? Therefore, industrial scientists tried to discredit the inventive "flash of genius" while characterizing their own R&D process as rational, continuous, and predictable.[100] For example, German chemist Carl Duisberg, director of Bayer's research laboratories, suggested that his lab's inventions had "*von Gedankenblitz keine Spur*"; that is, no trace of a flash of genius.[101] Even President Herbert Hoover—a pro-business Republican and supporter of industrial research—argued against the "Eureka!" conception of invention. In a 1929 radio address honoring Thomas Edison upon the fiftieth anniversary of his incandescent lighting system, Hoover suggested that inventions "do not spring full grown from the brains of men." Instead, Hoover emphasized that new inventions were "the products of long and arduous research" by "a host of men, great laboratories, [and] long, patient, scientific experiment."[102] Thanks to Edison, the image of a light bulb perched above a person's head would come to symbolize the "Eureka!" moment of invention. Yet, research directors—and even a sitting president—argued against the flash of genius to legitimize R&D's continuous, incremental progress as a credible form of invention.

Edison's career presented a rhetorical conundrum for the supporters of industrial research. These boosters hoped to trade on "The Wizard's" popularity as the nation's most iconic symbol of innovation while pointing to his Menlo Park complex as the prototype for their own corporate R&D labs. However, in order to legitimize their team-based, theoretical approach to invention, industrial scientists also needed to marginalize Edison, the archetype for the heroic, individual inventor they sought to displace.

During World War I, Edison actually questioned the effectiveness of science-based industrial research when he disagreed with plans to build the US Navy's research laboratories (see chapter 6). Yet, given Edison's popularity, the boosters of industrial research were quick to characterize their corporate R&D labs as direct descendants of his Menlo Park "invention factory." In his aforementioned speech in 1929, Hoover praised Edison for illustrating "the value of the modern method and system of invention, by which highly-equipped, definitely-organized laboratory research transforms the raw material of scientific knowledge into new tools for the hand of man."[103] General Electric prominently featured its heroic inventor-founder in its print advertisements and characterized its improved tungsten filament lamp—developed by William Coolidge and others at the GE Research Lab in Schenectady, New York—as the natural successor to the Edison incandescent lamp invented at Menlo Park.[104]

At the same time, the supporters of corporate R&D tried to marginalize Edison as an unsophisticated "cut and try" tinkerer. For industrial researchers, it was essential to discredit Edison—the most heroic independent inventor—if they hoped to legitimize their theoretical, team-based approach to R&D. When Edison died in 1931, several observers suggested that he symbolized the last of a dying breed. AT&T's Frank Jewett observed that Edison made few important achievements after the 1880s because the rapid advance of science "came to require a type of training which Edison did not possess." Likewise, MIT professor Karl Compton suggested that perhaps "the inventor of Mr. Edison's type has passed" because modern invention now demanded a "continually greater degree of specialization and scientific background."[105] In his eulogy, New York Times science editor Waldemar Kaempffert delivered the final coup de grâce for independent inventors. With Edison's death, wrote Kaempffert, "the heroic age of invention probably ends. The future belongs to the organized highly trained physicists and chemists of the corporation research laboratory."[106]

Independent inventors fought valiantly, but corporate scientists won a decisive victory in their war of words. Given their superior financial resources and PR acumen, industrial scientists spread their pro-research rhetoric to a much wider audience than the independents could. Consequently, industrial researchers succeeded brilliantly in establishing corporate R&D as a legitimate and credible form of invention while marginalizing independent inventors as antiquated, obsolete, and extinct.

Conclusion

In his memoirs, Duracell battery inventor Samuel Ruben recalled a newspaper's cynical reaction to one of his many awards:

> A research institute at George Washington University has just given its 1965 award for "Inventor of the Year" to Dr. Samuel Ruben. Dr. *Who*? . . . Last year's inventor of the year was Chester F. Carlson—remember? Somehow this information makes us a bit sorry for today's growing youngsters. Almost any American over fifty remembers how thrilled he was as a child to read about the experiments of Franklin, Edison, Fulton, Morse, Whitney, McCormick, Goodyear, Westinghouse and their peers. What's more, he can still hook up those names with inventions that have left their impact on his way of life. . . . [B]y this time next year Dr. Ruben's name will be as unfamiliar to most of us as that of Mr. Carlson. He invented the Xerox copying process.[107]

This wistful anecdote was emblematic of inventors' diminished fame relative to their heroic predecessors, and the growing belief that corporations had replaced them as the source of all new inventions. Although the miniature battery and photocopier clearly rated as important developments, their independent inventors—Ruben and Carlson—were scarcely recognized for their work. Instead, that recognition largely accrued to their corporate licensees, Duracell and Xerox.

During the first half of the twentieth century, a variety of factors conspired to render independent inventors less visible than their heroic forefathers and corporate contemporaries. Disenfranchised women and African American inventors were systematically underrepresented in the patent record, and many intentionally concealed their identities to avoid prejudice in the marketplace. Some independents struggled to escape the long shadow of Thomas Edison, who falsely claimed credit for their inventions. Inventors also faced a difficult trade-off: to commercialize their inventions, they often had to partner with large manufacturers, but those firms usually stipulated that their corporate brands would replace the inventors' names on their products and advertisements.

Meanwhile, sophisticated public relations efforts helped make industrial researchers more and more visible to the public. Firms such as AT&T, DuPont, and General Motors invested heavily in Atlantic City pavilions, traveling road shows, and extensive print advertising that celebrated their expansive research operations, even as those ads sometimes appropriated the familiar imagery of lone inventors. Through its government-sponsored program of radio broadcasts, popular lectures, how-to books, and laboratory inspection tours, the National Research Council likewise succeeded in "selling the research idea" to executives and the general public. With their superior financial resources and broader reach, industrial researchers won a decisive victory in their war of words with independent inventors. They convinced the public that independent inventors were obsolete and nearly extinct while portraying R&D labs as the wave of the future. In the process, industrial researchers replaced the myth of the heroic, independent inventor with the new, equally potent mythology of the corporate R&D lab.

The net effect of these activities was an emerging public perception that corporate R&D labs had supplanted independent inventors during the early twentieth century. However, the independents had not disappeared. In fact, as we'll see, the independents learned to survive by forging alliances with some of the same corporations that proclaimed their obsolescence.

3   Make, Sell, Ally

Am more anxious now to manufacture and market. . . . I am convinced the money is in making and selling it.

—Flexible straw inventor Joseph Friedman to his backer David Light,
July 16, 1941

Sold Patent First Stop, Sheaffer Company. Price Small But Am Satisfied. Obtaining Money Twenty Eighth December.

—Friedman to his wife, Marjorie, after selling his pen patent,
December 24, 1934

You will assist us in the development of the ideas included under this agreement and we will pay you at the rate of $100 per day; one half (1/2) of which payments are to be credited as advanced royalties.

—Clause 9 from Wadsworth Mount's standard client consulting
agreement, 1958

By 1850, the archetype of the struggling, impoverished inventor was already a well-established biographical trope. Even with their exaggerations, these rags-to-riches stories were popular because they were often true; invention was, and remains, a risky business.[1] An inventor might experiment for years and never earn a cent for his or her efforts. The pathways toward financial success became more complicated after 1900 with the emergence of corporate R&D labs. Independent inventors still had to compete against one another and contend with a fickle marketplace, but now they also had to compete with teams of industrial scientists from large, deep-pocketed firms. However, the big firms were not always competitors; sometimes they were clients, business partners, and potential sources of income. During the twentieth century, inventors balanced a number of considerations as they tried to earn a living in a complex economic environment.

For an independent inventor, the work of *invention* (creating technology) and *innovation* (commercializing technology) was fraught with costs and risks. During the invention stage, an inventor needed to pay for tools, materials, and sometimes the expertise of a skilled mechanic to build a prototype. Inventors also paid the attorneys' fees and application expenses associated with obtaining a patent. If not working from home, an inventor needed to rent a workspace; at all times, he or she needed a salary to cover basic living expenses, especially when supporting a family. Still, the work of *invention*—generating ideas, sketching, building a prototype, experimenting, and patenting—could often be achieved with only modest expenditures. For example, in 1927 the Munn & Company patent agency charged about $125 to prepare a patent application, which was a significant but surmountable expense for a manufacturing worker earning about $25 per week.[2]

A patent by itself, however, did not generate financial returns. The difficult follow-on work of bringing an invention to market—including manufacturing, shipping, advertising, sales, and customer service—entailed even greater costs and risks. This transition from invention to innovation forced an inventor to take stock of several factors—including their own personal finances and risk tolerance, and the state of the marketplace—before deciding the best approach to commercialization.

During the first half of the twentieth century, independent inventors pursued a series of commercialization strategies that reflected the heterogeneity of their profession and the variety of ways they interacted with firms (table 3.1). In order to profit from their ideas, independents could found a start-up to manufacture and market the invention (make), assign the patent to an established firm for a onetime payment (sell), or license the invention and earn royalties and consulting fees by entering into a cooperative alliance with a manufacturer (ally).

Established firms faced a parallel set of choices as they developed new technologies. Firms could produce inventions internally via a vertically integrated R&D lab (make), purchase individual patents (or acquire an inventor's firm) for a lump sum (buy), or license patents on a royalty basis via alliances and consulting agreements forged with external inventors (ally). Finally, both inventors and firms could pursue several of these approaches simultaneously in a mixed, hybrid economic strategy.

Table 3.1
Commercialization strategies of independent inventors and firms

Strategy	Independent inventors	Flow of intellectual property (IP) and technical information	Firms
Make	Inventor-entrepreneur founds a start-up to manufacture the invention	NA	Firm invests in a vertically integrated R&D lab and manufactures its inventions
Sell/buy	Inventor assigns (sells) IP for a lump sum	$\rightarrow$	Firm buys IP for a lump sum
Ally	Inventor grants licenses on IP, earns royalties and consulting fees	$\leftrightarrow$	Firm licenses external IP, pays royalties and consulting fees, shares its internal IP (for technical integration)
Mixed/hybrid	Both independents and firms pursue multiple innovation strategies simultaneously		

The two "make" approaches—the traditional Edisonian inventor-entrepreneur and the corporate R&D lab—have received the bulk of attention from historians.[3] More recently, historians have described a vigorous market for technology in which established firms acquired valuable, high-impact patents from outside independent inventors.[4] This chapter explores how a handful of independent inventors evaluated these strategies and commercialized their inventions. Collectively, their stories illustrate the advantages, disadvantages, and trade-offs of each approach—and the decisions inventors faced—as they strove to earn a living.

Make: Moonlighting Inventors, Reluctant Entrepreneurs, and Family Businesses

Much like Thomas Edison or Alexander Graham Bell of the earlier heroic era, twentieth-century inventors continued to launch entrepreneurial start-up firms to make and sell their own inventions. Joseph B. Friedman, inventor of the flexible drinking straw, provides a rich example of a small-scale inventor-entrepreneur. Friedman was born in Cleveland in 1900, the fifth of eight children of Jewish American immigrants. At age fourteen, Friedman conceived—but was unable to commercialize—his first invention, the battery-powered "pencilite" for writing in the dark. Friedman had no formal technical training, but he sketched continually and generated dozens of new product ideas, such as an automatic shutoff for a gas range, a portable movie screen, an improved ice cream scooper, and a reflective license plate.[5]

But inventing did not initially pay the bills. By the 1930s, Friedman had moved to San Francisco with his wife, Marjorie; daughters, Judith, Linda, and Pamela; and son, Robert. In the midst of the Great Depression, Friedman worked as an optometrist, insurance agent, and real estate broker. Though Friedman was only moonlighting as a part-time inventor, the 1930s proved to be his most prolific period, when he received six of his nine US patents, including one for his flexible straw.[6]

In 1936, Friedman sat with his daughter Judith at the counter of the Varsity Sweet Shop, his brother Albert's San Francisco soda fountain. He watched as the two-year-old struggled to sip from a glass sitting on the counter. Judith was not tall enough to reach the top of her straw, and when she bent it down, the resulting kink prevented the flow of her milkshake. Back at home, Friedman inserted a screw into a standard paper straw and wrapped dental floss tightly around the exterior. When he removed the floss and the screw, the straw was imprinted with accordion-like corrugations from the screw threads. These corrugations made the straw flexible; it could now bend without blocking the flow of the liquid. Friedman immediately filed a patent application for his "drinking tube" (figure 3.1) in November 1936.[7]

Friedman had cleverly improved on a quotidian but ubiquitous product, and he began to explore opportunities for its commercialization. He enlisted his childhood friend Irwin Geiger, a Washington, DC, attorney, to help him research the straw industry. Natural straws made from reeds and ryegrass had been around for centuries, but they often disintegrated in the beverage. Inventor Marvin Stone patented the wax-coated paper straw in 1888 and set up the Stone Straw Company to automate their mass production. In 1905, Stone's original patent expired, and several other firms entered the marketplace.[8] In May 1938, following research at the US Department of Commerce, Geiger reported that six major straw manufacturers supplied the $1 million annual straw trade. Straws had become a commodity item with rampant price cutting and thin profit margins across the industry. Friedman's flexible straw presented an opportunity for one of the

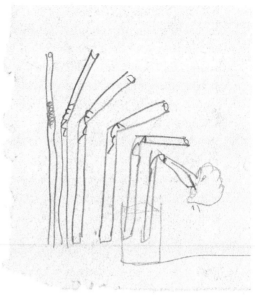

Figure 3.1
In 1936, inventor Joseph B. Friedman sketched the initial concept for his flexible straw. After failing to sell his patent, he reluctantly founded a family business to commercialize his invention. Joseph B. Friedman Papers, Archives Center, National Museum of American History, Smithsonian Institution, Washington, DC.

established companies (or a start-up) to disrupt the industry by introducing a new deluxe product with seventeen years of exclusive patent protection.[9]

Friedman initially looked to sell or license the flexible straw patent to one of the established firms; this had always been his inclination. Friedman had tried—but failed—to sell his pencilite (to the Sterling Corporation), ice cream scooper (to Hamilton Beach), portable movie screen (to Eastman Kodak), safety razor handle (to Gillette), and gas range safety device (to the Wedgewood Range Company). In 1934, Friedman did assign his improved fountain pen patent to the Sheaffer Pen Company for $3,000, and this successful transaction probably cemented his resolve to sell the straw patent as well.[10] Friedman was understandably reluctant to manage the risks of starting a new firm in a highly competitive industry at the height of the Great Depression, especially while supporting a young family.

In 1937–1938, Geiger approached several East Coast straw firms—including the Stone Straw Company, Lily-Tulip, Herz Straw Company, Glassips, Inc., and the Maryland Baking Company—to gauge their interest in buying Friedman's patent or taking a license. "All that stands in the way of effecting a satisfactory deal," wrote Geiger, is finding a "mechanical method for impressing the corrugation." This proved to be the catch. Glassips offered to take an exclusive license on the flexible straw patent and pay 2 percent royalties on future sales, contingent on Friedman's ability to provide a machine that could produce 1,000 corrugated straws per minute.[11] Geiger offered to sell the patent outright to the Stone Straw Company for $75,000, but President Landra B. Platt passed, citing the "impossibility" of constructing a workable machine. Instead, Platt proposed to license the straw but offered no advance on the royalties to construct the machine. Geiger told Platt to "jump in a lake."[12]

Friedman was at an impasse. As with his other inventions, he preferred to sell or license his flexible straw patent and eschew the risks of entrepreneurship, but no sale was possible without first developing an automated production machine. Friedman, now wary of partnering with the established straw firms, decided to "develop the machine myself" and take an "independent position" with respect to its commercialization.[13] Once the machine was completed, he could try again to sell or license the patent on more favorable terms or start a business to produce and sell the straws himself. Either way, he had to build the machine before the straw patent expired or risk losing any chance for profits. In 1939, Joseph Friedman became a reluctant entrepreneur and founded the Flexible Straw Corporation as a vehicle for commercializing his invention.[14]

From the beginning, the Flexible Straw Corporation was a family business. For start-up capital, Friedman turned to his sisters' husbands, Harry Zavin and David Light, who ran a furniture business in Portland, Oregon. Each brother-in-law invested $2,500 for a 12.5 percent equity stake in the business, initially capitalized at $20,000. Friedman retained the remaining 75 percent equity and initially drew a $250 monthly salary to work exclusively on building the corrugating machine.[15] However, this task proved much more difficult than expected. The initial $5,000 investment was soon expended, so Friedman gave up his salary and returned to "real estating" to "feed his family" while moonlighting on the straw project. With considerable time and his family's money invested in the project, Friedman was no longer willing to give up control by licensing or selling the straw patent. In 1941, he told Light that he was anxious "to manufacture and market" the straws and was convinced that "the money is in making and selling" them. However, Friedman struggled to automate the corrugation process, as restrictions on tools and materials during World War II hampered his progress.[16]

With the straw venture stalled and the war winding down in 1945, Friedman received another family assist from his sister, Betty Friedman, a career woman working in Cleveland. She had helped Bert and Milton Klein develop the Tarbonis Company, a successful local pharmaceutical firm that marketed an anti-inflammatory skin cream. Betty encouraged the Klein brothers to speak with her brother about his straw patent; in July 1945, they invested an additional $50,000 in the venture. The Kleins joined Friedman, Zavin, and Light as stakeholders in the original Flexible Straw Corporation, which would continue to own the straw patent and any subsequent improvements. Flexible, in turn, would grant exclusive licenses to a new Klein-backed entity, the Flex-Straw Corporation, which would manufacture, sell, and distribute the drinking straws.[17]

Friedman now devoted himself fully to the venture. In 1946, he closed his San Francisco real estate business and moved his family to Los Angeles to be nearer the Klein brothers, who were relocating some of their operations to Southern California. With an infusion of cash and the relaxation of wartime material restrictions, Friedman perfected his flexible straw and the automated machinery to produce it, while securing additional improvement patents.[18] By September 1947, Friedman had built two production machines, each capable of crimping and waxing approximately 50,000 straws per day. Drawing on Betty's connections in the medical supply industry, they sold the first Flex-Straws to a Cleveland-area hospital in October 1947. The Flex-Straws had several advantages over the glass straws typically used in hospitals: they were sanitary, disposable, shatterproof, and, of course, flexible, so they could be used by reclining, bedridden patients.[19]

As the flexible straw market expanded to grocery stores, drugstores, and restaurants, Friedman soon fielded and declined a handful of offers to sell his business to the established straw makers. Friedman was no longer a reluctant entrepreneur; in 1949, he wrote Betty that "I can make a fortune in the 5 years remaining of my patent protection."[20] Friedman, in fact, had two sources of income. First, as a 50 percent shareholder of the Flex-Straw Corporation, Friedman (alongside the Klein brothers) earned profits on the manufacture and sale of the straws. Second, Friedman (alongside Zavin, Light, and the Kleins) earned a 34 percent share of royalties, since the Flexible Straw Corporation licensed the straw patents to the Flex-Straw Corporation. In 1953, for example, Friedman earned approximately $55,000 from his 50 percent share of Flex-Straw's net profits and an additional $5,270 on his 34 percent share of Flexible's royalties.[21]

The Flex-Straw Corporation was essentially a small family business. As noted, Joseph Friedman's brothers-in-law, Harry Zavin and David Light, were his first backers. Joseph's wife, Marjorie, went on sales calls around Los Angeles; his father, Jacob, similarly pursued straw sales at hospitals in San Francisco and Oakland. Friedman's teenage daughter Judith—who inspired the flexible straw as a toddler—worked occasionally as a secretary in her dad's office. The correspondence among the family members was almost always warm and cordial, with paragraphs about royalty payments and sales calls interspersed with personal notes about birthdays, vacations, and other family gossip. These letters reveal Joseph Friedman as a man trying (often with some difficulty) to balance his ambitions as an inventor-entrepreneur with his responsibilities as a husband, father, son, brother, and in-law. He even admitted to some occasional "touchiness" with his family members, especially his sister Betty.[22]

Betty Friedman was crucial to Flex-Straw's success. While her brother focused on production in Los Angeles, Betty worked from Cleveland, where she split her duties between Tarbonis and Flex-Straw. She developed Flex-Straw's print and television advertising campaigns, directed the sales effort, and built a network of 3,000 distributors.[23] Betty and Joseph's relationship became temporarily strained in December 1950, when she sought a long-overdue equity stake in the business. "I built a company for you, Joe," she wrote, and "it was through me the thing got started in the first place. . . . I hate having to put it to you this way—I shouldn't have had to." By February 1951, the siblings had reconciled and Joe and the Kleins ended up giving Betty a 10 percent share in the Flex-Straw Corporation. Betty eventually moved to Los Angeles in 1954, becoming the company's general manger after Joe suffered a heart attack in September 1953.[24]

By the late 1960s, plastic flexible straws were beginning to cut into the market for waxed paper straws, Friedman's improvement patents were set to expire, and he was ready to retire. In 1969, the Maryland Cup Corporation acquired the Flex-Straw Corporation (the manufacturing company) for $320,000 and the Flexible Straw Corporation (the patent holding company) for an additional $60,000. Joseph Friedman—an optometrist, insurance agent, real estate broker, moonlighting inventor, reluctant entrepreneur, and family man—retired comfortably and died in 1982.[25]

Friedman's story illustrates some of the opportunities, risks, and trade-offs of pursuing the "make" strategy. Inventor-entrepreneurs retained their patents and an equity stake in their start-ups. Consequently, they maintained their autonomy and the potential to earn big profits if they could successfully commercialize their inventions. However, inventor-entrepreneurs endured financial uncertainty and time-sensitive pressure. They might work for years before making any income from an invention, and their window of opportunity closed once their patents expired. Given these risks, Friedman was a reluctant entrepreneur; he decided to manufacture his flexible straw only after failing to sell or license his patent. Once committed to the venture, Friedman enlisted financing and operational support from his family and then operated a successful small business for two decades before selling his firm to an established company.

Sell: Assigning Patents (and Facing Rejection) in the Market for Inventions

Friedman's first inclination—from the pencilite to the flexible straw—was to avoid the perils of entrepreneurship and sell his patents to established firms. This economic strategy had long been popular among independent inventors because it required minimal risk, minimal effort, and minimal expense. Patent assignments were often "arm's length" transactions—the established firm paid the inventor a onetime lump sum, the inventor transferred title to the patent, and the parties went their separate ways. The buyer could then integrate the newly acquired invention into its existing stable of products, resell the patent, use it as a bargaining chip in negotiations with other firms, or suppress it to prevent competitors from developing the idea.[26]

Historians have documented a robust market for technology throughout the nineteenth and early twentieth centuries. Through the 1930s, about half of all inventors eschewed entrepreneurship and elected to sell or license their patents to manufacturing firms. Corporate R&D labs, in turn, were often unable to supply all their firms' technology needs. Despite their disparaging rhetoric, they acquired approximately 20–25 percent of their patent portfolios from independent inventors to complement the inventions they produced in-house. To identify potential purchases, firms closely monitored the weekly notices of issued patents in the *Official Gazette of the United States Patent Office* and journals such as *Scientific American*. Intermediaries—such as inventors' professional organizations and patent attorneys—brokered connections between sellers and buyers. To better understand these transactions, we again turn to Joseph Friedman and another independent, Wadsworth Mount, to examine their negotiations with the Sheaffer Pen Company, Chrysler, and some other potential buyers.[27]

In April 1922—several years before he invented the flexible straw—Friedman earned his first patent for a modified fountain pen with a transparent barrel and ink reservoir. This feature allowed a writer to easily determine when the pen needed to be refilled with ink. Initially, Friedman was not proactive in marketing his pen patent, but he did respond to two firms that inquired about his invention. In December 1925, Friedman returned nearly identical letters to the Eisenstadt Manufacturing Company, a St. Louis jewelry maker, and the Sheaffer Pen Company of Fort Madison, Iowa. Friedman included a copy of his patent and a photograph of his modified pen prototype. The letter described his invention as a distinctive and marketable novelty for an otherwise commodity product. It eliminated the nuisance of "unexpectedly finding the pen dry" and the guesswork of refilling a typical "blind" pen. Friedman did not explicitly offer to sell the pen patent or name a price, but he agreed to answer any further questions.[28]

Apparently, nothing came of these first two leads. Then, eight years later, in 1933, Friedman received another inquiry about the pen patent, from W. Frank Wallace of Los Angeles. Friedman scratched out a handwritten reply: "If you are still interested in purchasing my fountain pen patent, I have decided to sell on your terms namely $100. You can draw up an assignment or bill of sale as you wish to have it. Mail same to me and I will sign." It is not clear whether Friedman or Wallace changed his mind, but Friedman ultimately retained the rights to his patent.[29] However, Friedman's three unsolicited inquiries—from Eisenstadt, Sheaffer, and Wallace—suggest that individuals and firms actively monitored the patent records to investigate potential purchases.[30]

The exchange with Wallace and some contemporary print advertising seemed to reinvigorate Friedman's efforts to dispose of his pen patent. Friedman clipped several print advertisements for Sheaffer's "Visulated" fountain pen, which featured a transparent ink window similar to Friedman's design. Friedman was determined to find out whether Sheaffer had stolen his idea after inquiring about his patent in 1925. In the December 1933 edition of *Thomas' Register*, Friedman circled the names and addresses of half a dozen firms in the "Pens: Fountain" sector.[31] In October and November 1934, he drafted nearly identical letters to several pen makers, including Sheaffer (Fort Madison, Iowa), the Wahl Company (Chicago), Waterman (New York), Conklin Pen Company (Toledo, Ohio), Parker Pen Company (Janesville, Wisconsin), and Carter's Ink Co. (Boston). Friedman included a copy of his pen patent and noted that it broadly covered "the use of an indicator or visible feature in connection with a collapsible reservoir pen." Friedman cleverly noted that he had "apparently been a bit early with this particular invention," given that the concept was now the subject of extensive advertising and sales efforts. Without using the word, Friedman intimated that any firm marketing a see-through pen might be guilty of infringement. Again, Friedman did not name a price but asked each recipient to consider purchasing or licensing his patent.[32]

Only Wahl and Sheaffer showed any interest in the patent. Wahl, for example, noted that the patent was set to expire in four years, leaving little time to develop the idea for the market. The firm declined to license the patent on a royalty basis but encouraged Friedman to "make an outright sale of it" and name his price. Sheaffer also declined to license the patent but would consider purchasing it outright.[33]

Friedman had correctly sensed a positive change in the market for his invention. In 1925, Sheaffer had shown some interest in Friedman's patent but ultimately passed on it. Now, in 1934, with a see-through pen in national print ads, Sheaffer appeared interested in purchasing Friedman's patent to avoid the risk of infringement litigation. From Wahl's point of view, purchasing Friedman's patent might help the firm develop its own see-through pen to compete directly with Sheaffer. With Friedman's patent in hand, Wahl might also use the threat of infringement to extract royalties from Sheaffer or force the firm to pull its pen from the market entirely. Friedman offered to sell his patent to Wahl for $11,250 and to Sheaffer for $10,000.[34]

To this point, Friedman had conducted all his sales efforts at arm's length, via correspondence, to a geographically broad spectrum of potential buyers. Friedman eventually decided to follow up on his leads by making a trip east from San Francisco to visit several pen manufacturers. Friedman's wife, Marjorie, sent him a telegram in Cleveland on Christmas Eve 1934 alerting him that Conklin and Waterman were no longer interested.

Friedman immediately sent a return telegram: "Sweet Little Chickie—Thought You Received Newspaper Describing Deal. Don't Wire Any More Pen Nonsense. Sold Patent First Stop Sheaffer Company. Price Small But Am Satisfied. Obtaining Money Twenty Eighth December."[35] On December 18, 1934, Sheaffer agreed to buy Friedman's pen patent for $3,000 to solidify its intellectual property position for the Visulated Pen.[36] With some simple market research in *Thomas' Registry*, a few solicitation letters to pen makers, and one face-to-face meeting, Friedman was able to sell his pen patent to Sheaffer for much less than the $10,000 he had requested but far more than the $100 he had been willing to accept in 1933.

We can also see how the "buy" strategy worked from Sheaffer's point of view. In 1925, the firm proactively contacted Friedman about his pen patent, suggesting that the firm regularly monitored, evaluated, and purchased patents from outside inventors. Friedman and Sheaffer eventually agreed to a sale in 1934, and the firm's standard assignment contract suggested the firm's frequent transactions with independent inventors. The boilerplate form had blanks where Friedman (or any other inventor) could fill in his name and place of residence, the patent number and issue date, and the date of the contract. Second, the language of the contract cleverly concealed the true purchase price of the patent. Patents were officially assigned to Sheaffer for $1.00 plus "other good, valuable, and sufficient considerations." Since all patent assignment contracts were registered publicly with the US Patent Office, any potential buyer or seller could determine what Sheaffer was paying for patents. However, by officially selling the patents for $1.00 and then privately settling on other "considerations," Sheaffer could maintain some secrecy in its frequent negotiations with inventors.[37]

Although Friedman had managed to sell one of his patents, not every inventor was so lucky. Inventor Wadsworth Mount struck out in his attempts to sell his ideas to Sheaffer and was stonewalled completely by the Chrysler Corporation. Mount (1907–1985) was born in Minnesota and raised in New Jersey. He studied economics and philosophy at Amherst College and then worked as a Wall Street stock salesman, a public utilities analyst, and a publisher of financial journals. During World War II, he developed his first major invention, a rocket-propelled antiaircraft weapon. Following the war, Mount became a full-time inventor and industrial consultant, working independently in the basement of his Summit, New Jersey, home. He was granted more than twenty US patents on a diverse range of technologies, including children's toys, military equipment, pipe-joining processes, and sailboat hardware.[38]

In June 1958, Mount received a letter from Sheaffer's patent counsel, W. K. Olson, who had seen him speak at a conference. Olson inquired whether Mount had "any inventions lying on the shelf" that he might offer "for sale to manufacturers." Independent inventors always worried that established firms would steal any ideas they shared while exploring a potential sale, so Olson enclosed a pamphlet describing Sheaffer's disclosure policies. Sheaffer agreed to maintain confidentiality while evaluating ideas, but its receipt of a submission did not imply any contractual relationship with the inventor. Furthermore, Sheaffer was not bound to divulge its reasons for rejecting an idea. This protected Sheaffer from leaking its own proprietary plans if an inventor's idea closely resembled a pending in-house patent application or future product release. By providing clear expectations, Sheaffer hoped to avoid false piracy accusations and maintain its good reputation among inventors.[39]

In July 1958, Mount sent his signed disclosure form to Olson along with a description of his idea: a retractable razor blade attachment for a mechanical pencil. Mount believed the invention would appeal to businessmen, who often took clippings from newspapers and magazines. In lieu of a drawing, Mount taped a small prototype of his razor blade attachment to the disclosure letter. Though he had no patent, Mount offered to license the invention to Sheaffer for a 5 percent royalty.[40]

Olson replied a month later to reject the idea. He explained that Mount's blade idea had been submitted many times before, but Sheaffer had determined that such a novelty would not appeal to a mass market. Olson encouraged Mount to keep sending additional ideas, since Sheaffer was trying to diversify beyond pens and pencils. Between 1958 and 1960, Mount disclosed several ideas, including a vacuum cleaner accessory, a diesel engine fuel injector, a traffic signal circuit, an electric baby-crib rocker, and several maritime devices (e.g., turnbuckles, snap shackles, and oarlocks). However, Sheaffer was uninterested.[41]

Mount could tolerate rejection as long as a firm like Sheaffer fairly evaluated his ideas. He discovered, however, that Chrysler and other automakers used their disclosure procedures to systematically reject the ideas submitted by

America's grassroots inventors. The automakers espoused the "not invented here" (NIH) attitude—a strong preference for their own ideas, combined with an inherent skepticism toward outsiders' ideas.[42]

The early auto industry had once been open to suggestions from independent inventors. Each model year brought new comfort and safety features such as bumpers, headlamps, and seat belts; moreover, car owners invented luggage racks, portable tents, and many other aftermarket accessories to aid their travels. Between 1920 and 1927, at the height of the Model T's production, Ford owners sent the company one hundred letters a day, including dozens of design suggestions from aspiring inventors. These "consumer-inventors" initially found many opportunities to sell their patents to the auto companies.[43]

However, as car designs stabilized and the "Big Three" automakers (Ford, General Motors, and Chrysler) began to develop their in-house R&D laboratories, the NIH attitude took hold in Detroit. *Business Week* described the automakers' conundrum: they could no longer "spare the time and effort of top executives to listen to all the inventors," but they also "didn't want to alienate any potential car buyers." In 1925, General Motors established its New Devices Section to evaluate—and mostly reject—the suggestions sent in by customers and outside inventors. Ford's Experimental Engineering Department and Chrysler's Engineering Improvements Committee emerged during the same period to serve essentially the same function. Between 1925 and 1949, GM's New Devices Section received more than 140,000 ideas from outsiders, a measure of the persistence and enthusiasm of independent inventors. However, "the chances of the average outside inventor ever making a fortune are slim," *Business Week* concluded, given the "trained, full-time auto-company engineers" on GM's in-house staff.[44]

Wadsworth Mount collided with this NIH attitude in February 1955 when he submitted two ideas to Chrysler. First, Mount suggested an adjustable rear-facing cowl on a car's hood that would allow drivers to direct hot engine air to defrost an icy windshield. Second, Mount suggested a redesigned glove compartment door that folded out to a level, horizontal surface for writing notes or holding a cup of coffee, plus an embedded mirror "for the ladies." With his letter, Mount enclosed a signed copy of Chrysler's Form 432–3, the "Submission Suggestion Agreement." Like the forms Mount had completed for the Sheaffer Pen Company, this disclosure agreement released

Chrysler from any liability associated with examining the suggestion. Two weeks later, Mount received a polite but cursory reply letter from J. R. Lemon, chairman of Chrysler's Engineering Improvements Committee. Lemon wrote that Chrysler was "familiar with both suggestions" but was not "in a position to take an active interest in them at the present time."[45]

Mount had received the NIH brush-off. He wrote Lemon a cranky reply letter and condemned Chrysler's unwillingness "to consider design changes from outsiders." One year after this disappointing exchange, Mount read an eye-opening account about Chrysler's Engineering Improvements Committee in the *Wall Street Journal*. "We got about 900 letters in February," Lemon reported, but "our main job is to keep the company out of trouble and send polite 'no thank you's' to people who send in ideas." Mount's suspicions had been confirmed. On paper, Chrysler's disclosure policy encouraged independent inventors to send in their new ideas. In reality, the automaker had institutionalized and bureaucratized the NIH attitude.[46]

Through the experiences of inventors Joseph Friedman and Wadsworth Mount, we can discern the pros, cons, and general characteristics of the "sell" strategy. Independent inventors could more easily afford, and tended to prefer, the work of *invention*—generating ideas, prototyping, and securing patents. However, many independents were unwilling to stomach the risks, costs, and complex tasks of *innovation and entrepreneurship*, such as raising venture capital, starting a firm, and presiding over manufacturing, marketing, sales, distribution, and customer support. In contrast, selling a patent to an established firm for a lump sum required minimal effort, minimal risk, and minimal expense, since these arm's-length transactions were generally negotiated via correspondence. Timing was everything; an independent like Friedman had to carefully read the changing market demand for his see-through pen patent in order to command the best price. But overall, many independents were content to specialize in the tasks of invention, extract immediate returns by selling their patents, apply those proceeds to the next invention, and leave the headaches and expense of commercialization to the acquiring firms.

There were a few risks, disadvantages, and regrets associated with the "sell" strategy. Inventors worried that unscrupulous firms might steal their ideas. They also had to contend with the "not invented here" attitude and the frustration of frequent rejections. Finally, by selling their patents, inventors relinquished control of their inventions. They also surrendered the potential for greater fame and fortune enjoyed by successful inventor-entrepreneurs and the satisfaction of seeing their own names on their inventions.

The analogous "buy" strategy also offered certain advantages for firms like the Sheaffer Pen Company. In contrast to the independents, established firms often excelled at manufacturing, sales, and distribution but sometimes struggled to generate new ideas beyond their existing product lines. By purchasing patents, firms could acquire creative new technologies to complement their in-house inventions while allowing outside inventors to assume much of the up-front risk and expense of new product development. Purchasing inventions at arm's length also allowed firms to catch up to rivals and protect themselves in competitive industries.[47] Firms developed sophisticated procedures to monitor and evaluate external inventions, protect themselves from piracy accusations, and conceal what they paid for patents.

There were also disadvantages to the corporate "buy" strategy. After relying on outside inventors during the industry's early decades, the Big Three automakers eventually became overwhelmed by the avalanche of unsolicited ideas and developed procedures to systematically reject inventors' suggestions. But the greatest disadvantage of the "buy" strategy was its inherent uncertainty. It was difficult to predict whether independent inventors could produce a steady stream of inventions in any technology sector. Firms also risked buying a "lemon" when negotiating via correspondence with unfamiliar inventors. Indeed, General Electric, AT&T, and DuPont established the first vertically integrated R&D labs precisely because they wanted to avoid the uncertainties of the external invention marketplace.[48]

Ally: Earning Royalties and Consulting Fees via Long-Term Partnerships

In a third economic strategy, independent inventors earned royalties and consulting fees by forging cooperative, long-term alliances with established firms. For inventors and firms, this "ally" model represented a middle path between the risks of the inventor-entrepreneur's start-up, the managerial structure of the corporate R&D lab, and the uncertainties of the technology marketplace.[49] To better understand this alliance model, we turn again to Wadsworth Mount.

Mount had failed to sell his ideas to the Sheaffer Pen Company, but he told patent counsel W. K. Olson that this was not his preferred way of doing business. Rather, Mount liked to "develop products that fit into a client's existing manufacturing and sales system" and then stay on as a consultant to "make it work." Working in this manner, Mount boasted, he had "succeeded in licking one problem after another" for his clients while retaining his independence. In short, Mount was neither a traditional inventor-entrepreneur nor a serial inventor who sold his patents. Rather, he was an inventor-consultant. He earned a living by contracting with about thirty-five firms over the course of four decades.[50]

Mount learned how to become an inventor-consultant while developing an antiaircraft weapon during World War II. As tensions grew around the world, the federal government established the National Inventors Council in August 1940 to solicit and evaluate the defense technologies submitted to the military by America's independent inventors (see chapter 6). While moonlighting from his job at the Commerce and Industry Association, Mount invented and prototyped a surface-to-air antiaircraft weapon consisting of a rocket-propelled grapnel hook attached to a long spool of steel cable. Ground-based soldiers could fire the projectile at a low-flying enemy airplane so that the grapnel hook and trailing metal rope disrupted the propellers and landing gear, causing a crash.[51]

Mount did not initially send his antiaircraft idea to the National Inventors Council.[52] Instead, he shared it with Neal Dow Becker, president of the Intertype Corporation. The two men had met through the Commerce and Industry Association, where Becker served as a board member and vice president. Intertype was based in Brooklyn and manufactured its eponymous typesetting machine, which competed with the Mergenthaler Linotype.[53] The specific details are unclear, but it appears that Mount approached Becker with his ordnance concept sometime in December 1941, perhaps in the days following the Pearl Harbor attack. On January 5, 1942, Mount entered into a contract to codevelop his antiaircraft idea with Intertype. Mount had impressed Intertype with his prototype, so the firm wanted him to construct some additional models and arrange for practical firing tests. If those tests were successful, Mount and Intertype would try to interest the various Allied militaries in adopting the device.[54]

The Mount-Intertype agreement was three pages long and included eight provisions. Mount assigned to Intertype all "manufacturing and selling rights in the device," including any of Mount's future modifications or improvements. Intertype would pay all costs related to the development of further models, the proposed firing tests, and, if warranted, the commercial manufacture of the device. Mount agreed to pursue both US and foreign patents on the device and any subsequent improvements, in his own name, and then assign exclusive rights to Intertype. Intertype agreed to handle all future legal expenses associated with securing, exploiting, or defending the patents. In these respects, Mount's retainer agreement was very similar to the employment contracts signed by the salaried employee-scientists at industrial research laboratories.[55]

Intertype agreed to pay Mount a 5 percent royalty up to and including $250,000 annual gross sales and a 2.5 percent royalty for receipts over $250,000. The same royalty structure applied if Intertype subcontracted with another firm to manufacture the patented items or sublicensed the patent to another firm. Intertype alone would determine the device's sales price and how best to market it. The two parties agreed to itemize and settle on expense reimbursements and royalties once a month. The agreement would remain effective for twenty years or "the life of any patent or patents" associated with the device. However, if Intertype eventually lost interest in the project, it could dissolve the contract at any time, with all patents, patent applications, design drawings, and other deliverables reassigned and returned to Mount. Thereafter, Intertype would be under no further obligations except to pay Mount anything he was owed in royalties and expenses.[56]

The agreement obligated Mount to collaborate on the "development, testing, and selling" of his antiaircraft weapon "to such extent as may be requested" by Intertype and so far as his "other obligations will permit." Within a month, Intertype came calling. With the United States now fully engaged in World War II, the firm suddenly found itself "overwhelmed in new and urgent armament work." In order to advance the ordnance project, Becker told Mount, "it will be necessary for you personally to take charge of the development work and the negotiations for introduction."[57]

Mount took his charge and ran with it. His potential to earn royalties relied on his ability to build—and sell—his antiaircraft device. Working with the Intertype engineering team, Mount built four new models, made contacts with Army and Navy ordnance officers, and arranged for a series of tests (figure 3.2) at the Army's Aberdeen Proving Grounds in 1942–1943. He applied for his first patent in February 1943 and then conducted additional tests at Florida's Eglin Field in May. In a demonstration before General Henry "Hap" Arnold, commander of the army air forces,

Figure 3.2
On December 1, 1943, inventor-consultant Wadsworth W. Mount tested his Mount-Intertype antiaircraft weapon at Aberdeen Proving Grounds in Maryland. Wadsworth W. Mount Papers, Archives Center, National Museum of American History, Smithsonian Institution, Washington, DC.

the Mount-Intertype antiaircraft weapon scored direct hits on seven out of seven attempts and completely demolished one drone.[58]

In June 1943, at the urging of the army and navy, Mount quit his job at the Commerce and Industry Association to work on his antiaircraft weapon full-time. In a letter to Mount's local draft board, Intertype vice president A. T. Mann vouched that "the work of designing, building, testing, and operating this weapon has been handled almost entirely by Mr. Mount" and that he was "the only person who can carry it on at the present time."[59] Mount received his deferment, earned two more patents in 1944 and 1945, and eventually sold thirty-seven arms contracts to the different branches of the military. In 1946, Mount received the Naval Ordnance Development Award for his wartime work.[60]

Intertype, via its contract with Mount, essentially outsourced the invention, development, and sales of a new product to an inventor-consultant. Mount—an independent inventor empowered by his alliance partner— acted as a semiautonomous agent for Intertype as he invented, codeveloped, and sold the Mount-Intertype device to the military. Mount's extant personal papers do not reveal his annual royalty receipts from Intertype. However, the alliance must have been mutually beneficial, since Intertype maintained it for thirteen years, long after the conclusion of World War II and the Korean Conflict. Intertype eventually terminated the agreement in 1955 and reassigned the ordnance patents back to Mount.[61]

Even while Mount was aligned with Intertype, he was still an independent inventor-consultant. As a free agent, he could—and did—work with multiple clients simultaneously. Like freelancers in any field, Mount had to hustle and network to generate new business. For example, in 1952 Mount sought consulting work from the vice president of the Roebling Company, which had supplied Intertype with the cable for his antiaircraft device. When soliciting new business, Mount advertised himself as "essentially an inventor . . . set up to work on a consulting basis" on "any special jobs on which my services might be of value." He included a résumé that read "Wadsworth W. Mount—Inventions to order for clients."[62]

Mount adopted and modified the language of the 1942 Intertype contract as his own standard retainer agreement for potential clients. First, Mount included flexible language that permitted him to develop multiple inventions with a client. Second, some new language outlined the two-way exchange of technical and marketing materials pertinent to each worthwhile idea. Mount offered his clients all relevant nonproprietary information from his independent inventive work and prior consulting engagements. In return, Mount would gain access to the client's own "records, manufacturing drawings, and other pertinent data" related to the project.

Third, Mount adopted similar royalty terms—5 percent on sales up to $1 million, 3 percent thereafter—but added provisions for a daily consulting fee. To the extent the client requested his services—and his other obligations permitted—Mount would assist the client in development work at the rate of $100 per day, half of which would be credited as advanced royalties on the future sales of Mount's patented items. The client also agreed to reimburse Mount's travel costs and other incidental expenses, provisions that were implied, but not explicit, in the original Intertype agreement. Finally, new language allowed Mount—not just the client—to remove an individual invention from the agreement or dissolve the contract entirely with sixty days' written notice, with all IP reassigned back to Mount. Collectively, these new provisions provided Mount with a more predictable, guaranteed income, better access to his clients' technical data, and more autonomy in his engagements.[63]

Mount subsequently entered into variations of this standard retainer agreement with several different firms, each of which modified the terms according to its needs. In 1950, while still allied with Intertype, Mount entered into a contract with Merriman Brothers, Inc., a Boston maritime equipment maker. Merriman hoped to adapt Mount's antiaircraft weapon into a maritime rescue system: a rocket-propelled flotation ring attached to a long cable. Mount helped Merriman negotiate a sublicense on his Intertype-assigned patents and agreed to help the firm develop, manufacture, and sell the concept. Merriman paid Mount on a graduated commission scale—10 percent of net sales up to $200,000, 5 percent between $200,000 and $500,000, and 4 percent for sales above $500,000—plus a $50 per day consulting fee as an advance against future commissions. Mount left no record of his project-related income, although Merriman later apologized that the deal did not "produce better results." Nevertheless, the parties remained allied for nineteen years before Merriman dissolved the agreement in 1969.[64]

Mount's alliance with the Raymond Concrete Pile Company was more of a pure consulting arrangement. Raymond was uninterested in developing any of Mount's prior inventions. Instead, the firm hired him to help solve their long-standing technical problems in the field of concrete piles, foundations, and cores. In lieu of a twenty-year agreement, Mount signed a one-year contract with Raymond in 1956, followed by a series of annual extensions. Mount observed Raymond's operations and submitted ideas for new inventions and improvements to Raymond's chief engineer, G. W. Bailey. If the ideas were useful, then Mount would help Raymond obtain the US and foreign patents, but that is where his work ended. Mount left the subsequent commercialization of those ideas to Raymond's engineering staff. There were no assignment or royalty clauses in the Raymond contract. Instead, as a pure consultant, Mount worked on a day-to-day basis, at the rate of $100 per day plus expenses. Between 1956 and 1964, Mount submitted approximately forty ideas. Raymond developed some of them and periodically reassigned others back to Mount per the stipulations in the contract.[65]

This alliance model held several advantages for independent inventors.[66] First, an independent like Mount could gain financial, legal, and material assistance from his partner firms and still retain his independence and autonomy. Firms like Intertype, Merriman, and Raymond allowed Mount to use their tools and workspaces, shared their technical information, paid all patenting costs, and generally covered all financial risks and up-front expenses of new product development. Second, Mount was not just selling his inventions; he was also selling his services. Under a standard licensing agreement (like Mount's first Intertype deal), independents had to wait months or even years before a product hit the market, and they earned royalties only if the product was a success. However, by selling his technical expertise as an inventor-consultant, Mount immediately earned a steady income of $50 or $100 per day while awaiting additional royalties. Third, Mount could tailor the deal to suit his clients; for example, he focused on generating ideas for Raymond but handled the full spectrum of invention, development, and commercialization for Intertype. Finally, his agreements were not exclusive. Mount contracted with dozens of firms over the course of his career, often working with multiple clients simultaneously.

Mount's clients also reaped several advantages from this alliance model. By outsourcing certain inventions, firms could transform R&D from a fixed cost into a variable one. With a vertically integrated laboratory, firms had to continually pay overhead for the lab itself plus the salaries of their employee-scientists, even if their ideas flopped. However, under retainer agreements, firms could pay a daily consulting fee (as needed) for an inventor-consultant's technical services. Plus, with patent licensing agreements, firms only paid additional royalties if the new product succeeded, since payments were computed as a percentage of actual sales. These payments were amortized over the life of the patent and ceased altogether when the patent expired. Finally, these contracts gave firms maximum flexibility. They could hire Mount to do a little or a lot, and if the arrangement did not work out, they could dissolve the agreement entirely with only sixty days' notice.

There were a few disadvantages to the alliance model. As with the "sell" strategy, independents assigned and licensed their patents to their clients, so they usually relinquished the satisfaction of seeing their own names associated with the end products. And while they earned a steady per diem and royalties, inventor-consultants abandoned the potential of the inventor-entrepreneur's bigger profits. For firms, the alliance model did not completely eliminate the uncertainties of licensing outsiders' patents, but it built in some helpful safeguards. Unlike a onetime patent purchase (in which all sales were final), Raymond could closely evaluate Mount's forty ideas and reassign any that were not promising. Similarly, firms might occasionally work with an idiosyncratic independent, but the daily interactions and long-term nature of these consulting agreements helped generate trust between the two parties. If an inventor remained incorrigible, the firm could simply terminate the agreement.

Mount's alliance strategy provided a third moneymaking alternative to running an entrepreneurial start-up or collecting lump-sum cash payments from patent sales in the technology marketplace. To maximize their chances for success, both independent inventors and established firms pursued multiple economic strategies simultaneously.

Mixed, Hybrid Innovation Strategies

During their careers, Joseph Friedman and Wadsworth Mount chose different economic strategies (i.e., make, sell, or ally) at different times as they encountered different situations. To choose the best strategy for a *single invention*, both men evaluated the specific technology in question, its potential market, and the resources and skills they could bring to bear on its commercialization. Looking across a *portfolio of inventions*, both Friedman and Mount adopted a hybrid mix of strategies while working simultaneously with a variety of firms.

Analogously, the vertically integrated R&D lab was not a firm's only path to innovation. Just like independent inventors, established firms could choose to make, buy, or ally, and they often pursued these different innovation strategies simultaneously. The long-term alliance between inventor Samuel Ruben and the P. R. Mallory Company provides a rich example of this mixed, hybrid approach.[67]

Samuel Ruben (1900–1988) was born in Harrison, New Jersey, and grew up in New York City. In his youth, Ruben performed his own self-directed electrical and chemical experiments and tinkered with radio sets, using available spare parts. During World War I, at age seventeen, Ruben worked as a technician with the Electrochemical Products Company, a Brooklyn start-up that experimented with high-frequency electrical discharges to extract atmospheric nitrogen for use in wartime munitions. The process never worked and the company floundered, but Ruben earned the respect of project consultant Bergen Davis, a physics professor at Columbia University. Ruben never earned a bachelor's degree, but he learned the basics of science and engineering by borrowing books from Davis's personal library and attending his lectures. In 1923, Davis convinced the start-up's primary investor, patent attorney Malcolm Clephane, to finance a private laboratory for Ruben using the equipment from the failed nitrogen project, first in lower Manhattan and then, after 1930, in New Rochelle, New York. Clephane paid Ruben a small stipend and covered all expenses in exchange for 50 percent of any future royalties.[68]

A serendipitous exchange helped Ruben strike his first licensing deal. In 1925, Ruben sent an assistant to Mallory's plant in Weehawken, New Jersey, to purchase some tungsten wire for an experiment. Philip Rogers Mallory had founded the company in 1916 as a supplier of tungsten filaments to General Electric and other light bulb manufacturers. By 1925, the firm was beginning to expand its product line to manufacture the resistors, capacitors, timers, and other components embedded in other companies' products.[69] Through the visiting assistant, Mallory's engineering staff learned about Ruben's new invention, a solid-state AC-to-DC rectifier, and arranged a meeting to negotiate a deal. Ruben granted Mallory an exclusive license on the rectifier for use in a new battery charger; in exchange, Ruben received advance royalty payments of $1,200 per month for six months while serving as Mallory's consultant. This arrangement was typical throughout Ruben's career. In lieu of manufacturing his inventions as an inventor-entrepreneur, Ruben preferred to work as an inventor-consultant. He licensed his inventions, collected royalties and retainer fees, and helped his clients solve their production problems, while developing new applications for his inventions.[70]

Ruben (figure 3.3) worked frequently, but not exclusively, with Mallory. In 1928, for example, Ruben licensed his quick-heating amplifier tube to

Figure 3.3
Inventor Samuel Ruben forged a long-term partnership with the P. R. Mallory Company. Samuel Ruben Photograph Collection, Science History Institute, Philadelphia, PA.

the Arcturus Radio Tube Company and subsequently visited their Harrison, New Jersey, lab "once a week as a consultant at a high fee" for the next two years. In fact, Ruben consulted with dozens of other electrical and radio firms, including Sprague, Emerson, and Grigsby-Grunow, even working with direct competitors in the same industry, such as battery makers Mallory and Ray-O-Vac. As an inventor-consultant, Ruben had no particular loyalties; in this respect, he was truly independent. However, Ruben enjoyed working with Mallory. He later recalled that their initial 1925 agreement was "the beginning of a long relationship with Philip Mallory, for I would be licensor to his companies both here and abroad on most of my inventions."[71]

Similarly, Mallory and its founder had a long history of working with independent inventors. Besides Ruben, his "prize example," Philip Rogers Mallory proudly noted the firm's collaborations with other inventors. During the 1930s, Paul Ware approached Mallory with a variable inductance television tuner that the company manufactured and sold to DuMont, Crosley, and other TV manufacturers. Mallory also worked with inventor Fred Hooven, who developed an interval bomb timer, a bomb release mechanism, and a bomb rack selector, inventions that became standard equipment on most American bombers during World War II.[72] Over the years, Mallory collaborated with a dozen other independent inventors whose inventions were not promising enough to add to the company's product line.[73] Nevertheless, independent inventors had a receptive audience at Mallory and remained an important part of the firm's innovation strategy.

Mallory and Ruben strengthened their relationship considerably during and after World War II. In 1942, the National Inventors Council asked Ruben to invent an improved battery that could withstand the heat and humidity of the Pacific theater. Ruben licensed the miniature mercury battery to Mallory, which supplied the batteries in great quantities to the Army Signal Corps for use in portable walkie-talkies (see chapter 6). Over the next fifteen years, Ruben and Mallory adapted these "button cells" for use in commercial applications, such as hearing aids, pacemakers, and electric watches. In 1961, Mallory introduced another Ruben invention: a new line of "copper top" alkaline batteries in standard AAA, AA, C, and D sizes for use in radios, flashlights, and cameras. In 1965, Mallory rebranded its booming battery business and evolved into the multibillion-dollar company we know today as Duracell.[74]

As Ruben and Mallory forged an alliance to commercialize Ruben's batteries, the firm began to invest in its own industrial research laboratories. After Bell Labs announced its transistor in 1948, Mallory formed its Central Research Division in 1953 at its Indianapolis headquarters to conduct research in solid-state physics that might lead to new components. In 1962, Mallory opened a new Laboratory for Physical Science in Burlington, Massachusetts. This new research center employed scientists from many disciplines to pursue "fundamental research in such areas as electrochemistry, materials, thin-film technology, and solid-state physics." Before long, Philip Mallory could proudly report that his employee-scientists had received twenty-three patents in 1965 and had filed an additional seventy-seven patent applications.[75]

Between 1951 and 1974, Mallory consistently increased its internal R&D investments, which ranged from 3 to 4 percent of annual sales. However, in 1966, Philip Mallory reiterated his firm's ongoing and valuable collaborations with the "creative inventors outside our own ranks." As a firm, Mallory was eclectic in its approach to innovation and rejected the "not invented here" attitude. Instead, Ruben reported, the firm was "never afraid to take on new products or to consider new ideas whether they emanated from within or from outside the Mallory Company."[76]

Mallory made a significant long-term commitment to Ruben when it relocated its battery operations in May 1946 from Indianapolis to Tarrytown, New York, just twenty miles from Ruben's laboratories in New Rochelle. This move, Philip Mallory explained, was explicitly undertaken to be "close to the inventor's laboratory."[77] In 1963, Mallory unveiled a new and improved Tarrytown Battery Research Center and dedicated the lab to Ruben in a public ceremony. Dr. Emanuel Piore, vice president of research and engineering at IBM, honored Ruben in a speech titled "The Independent Inventor in the Contemporary World." Piore noted that Ruben's inventions were "by one man who was without the resources represented by our large industrial and university laboratories," who nevertheless proved that "contrary to popular opinion, an individual can survive and be great in spite of big science, big engineering, and big industry."[78] Piore, a PhD physicist and director of IBM's expansive research operations, represented one end of the innovation spectrum. Ruben, the self-educated independent inventor, represented the other. Mallory's mixed innovation strategy occupied a position somewhere in the middle, embodied in its long-term relationship with Ruben and in an industrial research lab dedicated in his honor.

Altogether, Ruben partnered with Mallory and Duracell for fifty-eight years, from 1925 to 1983.[79] Ruben and Mallory's long-term relationship represents an extraordinary example of the alliance strategy and was a critical component of the firm's mixed, hybrid approach to innovation.[80] Certainly, an independent inventor like Ruben was free to work with multiple firms (e.g., Mallory, Arcturus, Sprague, Emerson), just as a firm like Mallory was free to contract with several different inventors (e.g., Ruben, Ware, Hooven). However, by choosing to work again and again with the same trusted ally, both Ruben and Mallory brought an additional degree of familiarity and certainty to an innovation process that was fundamentally unpredictable. Ruben benefited from the steady income, material resources, and manufacturing expertise supplied by his client, while Mallory benefited from Ruben's continuous supply of high-quality inventions and his skill at adapting them to new applications. Mallory eventually made concurrent investments in its own vertically integrated R&D labs, but the firm never abandoned its relationship with Ruben and in fact moved some of its research facilities from Indianapolis to New York to improve their coordination. With Mallory's mixed innovation strategy, the locus of invention resided both inside and outside the firm and crossed its permeable organizational boundaries.

Itinerant Careers and Permeable Organizational Boundaries

Some firms like Mallory invested simultaneously in multiyear inventor alliances and in their own R&D labs. Did these corporations try to hire inventor-consultants into their firms as salaried employee-inventors? And, if so, what were the considerations as inventors decided whether to "go corporate" or remain independent?

Historians have documented how the largest corporations occasionally hired independent inventors into their ranks. According to George Wise, General Electric president Charles Coffin recognized that hiring "outstanding technologists was preferable to merely buying or capturing their companies and leaving them free to compete again." Women and minority inventors were generally unwelcome within corporations like GE; however, Coffin hired White male inventors such as James J. Wood, Charles Vandepoele, Charles S. Bradley, and Walter H. Knight by making them "offers they couldn't refuse."[81] According to David Noble, individual inventors "flocked to corporate employment" when the dominant influence of big firms in certain industries made it difficult for the independents to compete. The "frustration of independent invention led the majority of inventors into the research laboratories of the large corporations," Noble argued, where they would be spared "the hardships of going it alone."[82] Most contemporary observers certainly believed that the corporate R&D labs had absorbed the independents into their ranks or wiped them out entirely (chapter 2).

But working inside a big corporation could be just as frustrating as competing against one. Employee-inventors admittedly enjoyed certain advantages, such as a steady paycheck and access to a firm's superior material, financial, and legal resources. However, employee-inventors often had little autonomy; they were told which problems to pursue and were required to assign their inventions to their employers. Even a prominent research director like Kodak's C. E. Kenneth Mees admitted that "many research workers are dissatisfied with the conditions attaching to their profession."[83]

There is no empirical data—only anecdotal evidence—on the movement of independent inventors into and out of employment. Certainly, some inventors were hired into the ranks of corporate R&D labs. However, other inventors chose to *migrate out* of industrial laboratories to attain or regain their independence. In fact, independent inventors often led itinerant, peripatetic careers. They alternated repeatedly between employment and independence and made multiple excursions across the permeable organizational boundaries of established firms.

For example, radio pioneer Lee De Forest (1873–1961) gained important experience as an employee-inventor at two firms before resigning in frustration to begin his career as an independent inventor. De Forest initially tried to find employment in the labs of other independent inventors as he completed his doctoral studies at Yale's Sheffield Scientific School. "I am writing for employment to Chas. Brush," De Forest wrote in a February 1899 diary entry, "and always hoping against hope for Tesla!" When those connections never materialized, De Forest took a job in the dynamo department at Western Electric, the Chicago-based equipment supplier for the Bell Telephone companies and AT&T. De Forest's diary entries record his frustration: "I work . . . from 7 a.m. to 5:15 with ¾ hr. for lunch at a lunch counter. I am learning a little bit, not very much of use—too much chasing of parts and mopping of grease—and all for $8 per week! I feel it is not all I might be worth in the proper place, & am sometimes discouraged, always impatient."

De Forest yearned to pursue his "cherished experimental work," and in October he received a requested transfer to Western Electric's Experimental Laboratory. "I can learn more here" in the presence of "refined and trained gentlemen," De Forest wrote in his diary, "Here, if ever, can I invent & have the ideas count for something." De Forest spent his days improving telephone cables at Western Electric and his evenings making "plans for wireless telegraphy."[84]

However, Western Electric never showed interest in De Forest's wireless experiments, so he joined the American Wireless Telegraph Company of Milwaukee in search of "more opportunities for recognition." However, when his wireless receiver experiments showed promise, De Forest balked at assigning his developments to the firm. "It is my invention," he told American Wireless's owner, and "I will not let it go into the hands of any company until that company is my own." De Forest resigned, returned to Chicago, worked part-time as an editor at the *Western Electrician*, and continued his independent wireless experiments with backing from his friend Ed Smythe. In December 1900, De Forest "crossed the Rubicon," quit his editor's job, and pursued invention full-time. In 1906, he invented the three-element "audion" vacuum tube, which could transmit, receive, and amplify wired and wireless electronic signals.[85]

Despite his earlier dissatisfaction as an employee-inventor at Western Electric, De Forest eventually entered into negotiations with its parent company; he sold his audion patent to AT&T in 1913 for $50,000. After AT&T established its own R&D lab in 1911, the firm continued to acquire and license the patents of outside inventors as part of a mixed innovation strategy. Using De Forest's audion as an amplifier alongside the patented loading coils acquired in 1900 from independent inventor Michael Pupin, AT&T achieved the holy grail of transcontinental, long-distance telephone service in 1915. AT&T's possession of the De Forest audion patents also helped the firm control the development of wireless (radio), which threatened the firm's massive investments in wires.[86]

Similarly, inventor Everett Bickley (1888–1972) alternated repeatedly between independence and salaried industrial research positions before licensing his most important invention to a former employer. Bickley graduated from Pittsburgh's Carnegie Institute of Technology in 1910 and took a job as the production manager for Detroit's EMF Automobile Company (later renamed Studebaker). While moonlighting from EMF, Bickley invented the "motograph," a now familiar electrical sign that displayed moving text on the sides of buildings and dirigibles. He founded the Bickley Manufacturing Company in 1912 to exploit the motograph and a few other inventions, but his ventures apparently earned little money. In 1914, Bickley joined Pittsburgh's H. J. Heinz Company as an efficiency engineer and eventually became the firm's chief engineer. As an employee-inventor, Bickley improved a can conveying system and invented an automated pickle-slicing machine that replaced twenty-seven workers. He also observed how hundreds of low-paid women stood before a moving conveyor belt of beans to remove stones, dirt, and rejects by hand.[87]

In 1918, Bickley resigned from Heinz because the firm "would not pay [him] adequately" considering the money he had saved them. He moved his family to the Philadelphia suburbs and became a home builder and real estate investor. He soon began conducting experiments on an automated bean sorter. Bickley's patented sorting machine used photoelectric cells to optically inspect the color of beans passing by on a conveyor; an attached vacuum tube sucked up and discarded any rejects or debris. When the real estate market dried up with the 1929 stock market crash, Bickley decided to demonstrate his bean sorter (figure 3.4) to Heinz, his former employer. He spent the next two years driving between Philadelphia and Pittsburgh, testing and improving his prototype as the eponymous inventor-entrepreneur of the revived Bickley Manufacturing Company. Meanwhile, Bickley simultaneously took positions as a salaried industrial researcher, first at the Radio

Figure 3.4
Inventor Everett Bickley standing outside his garage with his automated bean-sorting machine, circa 1931–1932. Everett H. Bickley Collection, Archives Center, National Museum of American History, Smithsonian Institution, Washington, DC.

Corporation of America (RCA) in nearby Camden, New Jersey, and then at Philadelphia's Philco Radio Corporation. With these jobs, Bickley earned a steady paycheck while learning more about electrical circuitry.[88]

Around 1931 or 1932, Heinz ordered four automated sorters and loaned Bickley the necessary capital to construct the machines as an advance on a leasing fee of $250 per machine per month. Bickley quit his job at Philco, built Heinz's sorting machines in his Philadelphia basement and garage, and remained an independent inventor-entrepreneur for the remainder of his career. He subsequently adapted his machine to sort different varieties of beans, rice, peanuts, coffee beans, and ball bearings. Over the next several years, he built, installed, and maintained ninety-three sorting machines for twenty-five firms, including thirty-one machines for Heinz.[89] During his itinerant career, Bickley alternated repeatedly between salaried employment as an industrial researcher (at EMF, Heinz, RCA, and Philco) and his work as the independent inventor-entrepreneur behind the Bickley Manufacturing Company. His greatest financial success came when he allied with Heinz, his former employer.

Similarly, Earl Tupper (1907–1983) pursued a range of employment options before developing "Tupperware" plastic containers as an independent inventor-entrepreneur. Tupper moonlighted as an inventor while working for his family's Massachusetts farm and greenhouse and later as a tree surgeon.[90] During the 1930s, Tupper invented a series of consumer novelty items, such as a "no-drip" ice cream cone, a combination tobacco case and pipe filler, and a comb affixed with a miniature mirror. Tupper also patented, but was unable to commercialize, an accessory that converted an automobile's rumble seat into covered trunk storage.[91]

In 1937, Tupper approached the DuPont Viscoloid Company factory in nearby Leominster, Massachusetts, to gauge their interest in manufacturing his combs and other novelties. The "Doyle Works" had pioneered the use of Viscoloid pyroxylin plastic in combs, toothbrushes, and toys. As Tupper recalled in his diary, he was working with the factory's staff to prototype a plastic watch case when DuPont offered him a job: "Mr. Dorr said I could begin work as soon as I wished at 60¢ to start. He said I'd get paid about 10% on any ways I found for the company to save money or make money. If the idea is meritorious, I get a batch of DuPont stock in addition. If I get up a patentable item they'll finance patent in my name assigned to DuPont.

And pay me a 2 to 5 percent royalty. . . . I passed my physical exam and start work Monday at 8:00am."[92]

In this mutually beneficial arrangement, Tupper gained access to DuPont's raw materials and factory equipment, which he was free to use on his own unpaid time after completing his assigned DuPont projects. DuPont, in turn, gained access to Tupper's ideas for new novelty products. Tupper clearly saw the position as a source of steady income during the Great Depression and a stepping-stone toward his own business. Recalling his first day as DuPont's "Class A sample maker and designer," Tupper noted that "the job is easy, and if it will be steady, so as to give me a dependable income, I'll be able to get a novelty business of my own going."[93]

In 1938, Tupper founded the Earl S. Tupper Company to sell his plastic novelties and advertise his freelance services as an "Industrial Inventor-Designer: Custom Inventing, Designing, and Manufacturing."[94] However, he maintained a tight connection with DuPont, which helped him learn more about the plastics trade and hone his skills at casting dies and molding new shapes. For example, Tupper molded plastic parts for gas masks and navy signal lamps while working as DuPont's subcontractor during World War II. Around 1942, DuPont provided Tupper with samples of polyethylene slag, a waste product from oil refinement. Tupper independently developed a chemical process to purify the slag into "poly-T," a flexible, durable, water-resistant plastic that could be manufactured in several different colors. Tupper saw the material's potential for consumer applications; he eventually patented his "burp" seal, an airtight lid for his Tupperware food containers. After experiencing disappointing sales at hardware and department stores, Tupper learned that veteran saleswoman Brownie Wise had enjoyed brisk sales by organizing in-home demonstrations. In 1951, Tupper recruited Wise to implement the home-party strategy nationwide, and sales took off.[95] Along the way, Tupper maintained a long and constantly evolving relationship with DuPont—as an employee, a subcontractor, and eventually as a bulk customer for the polyethylene slag he converted into Tupperware.

As the careers of Lee De Forest, Everett Bickley, and Earl Tupper demonstrate, large technical firms such as Western Electric, H. J. Heinz, and DuPont sometimes absorbed White male inventors into their ranks as salaried employee-inventors, at least temporarily. However, working as an inventor often meant alternating between periods of employment and

independence. Moonlighting inventors often supported themselves by taking jobs in related industries. These salaried positions provided employee-
inventors with a steady paycheck, an insider's knowledge of high-tech
industries, and access to machinery, production methods, and materials.
When they grew frustrated with their working conditions or balked at
assigning their inventions, ambitious employee-inventors often migrated
out of corporations to establish themselves as independent inventors and
entrepreneurs.[96] However, independents frequently sold and licensed their
inventions to their former employers or otherwise remained connected as
customers or contractors. In short, independent inventors led itinerant,
peripatetic careers as they migrated into and out of firms and forged cooperative alliances.

Conclusion

Independent inventors learned to adapt and survive in an era of corporate
R&D. To earn a living, they evaluated and executed a variety of commercial strategies. Inventors could either make, sell, or ally; that is, found a
start-up to manufacture and market an invention, assign the patent to an
established firm for a onetime payment, or license the patent to a manufacturing partner to earn royalties and consulting fees. Corporations faced
an analogous set of choices (make, buy, or ally) as they procured new technologies. Despite their disparaging rhetoric, corporations and their R&D
labs continued to purchase and license patents from individual inventors
to complement the inventions they developed internally.

  Independent inventors considered a variety of factors while weighing
their options for commercialization. Founding a start-up to manufacture
a new invention entailed significant costs and risks. However, inventor-
entrepreneurs retained their patents, their autonomy, and the potential for
substantial profits if they introduced a successful new product. Selling and
licensing patents to established firms allowed independents to specialize
in the act of invention while avoiding the uncertainties of entrepreneurship. However, by partnering with corporations to bring their inventions
to market, inventors relinquished the potential for huge profits and the
satisfaction of seeing their own names on commercial products. Inventors
were more likely to manufacture breakthrough inventions in emerging

industries with few competitors but sell or license patents in mature industries where corporations had already established a market.

To maximize their chances for success, independent inventors often pursued several of these strategies simultaneously in a mixed, hybrid approach to innovation. During the early twentieth century, women and minority inventors were generally unwelcome within corporations; however; itinerant White male inventors sometimes migrated into and out of established firms as they alternated between periods of employment and independence. When independent inventors and firms forged cooperative alliances, innovation occurred inside, outside, and across those firms' permeable organizational boundaries.

"People tend to think inventing is done by either lone individuals or corporations," observed historian George Wise, "In truth, it is as often a complex tangle involving both."[97] Corporate R&D labs did not wipe out independent inventors or absorb them permanently into their ranks. Rather, in their symbiotic relationship, inventors relied on the patronage of established firms for their economic survival, while corporations relied on outside, independent inventors as a crucial source of expertise and new inventions. To facilitate these interactions, inventors established a series of professional organizations to showcase their inventions, meet potential collaborators, and advance their interests.

4 Banding Together

It seems like we never will accomplish anything toward an organization of inventors.
—Independent inventor R. A. Summers, 1918

In 1907, a New York mathematician named G. W. Wishard wrote a letter to *Scientific American* and suggested that "some multi-millionaire should found a great institution to aid the worthy inventor." Any member could present an idea to a board of examiners. If the idea was promising, the institution would furnish the inventor with access to tools, materials, and its staff of mechanics to perfect the invention, plus an attorney to obtain a patent. The proposed institution would help its members exhibit their inventions at fairs and expositions with an eye toward licensing their patents to business partners who could manufacture and sell the goods. After paying the member his share of any profits, the institution would reinvest its share into libraries and evening schools so that its member-inventors would become "more intelligent, do better work, and produce still greater inventions." Wishard believed that these services "would greatly benefit the inventors, the institution, the manufacturers and the public, and wonderfully accelerate human progress."[1]

Wishard's imagined millionaire never came forward. Instead, independent inventors built their own ambitious organizations to confront the challenges they faced at the turn of the twentieth century. Individual inventors formed associations of like-minded colleagues who helped them evaluate, showcase, and commercialize their inventions. Independents also established political advocacy groups to reform an overwhelmed and underfunded patent system that was prone to the abuses of wealthy corporations. Women inventors, African American inventors, and eccentric independents each built supportive communities to garner recognition and respect from a prejudiced and skeptical public. Although independent inventors felt imperiled by these challenges, they were not helpless victims or passive bystanders. Rather, they developed a group consciousness, asserted their collective identity, and formed a series of informal communities and formal membership organizations to address their commercial, political, and communal needs.

Unfortunately, inventors' eagerness to join these kinds of groups made them vulnerable to exploitation by unscrupulous scammers. *Scientific American*, *Popular Mechanics*, and other magazines regularly warned readers to beware of predatory organizations that laid "traps for the unwary inventor."[2] These scams engendered fear, disillusionment, and distrust among inventors. As a result, even legitimate organizations struggled to retain enough members to stay in business and provide the practical support that inventors craved.

Scholars have generally paid little attention to inventors' professional organizations, especially when compared with the extensive literature devoted to American scientific and engineering societies.[3] Like their kindred scientists and engineers, American independent inventors banded together around the turn of the twentieth century to form a recognizable professional community. However, inventors' organizations were extremely fragile and short-lived. Each inventors' organization dissolved under its own specific circumstances, but the practical effect was always the same. Without the benefit of stable and durable professional organizations, independent inventors struggled to commercialize their inventions, achieve political reforms, and maintain their intellectual identity and authority in the face of competition from corporate scientists and engineers. When their professional organizations failed, independent inventors were left with fewer pathways for confronting the many challenges they faced at the turn of the twentieth century.

Banding Together for Commercial Support

"After an inventor has secured his patent, what is the next step for him to take in order to introduce it to the public," inventor "G. E." asked the editors of *Popular Mechanics*, "and at the same time get some slight returns out of its sale for himself?" Independent inventors hungered for this kind of practical advice. The editors of *Electrical Experimenter* apologized for being four months behind on their regular "Patent Advice" column, since they could print answers to only ten queries each month. Model makers and patent agencies wisely ran their ads alongside regular advice columns such as "Hints for Inventors," and offered to "assist inventors in perfecting their inventions, and in selling their patents."[4]

Independent inventors have never lacked ideas. Rather, they have traditionally struggled to assemble a network of technical, legal, and business partners to help them with commercialization. Inventors have often wondered: Is my invention any good? How can I learn more about the state of this particular art? How do I get a patent? How can I get my idea in front of financiers and manufacturers? Most of all, inventors searched for honest and trustworthy guides who were not trying to swindle them or steal their ideas.

Beyond the popular magazines and patent agents, American inventors sought this kind of advice in a handful of local inventors' leagues, unions, and associations that emerged in New York, Boston, and Baltimore during the nineteenth century. These groups—such as the New England Association of Inventors and Patrons of Useful Arts and the Inventors' National Institute of Baltimore—provided forums where local inventors could demonstrate their new technologies, gather advice from fellow inventors, and make connections with patent attorneys, financial backers, and manufacturers.[5]

Philadelphia's Franklin Institute remains the oldest and most venerable of these nineteenth-century technical institutions.[6] It was founded in 1824, at a time when the United States had enjoyed a half century of political independence but remained economically dependent on Great Britain for imported technologies and manufactured goods. To stimulate homegrown American industries, the charter members of the Franklin Institute incorporated it for "the Promotion and Encouragement of Manufactures and the Mechanic and Useful Arts." Its founders were Philadelphians Samuel Vaughan Merrick, owner of a small manufacturing business, and William H. Keating, a chemistry professor at the University of Pennsylvania. It was a hybrid of two established associational formats—the mechanics' institute, which appealed to artisans and inventors, and the salon or philosophical society, which appealed to a more elite, learned class of patrons. The egalitarian spirit of the Franklin Institute would have pleased its namesake, Benjamin Franklin, and was reflected in its motto: "Science with Practice; Practice with Science." It proposed to offer public lectures on the arts and sciences, establish a library and reading room, examine and evaluate new inventions, host exhibitions and award medals, and publish a journal.[7] The annual dues were $8 in 1895, which attracted a wide spectrum of inventors, mechanics, industrialists, and scholars, who took advantage of the institute's many services.[8] By the institute's centennial in 1924, it boasted 1,378 members, including 563 members from outside the Philadelphia area.[9]

In an era before the widespread establishment of formal universities in the United States, the Franklin Institute offered remarkable opportunities for technical education. It opened a School of Mechanical Drawing in 1824 and soon supplemented its drafting classes with lessons in chemistry, physics, machine design, naval architecture, and mathematics. It also hosted weekly Friday night lectures, where members and invited scientists and inventors presented their new discoveries and inventions. These lectures were reprinted and disseminated widely via the *Journal of the Franklin Institute*.[10]

Inventors, patent attorneys, and manufacturers also took advantage of the Franklin Institute's library, arguably the best technical library in the United States. By 1888, the library subscribed to 200 technical periodicals and, through an arrangement with the US Patent Office, held a nearly complete collection of reprinted US and foreign patents. Through the late 1800s, the library was free and open to nonmembers and attracted 150 readers per day. The Franklin Institute did not help its members obtain patents. However, through its vocational courses, lectures, journal, and library, inventors could educate themselves on the state of the various arts.[11]

The Franklin Institute also offered free evaluations of new inventions through its Committee on Science and the Arts (CSA). America's inventors enthusiastically sent their written descriptions and models to Philadelphia. Between 1824 and 1900, the CSA formed 2,200 subcommittees to investigate new inventions and issued detailed reports on 1,600 cases. A who's who of nineteenth-century inventors submitted their ideas to the CSA, including Thomas Edison, Nikola Tesla, George Westinghouse, and Rudolf Diesel.[12] As institute secretary William Wahl observed, the sixty-member CSA was a "competent, trustworthy and impartial body" that provided inventors with an informed opinion on "the usefulness of their inventions." If an idea was impractical or already in widespread use, the inventor received a polite but unfavorable report that saved him time, money, and unnecessary embarrassment. However, promising inventions received laudatory committee reports that were subsequently published in the *Journal of the Franklin Institute*. The most meritorious inventions also received medals and cash prizes. Inventors leveraged the CSA's trusted seal of approval to attract investors and business partners. As *Scientific American* observed in 1920, the CSA's published reports drew the attention of "every large corporation in the world," so that the majority of recommended inventions had "come into wide commercial usage within a relatively short time."[13]

The Franklin Institute also organized technology fairs and exhibitions to help inventors showcase their inventions. In October 1824, it hosted the first exhibition of American manufactures held in the United States, and it continued to host fairs either annually or biennially until 1858 and then less frequently through 1899. These exhibitions provided a forum for connecting inventors, manufacturers, and consumers, so they helped stimulate the development of new technology-based industries. At the fairs, the institute's judges awarded medals and cash prizes to outstanding inventions. Like the CSA's awards, these exhibition prizes bolstered the reputations of award-winning inventors, who often reproduced renderings of the Franklin Institute medals on their letterhead, product packaging, and printed advertisements.[14] The exhibitions also provided tremendous mass-marketing opportunities in an era before national magazines and broadcast media. The institute's fiftieth-anniversary exhibition in 1874 attracted 1,251 exhibitors and 267,638 paying visitors and set the stage for Philadelphia's successful 1876 Centennial Exhibition. The institute's International Electrical Exhibition in 1884 drew nearly 300,000 visitors to exhibits by Thomas Edison, Alexander Graham Bell, Charles Brush, and Franklin Institute member Elihu Thomson.[15]

Thomson's career underscores the Franklin Institute's significant influence among American independent inventors. Thomson was born in Manchester, England, in 1853 and immigrated to Philadelphia with his family in 1858. He attended Central High School, where his professors—including future business partner Edwin J. Houston—encouraged his explorations of chemistry, photography, and electricity. Following his graduation in 1870, Thomson taught chemistry at his alma mater and conducted electrical experiments at home in his attic laboratory. With Houston's nomination, Thomson joined the Franklin Institute in June 1874. His membership helped him gain expertise in the electrical field, attain standing in Philadelphia's technical community, and forge the personal connections that launched his inventive career. Thomson eventually became one of the most prolific inventor-entrepreneurs in American history, with 696 patents covering arc and incandescent lighting systems, alternating current motors and transformers, and electric welding equipment. In 1892, the Thomson-Houston Company merged with Edison's interests to form General Electric. Thomson—the onetime independent inventor in the attic—became a pioneer of corporate R&D when he, Charles Steinmetz, and Willis R. Whitney established the General Electric Research Laboratory in 1900.[16]

In September 1924, the Franklin Institute celebrated its centenary and invited the seventy-one-year-old Thomson to deliver an address describing the organization's substantial impact on his career. Thomson recalled that he regularly attended the institute's Friday night lectures and made extensive use of its library to teach himself the electrical arts. In 1876–1877, Thomson himself presented a series of five demonstration lectures (figure 4.1) on electricity, using a small dynamo and arc light he had constructed in his attic.

Figure 4.1
Inventor Elihu Thomson (center) prepares for an electrical demonstration in the Franklin Institute's lecture hall in 1913. Photo by William N. Jennings, from the Collections of the Franklin Institute, Philadelphia, PA (cat. no. 6930-1).

During the fifth lecture, Thomson accidentally melted two wires together and serendipitously discovered the principle underlying his resistance welding patents. In 1877–1878, Thomson and Houston served on a special institute committee charged with evaluating the commercial dynamos then offered by Charles Brush and his competitors; they subsequently published their results in the *Journal of the Franklin Institute*. Thomson recalled that observing the state of the art during the dynamo tests was "most inspiring and valuable as a foundation for my future efforts." He and Houston soon demonstrated their own improved dynamo and arc lighting system during an institute lecture in January 1879. Philadelphia businessmen Thomas McCollin and George S. Garrett were in attendance and offered to underwrite the inventors. At the institute's International Electrical Exhibition in 1884, the Thomson-Houston Electric Company proudly displayed its commercial dynamos, lamps, and motors alongside products from other leading inventors. As he recounted these experiences during his celebratory speech, Thomson concluded that the Franklin Institute had provided important "encouragement to invention and discovery, and to inventors themselves."[17]

The Franklin Institute's vocational courses, scientific lectures, technical evaluations, and exhibitions helped inventors like Thomson learn their craft, stay abreast of new developments, and commercialize their inventions. However, around 1900, the institute began to change in ways that were detrimental to independent inventors. Faced with a financial crisis, it gradually abandoned its technical services and refashioned itself as a popular science museum. As the Franklin Institute evolved from a professional society for adults into a science museum for children, American inventors lost an important source of institutional support.

The first change occurred over several decades, as the Committee on Science and the Arts slowly relinquished its role as an evaluator of emerging technologies. In 1886, institute president Charles Banes proposed that, to avoid becoming embroiled in priority disputes or damaging litigation, the CSA should evaluate only patented inventions. The CSA resisted this proposal for decades, but by 1920 it had adopted this more conservative approach. The "patented inventions only" policy protected the institute from legal entanglements, but it also weakened the CSA. When the committee evaluated unpatented, cutting-edge ideas, it could influence the trajectory of emerging technologies. Instead, the CSA adopted a diminished role as a cheerleader and promoter of more proven, patented technologies.[18]

A proliferation of new endowed prizes and medals also provoked a major shift in the character of the CSA's evaluations. Thanks to generous donors, the four medals and certificates that the CSA offered in 1893 gradually ballooned to thirteen by 1958. However, the CSA could no longer rely on the volume or quality of external submissions to find suitable recipients for its annual awards. In 1909, the CSA established a standing Subcommittee on New Subjects and Publicity to internally suggest new inventions worthy of study.[19] Gradually, the New Subjects subcommittee became less concerned with recommending new *inventions* for evaluation and more concerned with nominating appropriate *inventors* for the institute's growing slate of medals.

This change is clearly reflected by comparing two of Elihu Thomson's CSA case files. In 1900–1901, the CSA investigated Thomson's "Constant Current Arc Light Transformer." The case file included measurements and performance graphs from tests the CSA investigators conducted on the transformer, as well as endorsement letters from cities that had successfully employed the transformer for their municipal lighting. Thomson was duly awarded the institute's John Scott Legacy Medal. In contrast, Thomson's 1912 CSA case was not a technical investigation. Rather, the report merely recommended awarding Thomson the Elliott Cresson Medal for his "leading and distinguished work in the industrial applications of electricity."[20] Eventually, the CSA stopped evaluating technologies altogether and became an honorary society. In the process, independent inventors lost a crucial professional service.

The Franklin Institute also abandoned its long tradition of hosting inventors' fairs and exhibitions. The equipment and infrastructure required to host the exhibitions had become financially burdensome for the institute, which had a history of precarious finances. For example, the 1884 Electrical Exhibition ran a deficit of $6,910 and the 1885 Novelties Exhibition lost $9,125; these consecutive shortfalls nearly bankrupted the institute.[21] By 1895, the Franklin Institute's secretary, William Wahl, had concluded that it could no longer present the breadth and depth of America's inventive talent without some financial assistance. Instead, Wahl believed that the government-sponsored exhibitions in New Orleans (1884), Chicago (1893), and Atlanta (1895) had become the premier destinations for independent inventors and manufacturers.[22] After mounting thirty exhibitions, the institute hosted its final event in 1899, and independent inventors lost an important showcase for their inventions.

The Franklin Institute rebuilt its finances by raising money to construct a new museum and recasting itself as a purveyor of popular science education. On January 1, 1934, it moved from its downtown Philadelphia headquarters to a magnificent new museum and planetarium on the Benjamin Franklin Parkway that was hailed as "A Wonderland of Science." The institute, which had long served an adult population, eventually became a popular destination for Philadelphia's schoolchildren. The inventors' fairs had once been showcases for commercializing emerging technologies. However, the institute's permanent historical exhibits now displayed technologies so well established—such as Linus Yale's pin-tumbler lock and Matthias Baldwin's full-sized locomotive—that they belonged in a museum.[23]

At the Franklin Institute's final meeting in the old building, Philadelphia inventor Samuel Vauclain wistfully noted the change. He remembered nervously submitting his design for an improved locomotive to the CSA—"everyone said it would not work, that I was wasting my time"—and his vindication at earning the Elliot Cresson Medal. "An inventor usually has a hard road to travel," Vauclain told the members, but the Franklin Institute had encouraged him and thousands of his brethren. "When you leave this old building . . . and go into that mansion on the Parkway," Vauclain predicted, "you are going to miss this."[24] As the Franklin Institute gradually abandoned its technical services, it became more difficult for independent inventors to commercialize their inventions.

* * *

"Is there anyone here," Thomas Howard asked rhetorically, "that knows of a reliable patent promoter?" Howard, the executive chairman of the National Institute of Inventors (NII), received no answer from the 500 members who had gathered in June 1919 for the group's annual meeting. As they strove to commercialize their inventions, inventors were constantly on guard against predatory scammers. The assembled inventors had joined the NII in part because the organization promised to protect them. Howard boasted that the institute had worked with Post Office fraud inspectors to put six crooked patent promoters out of business. Howard had also helped NII members recover fees paid to unscrupulous model makers and patent attorneys. "The Institute stands on the watch," Howard pledged, "for that particular class of frauds."[25]

Howard was a Brooklyn-based independent inventor who had previously worked in the motion picture industry. He founded the NII in 1914 as

Figure 4.2
Letterhead from the National Institute of Inventors, 1917. Orville and Wilbur Wright
Papers, Manuscript Division, Library of Congress, Washington, DC.

an organization dedicated to the "mutual aid, betterment, and protection"
of his fellow inventors. Based in New York City, the NII (figure 4.2) was a
cooperative in which dues-paying members could receive impartial advice
on their new ideas, legal and financial assistance for commercializing their
patents, and protection from con artists. Unfortunately, the NII itself was
a charade. Howard simply pocketed the membership dues and embezzled
thousands of dollars from America's unsuspecting inventors.[26]

At the most basic membership level, inventors could join the NII for an
initial fee of $25 with annual dues of $5. By 1919, the NII had grown to 2,262
members, including high-profile honorary members such as radio pioneer
Guglielmo Marconi and airplane coinventor Orville Wright. According to
its literature, the NII offered a range of technical, legal, and marketing services. It boasted twenty technical committees on subjects such as mechanics, electricity, and chemistry, in which five elected members evaluated the
"practicability, marketability, adaptability and commercial value" of the
ideas submitted by their fellow members. The Relief Committee could dispense up to $250 to a worthy inventor to aid the further development of
any meritorious ideas. The NII also covered the first $25 of billable consulta-
tions with its staff of patent attorneys, with members paying the balance. In
a bitter irony, the NII's Grievance Committee promised to aid any members
who had been defrauded by predatory patent agents or other "exploiters."[27]

In April 1918, the NII began to attract suspicion when Howard mailed
more than 1,000 inventors and scientists his proposal to build a "National

Laboratory for Invention and Research" to stimulate technical mobilization during World War I. The privately endowed philanthropic institution would be equipped with a modern research laboratory, machine shop, and patent library, with operations overseen by a corps of engineers, chemists, and mechanics. The lab would assist independent inventors in developing and marketing their military inventions, then take a share of any profits after the inventions were adopted by the US and allied governments. The proposed laboratory also promised to "help America, through invention, maintain supremacy in the commercial markets of the world after the war." A second circular urged each NII member to donate $100 to establish the laboratory's endowment.[28]

Orville Wright had accepted Howard's invitation to become an honorary NII member in 1917. Soon after Howard's national laboratory solicitation, he began to receive inquiries from correspondents who were anxious for his opinion of the organization. Omar M. Highley of Converse, Indiana, granted that the NII's proposed lab could be beneficial for inventors, but added that he was hesitant "to part with money under any circumstances." Wright replied that he was not yet well acquainted with the NII but believed its purposes were noble.[29] Charles A. Munn, publisher of *Scientific American*, asked for Wright's opinion after the NII sought the journal's endorsement for the proposed laboratory. Munn supported the concept but believed the lab should be operated "by the Federal Government and not by some private and money-making enterprise." Munn confessed that he had some "grave doubts" about the NII and its plans.[30]

Before Wright could reply to Munn's inquiry, he received a searing letter from Samuel Biddison, a Los Angeles inventor. After Biddison met with the NII's officers in New York City to develop a marketing plan for his patented gas generator, he sent Wright a warning:

> Dear Sir, I see you are a member of the National Institute of Inventors. . . . I assume of course you are not aware that this is a swindling concern, composed of one Thomas Howard or Cohn [sic]. . . . They are using your name with that of Marconi and Thomas A. Edison to rake in the dollars. . . . On the strength of some of the names on their literature I was induced to become a member and came here from California with a valuable patent. They tried to get it away from me for nothing, but failed. Mr. Edisons [sic] is not a member and was not aware his name was being used. Both the Department of Justice and the Attorney General are after them and they are liable to be locked up any day.

Biddison had accurately surmised several aspects of the NII's racket. "As soon as I saw the bunch," Biddison concluded, "I demanded my name be stricken from their books."[31]

Wright had heard enough. He resigned from the NII and asked Howard to remove his name from its promotional literature. He then sent follow-up letters to several inventors, retracting his earlier endorsement. Finally, Wright replied to *Scientific American*'s Charles Munn and confirmed the editor's suspicion that the NII was likely a scam. Munn replied that it appeared to be a fraud in which "poor inventors" paid their dues to access the services advertised in the NII's "pretentious literature" but received "absolutely no return."[32]

A few months later, Wright agreed to cooperate with a Post Office fraud investigation involving numerous complaints against the NII. He sent the inspector copies of his correspondence with Howard, Munn, and several inquiring inventors. In his defense, Howard countered that vengeful, disreputable patent promoters had made the spurious fraud allegations after the NII tried to shut them down. Additional fraud accusations, Howard reported, had also come from some jilted members after the NII declined to develop their impractical ideas.[33]

Meanwhile, the National Research Council (NRC) launched its own investigation into the NII. Several scientists had received Howard's laboratory solicitation and asked the NRC about its legitimacy. John Johnston, the NRC's executive secretary, visited Howard and his associates at NII's headquarters in New York City. They seemed honest, Johnston reported, but "not quite the kind of people we should like to see in charge of their so-called 'National Laboratory for Invention and Research.'" He urged an Ohio State physics professor to "proceed with them rather cautiously."[34]

To learn more about Howard and the NII, Johnston contacted Proudfoot's Commercial Agency, a private investigator in New York City. The news was not good. As a result of other complaints, the agency had been tracking the NII since 1914, and it forwarded two reports to the NRC. In the first report, from October 1917, Proudfoot found that several inventors and patent attorneys had been listed—without their knowledge or permission—as members of the NII's technical committees. A second report, from January 1920, focused on an upcoming 600-person banquet at the Hotel Astor to raise funds for the proposed national laboratory. Reporting just four days before the event, Proudfoot found that the NII had not yet paid a deposit for the $1,800 dinner. The agency predicted, of course, that the banquet would never occur and that the laboratory would never be built. In summary, Proudfoot believed that Howard was "crooked" and had founded the NII as a scheme to "make some easy money out of some unsuspecting people." Proudfoot urged the NRC to give the NII a wide berth and predicted that Howard would soon find himself in "serious trouble with the authorities."[35]

Of course, that is exactly what happened. Howard filed for bankruptcy in June 1920, and an investigation performed by his creditors uncovered new details about his fraudulent activities. As inventor Samuel Biddison had correctly surmised, Howard had used the alias Henry Cohen to engage in shady real estate deals, resulting in two felony convictions in Pennsylvania. Moreover, Howard had allegedly embezzled between $50,000 and $100,000 by transferring the NII's funds into his wife Rose's personal bank account. Howard was indicted on federal fraud charges, but he died suddenly in June 1922 before the case could be resolved. Howard's federal charges were dismissed, but postal inspectors officially put the NII out of business by dissolving its articles of incorporation.[36]

Howard had conned inventors for eight years (1914–1922), and his death brought an ignominious end to the National Institute of Inventors. His scam had exploited inventors' eagerness to find practical solutions in professional organizations. However, as the NII dissolved and the Franklin Institute shifted its focus, inventors were left with fewer (legitimate) institutions that could help them commercialize their ideas.

Banding Together for Patent Reform

At the turn of the twentieth century, inventors had become gravely concerned about the deteriorating conditions at the US Patent Office. The American patent system was established in 1790, and a century later it had become overwhelmed by what *Scientific American* called "a gigantic tidal wave of human ingenuity." R. H. Thurston, dean of Cornell University's Sibley College of Engineering and inventor of materials testing equipment, warned the journal's editors that the shorthanded examiner corps was struggling with a heavy workload in unhealthy, overcrowded conditions. The Patent Office received 135 applications per day, and a growing backlog of 5,000 pending cases forced inventors to wait at least eight or nine months for their patents. Inventors, Thurston complained, had paid for their examinations but were not receiving them in a timely manner. With the influx of application fees, the Patent Office had transferred a \$4 million surplus to the US Treasury, yet Congress sat idle instead of appropriating the funds to remedy the situation. This "barbarism" had persisted for half a century, Thurston argued, yet "the appeals of generations of commissioners, of thousands of inventors," had gone unheeded.[37]

In addition, wealthy corporations were exploiting systemic inequities in the patent system to harass independent inventors. In 1902, an anonymous *Popular Mechanics* subscriber described how he had sued an "unscrupulous corporation" for deliberate infringement but was forced to drop the case as his legal expenses mounted. "The courts only guarantee the inventor such justice in the protection of his patent as he has money to pay for," he concluded, adding that "when a poor man has an invention that proves attractive to a wealthy corporation, it is foolish for him to resist." In a legal battle, wealthy corporations could use their superior financial resources and aggressive legal tactics to make poor inventors capitulate. Observers worried that these abuses would eventually wipe out independent inventors, stifle meaningful competition, and bring all patented technologies under the control of the largest corporations.[38]

With the patent system in disarray, inventors mobilized to campaign for reform. At the turn of the twentieth century, inventors founded two political advocacy groups—the American Association of Inventors and Manufacturers and the Inventors' Guild—to bring their collective influence to bear on Congress.

*   *   *

The American Association of Inventors and Manufacturers (AAIM) emerged in conjunction with celebrations marking the one hundredth anniversary of the US patent system. Representatives from the Patent Office, the Smithsonian's US National Museum, and the Washington Board of Trade circulated a press release announcing a three-day celebration to be held in Washington, DC, in April 1891. Beyond the celebratory parades and speeches, the organizers also announced plans to establish a "National Association of Inventors and Manufacturers of Patented Articles." The organizers believed the celebration would attract hundreds of inventors to the capital and draw attention to the needs of the Patent Office.[39]

In February 1891, a second circular specifically encouraged inventors to "unite in the formation of this Association." It described how overworked and underpaid examiners toiled in overcrowded conditions at the Patent Office and faced a growing backlog of applications. The long pendency of patent cases in turn delayed inventors' efforts to capitalize on their inventions. Both inventors and manufacturers worried that hastily examined applications (and the resulting weak patents) would encourage willful infringement, excessive litigation, and the invalidation of those patents in the courts. The group's organizers believed the situation could be resolved if Congress would simply appropriate the accumulated surplus of patent application fees to acquire more office space and hire more examiners. In pursuit of these goals, the association promised to bring "concerted effort to bear upon their representatives in Congress."[40]

On April 8, 1891, 700 delegates convened in Washington, DC, to celebrate the centenary of the US patent system. The delegates were treated to speeches by President Benjamin Harrison and other dignitaries, a White House reception, a military parade, and a steamboat excursion to Mount Vernon. On Anniversary Day, April 10, a group of delegates met to adopt a preliminary constitution and bylaws, and the American Association of Inventors and Manufacturers was born.[41]

The AAIM's officers and directors included several prominent inventors, industrialists, and politicians, including automatic gun inventor Richard J. Gatling (president); paper straw inventor Marvin Stone (treasurer); Bell Telephone president Gardiner Hubbard (first vice president); Ohio Republican and chairman of the House Committee on Patents Benjamin Butterworth (fourth vice president); and arc lamp inventor Charles Brush (director).[42] Annual dues in 1891 were $5, and membership was open to anyone with one member's endorsement and approval from two-thirds of the officers. In its first year, the AAIM recruited several notable inventors to its ranks, including Alexander Graham Bell, Emile Berliner, Octave Chanute, and Elihu Thomson, whose AAIM correspondence provides a window into the group's activities.[43]

The AAIM held its first annual meeting on January 19, 1892, in Washington, DC, and began pressing its agenda to improve the conditions at the Patent Office (figure 4.3). An AAIM circular explained that the Patent Office had been forced to share space in its building with its Interior Department colleagues from the General Land Office. It described examiners working

Figure 4.3
The AAIM sought to improve examiners' pay, relieve overcrowded conditions, and
eliminate the backlog of pending applications at the US Patent Office.
Source: "The Crowded Condition of the Patent Office," *Scientific American* 66, no. 25
(June 18, 1892): 391.

in makeshift basement offices and hallways "lined on both sides with unsightly cases of papers," conditions that interfered with "the health of the force" and slowed their work. The AAIM urged members to write their congressmen to request an appropriation bill so that the General Land Office might rent space elsewhere and give the Patent Office some breathing room. The journal *Science* praised the AAIM's lobbying efforts and urged Congress to provide the Patent Office with "respectable quarters and [a] sufficient working force."[44]

The AAIM also tracked patent legislation in Congress. For example, the AAIM opposed an amendment to a bill in the House of Representatives that would have abolished all interference proceedings. These administrative hearings determined the priority of invention when two or more pending applications bore on the same subject matter. The AAIM also supported a proposal that would permit patent holders to sue manufacturers and sellers for infringement but prevent cases against consumers. According to the AAIM, several farmers had been victimized by "*sharpers*, who go around demanding royalties" on the pretense that they had infringed some patent on their plows and reapers. Moreover, these frauds had brought "*the patent system into disrepute*" and motivated several reactionary bills designed to weaken it. In October 1893, a group of midwestern congressmen introduced a proposal to reduce patent terms from seventeen to seven years and another proposal to nullify a patent entirely once the inventor had earned $25,000 from the invention.[45] The AAIM helped thwart these misguided bills and promised to remain vigilant against any attempts to weaken the patent system.

However, a public relations effort revealed the AAIM's precarious finances and limited influence. In 1897, the AAIM's leaders made plans to publish a volume of member testimonials in favor of a strong patent system. The group assembled several statements but lacked the $250 necessary to publish them. Secretary-treasurer Arthur Steuart, a patent attorney, solicited the membership for $25 donations and succeeded in raising the money, but the volume apparently lacked a wide readership. Steuart asked Elihu Thomson to forward the testimonials to various trade journals where he had some influence as a frequent advertiser. Admitting the AAIM's lack of notoriety, Steuart hoped that the reprinted testimonials might "bring our association to the knowledge of a large number of people who do not know anything of it."[46]

It is unclear whether Thomson followed through on Steuart's request. However, on January 11, 1900, Thomson wrote Steuart, paid his $5 dues for

1899, and resigned from the AAIM, saying only that he was "of little use to the Association." Steuart disagreed and asked Thomson to remain on board as an honorary member, since the AAIM needed the legitimacy conferred by its most successful members. Thomson acquiesced and continued to receive the AAIM's mailings and bulletins. In his final correspondence, Thomson declined an invitation to attend an AAIM meeting in March 1902.[47]

Here the evidentiary trail grows cold and one can only speculate regarding the AAIM's demise. One possible explanation is a degradation in the group's leadership. The AAIM initially counted several notable inventors and manufacturers among its ranks, including Richard Gatling, Elihu Thomson, and Alexander Graham Bell. However, as these prominent inventors resigned from the AAIM, rotated out of leadership positions, or died, less prominent— and perhaps less capable—men took the helm. At the turn of the century, the AAIM's leadership consisted of relative unknowns such as president Francis H. Richards (inventor of an envelope-stuffing machine), Theodore N. Ely (chief of motive power, Pennsylvania Railroad), and secretary-treasurer Steuart (a patent attorney). Absent the legitimacy conferred by inventors with higher profiles, the AAIM was presumably unable to draw ongoing press attention or recruit new dues-paying members to sustain its agenda.[48] Sometime after 1902—a mere decade after its founding—the AAIM faded from existence.

*   *   *

The AAIM had failed to mobilize Congress on patent reform, but a new inventors' advocacy group soon emerged to pick up the baton. On December 13, 1910, the *New York Times* reported the formation of the Inventors' Guild in New York City. Like the AAIM, the guild hoped to remedy the deteriorating conditions at the Patent Office, an agenda that appealed to both independents and corporations alike. The guild, however, sought additional populist reforms designed to "defend poor inventors against 'rich corporations.'" The "inventor knows that there is something wrong with the laws and the Patent Office," an unnamed spokesman told the *Times*, but "he doesn't know what the remedies are and how he can defend himself or improve conditions." Therefore, the guild planned to consult with legal experts to determine "exactly what is wrong and exactly how to make the wrongs right." Unlike the AAIM, which sought a wide membership, the Inventors' Guild preferred an exclusive organization capped at fifty elite members who were "capable of exerting some influence." The guild's twenty-nine charter members included electrical giant Thomas Edison, plastics pioneer Leo H. Baekeland, and mercury lamp inventor Peter Cooper Hewitt.[49]

After taking some time to study the various defects in the patent system, the Inventors' Guild published its reform agenda in *Scientific American*. First, it echoed the AAIM in calling for more space and better pay for Patent Office examiners. These improved conditions would result in strong, well-examined patents that were less likely to be disputed. Second, the guild sought to bring international patent laws in line with US standards, so that American inventors abroad would receive the same protections as foreign inventors in the United States. Third, the guild suggested a series of reforms to protect independent inventors against wealthier rivals when their patents were disputed, either in interference hearings—in which the Patent Office determined the priority of competing applications—or infringement lawsuits, which were tried in the federal courts.

In patent disputes, wealthy corporations could compel excessive testimony, introduce delaying motions, and purposefully prolong their cases to bankrupt poor rivals until they capitulated. Wealthy plaintiffs could also sue poor defendants on the same patents across multiple jurisdictions until they received a favorable ruling. To remedy these inequities, the Inventors' Guild proposed streamlined discovery processes, limits on depositions, and other procedural reforms to expedite both interference and infringement cases. The guild also called for a single Court of Patent Appeals with nationwide jurisdiction to standardize appellate rulings and limit excessive litigation. To discourage "hasty and ineffective legislation" that might serve narrow interests, the guild urged all stakeholders to carefully study the proposed reforms.[50]

At the same time, antitrust watchdogs were concerned about two additional abuses—patent suppression and restrictive licenses—that seemed to subvert the constitutional aim of fostering innovation. As the *New York Times* explained, corporations took out "defensive" patents on ideas they never intended to use and sometimes shelved the patents they purchased from independent inventors. By suppressing these patents, corporations protected their own technologies and prevented rivals from bringing forth competing ideas.[51] Critics also worried that restrictive patent licensing agreements had helped patent holders sustain illegal monopolies by fixing prices and restraining trade. For example, the Supreme Court was considering whether a patent holder could legally stipulate the price at which a licensee could resell the patented article. In another case, the court was considering whether the A. B. Dick Company could force licensees of its patented rotary mimeograph machine to buy unpatented Dick-brand paper and ink.[52] Accusations of patent suppression and the pending licensing cases had stirred much activity among the House and Senate Patent Committees, leading the *New York Times* to worry about misguided legislation, "ill expressed in the form of dangerous and radical bills." The guild, aware that Patent Commissioner Edward Moore would soon recommend new patent reform legislation, had likewise petitioned President William Howard Taft and Congress to carefully study the various defects of the system before enacting any legislation.[53]

On April 17, 1912, William A. Oldfield (D-AR), chairman of the House Committee on Patents, introduced H.R. 23417, a sweeping revision of the patent laws. The bill, proposed by Commissioner Moore, was designed to curb the "power of large corporations." The language echoed the Inventors' Guild's calls for better facilities and higher salaries for the Patent Office and provided sensible suggestions for streamlining interference proceedings. However, some of the bill's proposed remedies—including compulsory licensing provisions and severe limits on restrictive licenses—were arguably worse than the original ailments.[54] The guild had proposed a series of moderate reforms and asked for a deliberate, careful study of the patent system. As the House Patent Committee convened hearings on the Oldfield bill, the guild's members found themselves in the awkward position of testifying *against* a series of reforms they had helped inspire but nevertheless considered too radical.

Guild member Leo H. Baekeland took the lead in advocating for sensible reforms. Baekeland was born in Belgium and earned his PhD in chemistry at the University of Ghent. In 1889, he traveled to the United States on a scholarship, with plans to pursue a career as a consulting chemist and independent inventor (figure 4.4). Baekeland invented a new photographic paper, cofounded the Nepera Chemical Company to market it as Velox, and eventually sold the firm to Eastman Kodak in 1899. In 1909, Baekeland patented Bakelite, the first true synthetic plastic, a heat- and chemical-resistant material that would eventually be molded into billiard balls, jewelry, and the casings for electrical appliances.[55]

Figure 4.4

Leo H. Baekeland in his Yonkers, New York, laboratory around 1910. As a charter member of the Inventors' Guild, Baekeland advocated for patent reforms that would curb corporate abuses against independent inventors. amsterdambakelitecollection .com/Courtesy of Reindert Groot.

During his congressional testimony in 1912, Baekeland introduced himself to the House Committee on Patents as a chemist, independent inventor, businessman, and "charter member of the Inventors' Guild." His broad experience made him a valuable expert witness. Baekeland had earned patents in the United States and across Europe, so he could compare the merits of different patent systems. He also testified to his experience as both a "poor inventor" and a "comfortable inventor." So, the committee inquired, was the US patent system equitably serving both corporations and independent inventors? Baekeland did not mince words: the US patent system was "a godsend for large corporations, for lawyers, and for experts" and "a very poor thing for an inventor who is not a wealthy man." Baekeland testified that large corporations could bluff the public "with their worthless patents" and willfully infringe the good patents of a "poor competitor." He believed that corporate abuses had turned the patent system into a rigged game of "Heads I win, tails you lose."[56]

Baekeland and his fellow guild members then testified on various provisions of the Oldfield bill. To prevent purposeful delays and encourage expedited interference hearings, Section 17 proposed the "nineteen-year rule." In lieu of a seventeen-year patent term beginning at the *date of issue*, the provision called for a nineteen-year patent term beginning at the *date of filing*. Inventors or firms that employed dilatory tactics would only penalize themselves by diminishing their remaining patent term after issuance. The guild's H. Ward Leonard, an inventor of electric control systems, testified that he had spent approximately $75,000 fighting thirty-four different interferences concerning his one hundred patents. Interference proceedings were only necessary because the United States was peculiar among the world's patent systems in awarding priority to the inventor with the earliest proof of conception (first to invent) instead of the first inventor to file an application (first to file). In lieu of the nineteen-year rule, both Leonard and Baekeland proposed converting the United States to a first-to-file system, awarding the patent to "the first man who discloses it" and eliminating interference proceedings entirely.[57]

Guild members also testified against Section 17's compulsory licensing provision, which aimed to curb the suppression of patents. If a petitioner could prove that a patent had been undeveloped for four years, a federal court could compel the patent holder to grant a license for an equitable royalty. Guild members testified that concerns about patent suppression had been overblown. Thomas Edison described cases in which a firm had "bought up an invention, introduced it, and afterwards bought up an improvement and ceased using the first patent—suppressed it, in fact." Edison believed there was nothing malicious in the public getting "the latest improvement." Leonard testified that the compulsory licensing provision could actually work against independent inventors by compelling them to grant licenses to wealthy competitors. It would be unfair, Leonard testified, if General Electric petitioned for a compulsory license on one of his control system designs and then used Leonard's own patent to compete against him.[58]

Finally, guild members objected to Section 32, which would prohibit the restrictive patent licenses that monopolies had used to illegally fix prices and restrain trade. Leonard's objection was practical: to ensure quality. If a patented device were used improperly, in ways not stipulated in a license, then "the reputation of that machine would be destroyed." F. L. O. Wadsworth's objection was over legal jurisdiction. If a licensee used the wrong ink in a mimeograph machine, A. B. Dick should sue for breach of contract, not infringement of the patent. Baekeland's objection was fundamental. Both compulsory licenses and licensing restrictions undermined an inventor's exclusive constitutional right to dispose (or not dispose) of the patent in any way he saw fit. Baekeland believed that Congress should not meddle with an inventor's temporary patent monopoly, since the technology eventually became available to everyone after seventeen years.[59]

On August 8, 1912, the House Committee on Patents issued a favorable report and sent the Oldfield bill to the House floor. However, the August revision contained twelve pages of antitrust provisions—amendments to the Sherman Act—which had barely been discussed during the twenty-seven public hearings. The Oldfield bill should be discredited, argued the *Yale Law Review*; it had been "illegitimately amended" and sought to make radical changes to the antitrust laws without "any opportunity for public discussion or public hearings."[60]

Baekeland agreed that the Oldfield bill did not address the real concerns facing independent inventors. In his December 1912 presidential address before the American Institute of Chemical Engineers (AIChE), Baekeland observed that "individual patentees of slender means" operated "under smothering disadvantage when opposing rich antagonists." The broken patent system, he continued, "simply reinforces the rich and big industrial enterprise, and discourages the individual inventor unprovided with a liberal bank account." Baekeland thought Oldfield's bill was a well-intentioned but misguided attempt "to hit the trusts" and only succeeded in weakening the patent laws. He believed the bill's key provisions—compulsory licenses and restrictions on licensing terms—would "practically annihilate the advantage of exclusive ownership" conferred by a patent. Without that protection, Baekeland argued, inventors and their backers would be unwilling to take the risks necessary to commercialize new inventions.[61]

Following the uproar over Oldfield's bill, Congress passed a joint resolution in August 1912 appropriating $10,000 for a comprehensive investigation of the patent system. This is what the Inventors' Guild had wanted all along—a thorough study, not ill-advised legislation. The commission's 624-page report echoed many of the guild's original recommendations, such as a single court of patent appeals and streamlined interference procedures. Meanwhile, objections from the Inventors' Guild, the AIChE, and the Patent Law Association doomed the Oldfield bill, which languished in the House of Representatives. Oldfield introduced another version of his bill (H.R. 1700) in April 1913, but the guild quickly condemned it and the bill again went nowhere. Even with the commission's robust recommendations, the House Committee on Patents was understandably hesitant to draft any new legislation after so much scrutiny.[62] In the end, the Inventors' Guild defeated the Oldfield bill's harmful proposals, but it also failed to achieve the reforms it really wanted.

The Inventors' Guild receded from the public arena for two years before adopting a new cause. Between 1915 and 1918, it helped mobilize independent inventors for World War I, as Edison, Baekeland, and several other members played prominent roles on the Naval Consulting Board (chapter 6). However, the guild never reengaged in policy debates after the 1918 armistice and was essentially dormant by 1920. Like the AAIM, the Inventors' Guild was active for only about ten years.

Both the AAIM and the Inventors' Guild sought reforms to an overburdened patent system that was increasingly prone to abuse. Both groups were short-lived, and their lobbying was ultimately ineffective. As a result, independent inventors continued to struggle with a patent system whose structural inequities favored the wealthy corporations.

Banding Together for Recognition and Respect

Beyond their commercial and political goals, independent inventors pursued something even more elusive: the respect of a skeptical public. At the turn of the twentieth century, women and African Americans challenged the pernicious assumption that they were intellectually inferior and incapable of being inventors. Independents of all stripes confronted negative stereotypes about "crackpot" inventors as they asserted their relevance in an era of corporate R&D.

These challenges motivated socially marginalized inventors to band together to secure recognition and mutual support. Informal networks of women inventors and African American inventors advocated on behalf of their communities and publicized their achievements during a time when many Americans assumed they did not exist. Likewise, the National Inventors Congress offered a welcoming membership and proud sense of identity for eccentric independent inventors in the face of public derision. These supportive communities formed—but soon dissolved—around charismatic leaders who were unable to sustain them. As a result, inventors on the social margins were left to face their struggles without the support of sympathetic colleagues.

*   *   *

In 1883, suffragist Matilda Joslyn Gage noted that every woman had confronted the assumption "that she possesses no inventive or mechanical genius." There was ample anecdotal evidence of women's ingenuity—such as Margaret Knight's machinery for producing paper bags—yet society had not encouraged the full "exercise of woman's inventive powers."[63] In an 1887 essay, journalist Ida Tarbell examined the various barriers to women's inventive progress. Women, she found, had few opportunities for technical education. They were systematically excluded from the artisanal trades, technical schools, and professional engineering societies. Moreover, some elites considered it "unladylike" for a woman to perform technical work, so many women avoided mechanical pursuits to avoid social stigma. Given these conditions, women sometimes hesitated to commercialize a good idea because of "ignorance about securing patents, doubt about her model, the expense to be incurred, the skeptical remarks of friends, or the demands of her daily life." Mrs. A. D. T. Whitney, for example, told Tarbell that her patent for educational alphabet blocks had "lain almost idle" for want of some trustworthy "channel or agency" to help bring it to market. Finally, common law in many states prohibited married women from owning or controlling any property. Therefore, a married woman could earn a patent, but only her husband could sell it, license it, or sue an infringer. These laws deterred many women from ever becoming inventors or compelled them to apply for patents in the names of their male relatives and attorneys. Overall, the restrictive laws and conservative social norms of the period made it difficult for women to pursue invention.[64]

To confront these challenges and draw attention to their cause, women inventors—led by activist Charlotte Smith—banded together to establish an informal supportive community. Smith was born Charlotte Odlum in 1840 in upstate New York, the daughter of Irish Catholic immigrants. During her twenties, Smith worked as a hatmaker, grocer, and bookseller while moving between Memphis, Mobile, and Chicago. She eventually married Edward Smith in Philadelphia and had two sons before settling in St. Louis. Between 1872 and 1878, Smith published a regional general-interest magazine, *The Inland Monthly*, which was written, typeset, and printed entirely by women. Smith sold the magazine for $75,000, moved to Washington, DC, in 1879, and spent the next fifteen years as an advocate for working women. In 1882, Smith founded the Women's National Industrial League (WNIL) to gain better pay and working conditions for women who were barred from joining traditional unions. In an era before she could vote, Smith regularly testified before Congress and helped pass laws protecting the safety of foods, cosmetics, and medicines.[65]

Although she was not an inventor herself, Smith began to champion women inventors after meeting Mary S., the pseudonym for the twenty-four-year-old daughter of a talented but feckless inventor who sold each of his inventions "for the paltry sum of five or ten dollars." Mary tried her own hand at inventing to supplement the family's income and eventually completed thirty-seven inventions initiated by her father, plus another sixteen inventions of her own. However, like her father, Mary sold her inventions to patent agents in New York, Washington, and St. Louis for only small sums. In particular, she patented one invention under a male lawyer's name, sold it for just five dollars, and then watched as the attorney turned the invention into a windfall. Smith lamented that her friend "furnished the brains for man to create wealth, when he himself had not the genius to invent" but only "the craftiness and cunning to take away the products of that woman's inventive labor." Mary S. died a pauper in 1880, and her story inspired Smith to seek justice for women inventors.[66]

Smith naturally wondered how many women inventors like Mary S. might be suffering similar injustices, so after moving to Washington, DC, in 1879, she began her crusade by convincing the Patent Office to compile an official comprehensive list of female patentees. Smith understood that women inventors served as important role models. When they earned patents—objective, government-issued proof of their inventions—sexist arguments about their intellectual inferiority were difficult to sustain. By extension, a comprehensive list of women patentees, prepared and published with the imprimatur of the Patent Office, would serve as a powerful inspiration for all women. Smith spent the next several years doggedly calling on four successive patent commissioners, including seventeen visits to Commissioner Benton Jay Hall. The commissioners' polite excuse was always a shortage of funds and clerical time, so Smith testified before Congress and obtained the necessary $300 appropriation.[67]

Four clerks spent about ten days compiling what sounded like women's names from the cumulative list of half a million patents. In 1888, the Patent Office published the official list, *Women Inventors to Whom Patents Have Been Granted by the United States Government, 1790 to July 1, 1888*. The booklet was a simple chronological listing of 2,297 women's patents that included the patent number, title, and issue date, plus the inventor's name and address. Scholars have since identified several omissions, plus the accidental inclusion of some male patentees. Nevertheless, the 1888 list—and two subsequent editions, published in 1892 and 1895—provided Smith and other advocates with an unassailable, government-sponsored record of women's inventive activity.[68]

Using the names and addresses from the list, Smith spent the next three years corresponding with women inventors, gathering their stories, and developing her political agenda. As delegates arrived in Washington, DC, to celebrate the patent system centennial in 1891, Smith handed out copies of *The Woman Inventor* (figure 4.5), which she advertised as the first periodical "devoted to the cause of women inventors of this country." The rendering on the cover featured about ten women inventors confronting a larger gathering of their male counterparts in the halls of the US Patent Office. The women held a banner that read: "We ask for Recognition in the Industrial and Inventive Worlds."[69]

The journal celebrated America's women inventors, both past and present. Smith described how in 1809 Mary Kies became the first woman to earn a US patent, after inventing a method for weaving straw for hats. Smith, echoing Matilda Joslyn Gage, provocatively suggested that Catharine Greene allowed Eli Whitney to patent her idea for the cotton gin

Figure 4.5
Charlotte Smith published the first issue of *The Woman Inventor* in April 1891 (not April 1890, as erroneously printed). Smithsonian Institution Libraries, Washington, DC.

because she feared "ridicule because of her social position." Smith also celebrated Philadelphia's Helen Blanchard, who in 1873 patented a zig-zag
overstitch for sewing machines and earned a substantial income as founder
of the Blanchard Over-Seam Sewing Machine Company.[70]

The Woman Inventor also announced Smith's political agenda. Echoing
the AAIM's platform, Smith proposed the eviction of the Department of
the Interior from the overcrowded Patent Office building. She also sought
reduced filing fees and cash subsidies so inventors could perfect their
inventions. To aid women inventors specifically, Smith called for the states
to reform their married property statutes and urged the federal government
to subsidize women inventors' infringement cases at taxpayers' expense.
Finally, Smith requested a spacious hall in the Patent Office, managed
entirely by women, to exhibit women's inventions.[71]

In the second issue of *The Woman Inventor*, Smith reprinted several letters from women inventors. Eliza Wood, a farmer's wife from New York,
sought Smith's advice. Wood had no funds to market her patented mop
pail and wringer, but she was wary of engaging with crooked patent agents.
Smith replied that she was in the process of establishing the Woman Inventors Mutual Aid and Protective Association, a trustworthy patent agency
dedicated to helping women inventors like Wood. The agency would be
administered entirely by women, with Smith as president and Martha Coston, coinventor of nighttime signaling flares, serving as vice president. For
a $5 annual fee, members would receive one copy of *The Woman Inventor*,
information on US and foreign patenting procedures, and access to low-cost
model making and drafting services. When presented with a woman's promising idea, the association would advance her Patent Office fees and help
secure the patent. Then, in exchange for a moderate royalty, the association would either negotiate the sale and licensing of the patent or help the
woman inventor set up a manufacturing business. While Smith waited for
the Patent Office to establish her proposed women's gallery, the association's
Washington offices would exhibit women's patent models, dispense advice,
and generally serve as a hub for the community of women inventors.[72]

Smith was unable to carry out her ambitious plans. She persuaded the
AAIM to admit female members, but she undercut that concession by
announcing her own political agenda. Smith convinced the Patent Office
to update the list of women patentees in 1892 and 1895, but she never succeeded in reducing patent fees, subsidizing women's infringement lawsuits,

or securing a special gallery to display women's inventions. *The Woman Inventor* gave its constituency a powerful new voice, but it ceased publication in 1891 after only two issues. Smith also failed to gain any traction for her proposed women's patent agency. Smith joined an effort to establish the Woman's Building at Chicago's 1893 World's Columbian Exposition, but she clashed with the pavilion's organizers and ultimately played little role.

Smith was unable to sustain her efforts on behalf of women inventors, so she shifted her focus. She moved to Boston in 1892 and operated a shelter for battered and homeless women until her death in 1917.[73] However, in a zealous burst of activity from 1879 to 1892, Smith galvanized women inventors into a recognizable professional community at a time when few people scarcely realized they existed.

*   *   *

Much like women inventors, African American inventors faced racist assumptions about their intellectual inferiority and technical incompetence. Although born enslaved, Congressman George Washington Murray (R-SC, figure 4.6) had pursued his education, won election as the only Black representative in the 53rd and 54th congresses, and earned eight patents for improved farm equipment, such as a "cotton chopper" and a "fertilizer distributer." During a floor debate in 1894, Murray excoriated the "slanderers of our race" who claimed "that we are not equal to others because we have failed to produce inventors." African Americans, Murray argued, had "taken out patents upon almost everything," and, to prove his point, he entered the names and inventions of ninety-two Black patentees into the *Congressional Record*, including his own.[74]

Despite this demonstrated record of achievement, African American inventors struggled to overcome the persistent disenfranchisement, segregation, and discrimination of the "Jim Crow" era.[75] Even with the postbellum establishment of Black colleges and technical schools—such as Alabama's Tuskegee Institute, founded in 1881—African Americans generally had few opportunities for technical training. They were systematically excluded from the professional engineering societies and initially barred from attending the lectures and technical courses at Philadelphia's Franklin Institute. With limited access to wealthy backers and mainstream banks, they were forced to develop alternative networks to access the start-up capital and expertise necessary to commercialize their inventions. Overall,

Figure 4.6
Inventor and congressman George Washington Murray (R-SC) earned eight patents for improvements in farm equipment. He supported federal funding for exhibitions featuring African American inventors. C. M. Bell Studio Collection, Prints and Photographs Division, Library of Congress, Washington, DC.

African Americans had to battle what historian Bruce Sinclair has called the "deeply ingrained and long perpetuated myth of black disingenuity."[76]

Murray confronted these challenges directly. After Reconstruction, Murray and other African American leaders sought to compile a comprehensive list of Black patentees who might serve as inspirational heroes for their communities. The Patent Office always resisted, noting that patent records did not indicate racial heritage. However, as the federal government and individual states began to underwrite major fairs and expositions, elected officials began to demand a list of Black patentees to aid the development of displays featuring African American inventors. The Patent Office obliged by compiling partial, unofficial lists for the Cotton Centennial in New Orleans (1884), the World's Columbian Exposition in Chicago (1893), and the Cotton States and International Exposition in Atlanta (1895). Murray, in fact, had shared his list of ninety-two Black inventors during a floor debate over funding for the Atlanta exposition.[77]

Murray's roster of Black inventors was compiled by patent examiner Henry E. Baker. Baker (1857–1928) was born in Columbus, Mississippi, just before the Civil War. In 1874, he became the third Black midshipman appointed to the United States Naval Academy, but he withdrew after enduring two years of racist insults and harassment. The secretary of the navy offered to reinstate him, but Baker opted to take an entry-level position as a copyist at the US Patent Office. Meanwhile, he completed his undergraduate studies at Washington's Ben-Hyde Benton School of Technology and then graduated from Howard University's law school in 1881 at the top of his class. Baker was promoted from copyist to clerk and then to assistant examiner and became a leader in Washington's African American community, where he served as an officer for two banks and several social clubs. Most of what we know about early African American inventors derives from Baker's efforts to compile and publicize their contributions.[78]

Patent Commissioner Charles Duell directed Baker to compile the first comprehensive list of Black patentees in preparation for W. E. B. Du Bois's planned "Negro Exhibit" at the 1900 Paris Exposition. Baker mailed a questionnaire to thousands of registered patent agents and attorneys seeking information about "colored inventors." Some racist respondents believed Baker's inquiry was a joke. However, Baker identified 370 patents by Black inventors and assembled reprints of their specifications into four leather-bound booklets for display at the Paris Exposition.[79] Baker subsequently published his list and highlighted the achievements of Black inventors. Jan Matzeliger, an immigrant from Dutch Guiana, had earned four patents for his automated shoe-lasting machine. Detroit's Elijah McCoy held twenty-eight patents and was best known for the "real McCoy," his automated lubrication system for locomotives. Cincinnati's Granville T. Woods was hailed as the "Black Edison" for his twenty-two patented improvements in telegraphy, telephony, and electrical circuitry. Baker conducted an expanded search in 1913 to mark the fiftieth anniversary of the Emancipation Proclamation. This second list, published as *The Colored Inventor*, listed nearly 800 patents. These compilations, Baker concluded, provided irrefutable evidence to anyone who would erroneously "assert that the Negro has made no contribution to the civilization of the age."[80]

Baker's lists emerged from the widespread interest—from Atlanta to Paris—in displaying the inventive genius of African Americans. Richmond attorney Giles B. Jackson (figure 4.7) founded the Negro Development and Exposition Company (NDEC) to capitalize on that interest. Jackson (1853–1924) was born enslaved in Goochland County, Virginia. After the Civil War, he worked as a laborer and then in the law offices of William H. Beveridge. Beveridge tutored Jackson, who became the first Black attorney

Figure 4.7
Giles B. Jackson founded the Negro Development and Exposition Company to show-case the achievements of African Americans at the Jamestown Tercentennial Exposi-tion of 1907 and other fairs. Virginia Museum of History and Culture, Richmond, VA (Object No. 1991.1.14743).

certified to practice law in Virginia. In 1888, Jackson helped incorporate the True Reformers Bank, one of the first Black-owned and operated banks in the United States. The bank and other flourishing Black businesses in Richmond's "Jackson Ward" drew the attention of Booker T. Washington, who asked Jackson to serve as vice president of his National Negro Business League.[81]

In 1901, the Virginia legislature authorized plans for the Jamestown Tercentennial Exposition of 1907 to mark the three hundredth anniversary of America's first permanent English settlement. Jackson began promoting the idea of erecting a "Negro Building" with exhibitions by and about African Americans. He secured the endorsement of the fair's commissioners and in 1903 organized the NDEC, with its headquarters in Richmond.[82] Critics—including Jackson's mentor, Booker T. Washington—suggested that a separate Negro Building would be "a Jim Crow affair," which would only emphasize African Americans' marginal status in a segregated society. However, in an October 1905 appearance in Richmond, President Theodore Roosevelt defended Jackson's efforts to make a "creditable exhibit of the achievements of your race." To build the pavilion, Jackson raised $50,000 from individual donors and secured an additional $100,000 from Congress in 1906. A dozen NDEC field agents crisscrossed the country to publicize the Negro Building and recruit exhibitors.[83]

The NDEC overcame construction delays and assembled a series of exhibits that demonstrated the full range of Black progress. The Negro Building showcased sculpture and handicrafts by Black artisans, books by Black authors, recitals by Black musicians, a hospital exhibit featuring Black physicians and nurses, and an operating branch of the True Reformers Bank. Notably, a 1,200-square-foot Inventions Section displayed fifty prototypes and 351 US patent specifications contributed by African American inventors. For example, Aiken C. Taylor of Charleston, South Carolina, presented three inventions: a combined cotton planter and fertilizer distributor, an extension stepladder, and an "invalid's bed" that converted into an easy chair at the turn of a crank. G. F. Carr exhibited a two-sided, heated cabinet that served as both a bread oven and an incubator for baby chicks (figure 4.8). Baltimore's Samuel T. Crawford won the gold medal in the Inventions Section for a hand-cranked boat propeller that attached to the stern of a small watercraft.[84]

THE INCUBATOR AND BREAD RAISER by G. F. Carr,
is an arrangement whereby the same machine may be used as an
incubator on one side and a bread raiser on the other. As long as
the little chicks and the dough are not in the same compartment at
the same time it would appear that the scheme ought to work well.

Figure 4.8
G. F. Carr's combined incubator and bread oven was one of fifty prototypes displayed
by African American inventors in the Negro Building at the Jamestown Tercentennial
Exposition of 1907.
Source: Giles B. Jackson and D. Webster Davis, *The Industrial History of the Negro Race
of the United States* (Richmond: Virginia Press, 1908), 311.

The Negro Building was a bright spot in an otherwise unsuccessful Tercentennial Exposition. The fair, which the *New York Times* deemed a "colossal failure," had drawn only half of the six million expected visitors and incurred a $2.5 million debt. The Negro Building had drawn between 3,000 and 12,000 visitors per day between April and November 1907, but Jackson estimated that less than 1 percent of African Americans living in the United States had seen the exhibition. He therefore started a campaign to relocate the Negro Building to Richmond and convert its displays into a permanent "National Museum for Colored People." In August 1908, the NDEC paid a $5,000 bond to the US Treasury to claim possession of the federally funded building, but the effort apparently faced local opposition in Richmond and was abandoned.[85]

Despite this setback, the example set by Jackson and the NDEC inspired African Americans to host additional expositions highlighting the work of Black artisans and inventors, such as the Negro National Fair in Mobile, Alabama (1908) and fiftieth-anniversary "Emancipation Expositions" in Philadelphia, New York, Chicago, and Washington, DC, held between 1913 and 1915. Jackson and the NDEC stayed active as well, organizing the Second National Negro Exposition in Richmond in July 1915. However, with the completion of that exposition, the NDEC apparently ceased operations, after a twelve-year run.[86]

Jackson spent the remainder of his life advocating for fair labor practices on behalf of African Americans. He died in 1924—followed closely by George Washington Murray in 1926 and Henry Baker in 1928—and so ended the first generation of advocacy on behalf of African American inventors. Over a twenty-year period, these leaders documented the achievements of African American inventors and convened exhibitions where they could meet and support one another. However, when these champions died, their informal network died with them.[87]

*   *   *

The American public should show "more respect for the work of inventors," declared Albert G. Burns, president of the National Inventors Congress (NIC). Speaking in a radio address before the NIC's 1937 convention, Burns lamented that the inventor is "'a forgotten man' who is laughed at by some people as 'a nut.'"[88] Burns estimated that only about 6 percent of his 32,000 members were "professionals" who made a living from invention. Rather, the vast majority were amateur, "happenstance" inventors, who had stumbled onto a bright idea. As *Time* magazine elaborated, NIC members were not "the bigwigs of industrial and academic laboratories" but rather the "indefatigable purveyors of small ingenuities," the "humble rank & file of U. S. idea men."[89] Consequently, the NIC's amateur inventors endured their fair share of condescension.

Burns and his members were representatives of a uniquely American archetype—the amateur scientist, the do-it-yourself garage tinkerer—that has permeated the culture since Benjamin Franklin's kite experiments. Skeptics derisively characterized these eccentrics as "cranks" or "crackpots" who toiled in obscurity on their pseudoscientific gadgets.[90] However, the NIC served as a safe space for amateur independents amid the cynicism of elite industrial scientists and a skeptical public. It provided a community of mutual support and acceptance that embraced and celebrated the quirkiness of independent inventors in an era of corporate R&D.

The NIC was founded in 1924 as an "impoverished inventors' cooperative" dedicated to hosting commercial fairs in the tradition of the Franklin Institute and the NDEC. Cash E. Naylor, an inventor from Grands Rapids, Michigan, recalled that a group of inventors "decided it would be better to have a convention and invite manufacturers to attend" rather than "cool their individual heels at the doors of prospects' factories." The NIC grew under the leadership of Burns, who joined the group in 1928 and became its president in 1931. Like many moonlighting independents, Burns had found employment in a series of day jobs—working at a coffee shop, directing a chamber of commerce, and managing a sanitarium—before earning some success as an inventor. Burns sold 800,000 of his patented auto locks for Model T Fords and later marketed an improved bread knife that earned him "quite a nice little sum." The mischievous Burns referred to himself as the "Nation's Gadget Chief," the "Chief Nut," or "Doodlebug No. 1."[91]

Burns and his sons Roy and Albert Jr. ran the NIC from Chicago when they were not on the road hosting an exposition. In exchange for the $5 annual membership dues, members received a copy of the group's monthly newsletter, *The Inventor*, and the opportunity to display their inventions for manufacturers and a curious public at the NIC's regional conventions. Whereas the Franklin Institute had attracted nearly 300,000 visitors to large fairs held once every two years, the NIC hosted three smaller conventions per year—one each in the East, the Midwest, and the West. For example,

New York City's June 1937 expo attracted 1,000 delegates displaying 500 inventions for 25,000 visitors. Burns estimated that one in five inventors disposed of an invention at each convention; a 1932 congress in St. Louis (figure 4.9) "did $700,000 in business." Delegates also attended topical lectures (e.g., "How to Market Your Idea") and a concluding awards luncheon honoring each expo's top inventions. Some members exhibited at multiple

Figure 4.9
Inventor Florence Noonan Gemmer (standing) displays her safety seesaw at a National Inventors Congress expo in St. Louis, Missouri, on November 26, 1932. Sitting opposite an unidentified girl is the inventor's daughter, Virginia Gemmer, wearing her tiara as Queen of the Inventors' Congress. Photo by Harold Sneckner, Sievers Studio Collection, Missouri History Museum, St. Louis.

consecutive expositions, so a congenial community of inventors coalesced around Burns's traveling road show.[92]

Newspaper reporters loved to write patronizing stories about the odd-ball amateur inventors who peddled mind-reading headgear and weather modification schemes at the NIC's conventions. "Of course there are 'crackpot' inventors," Burns admitted, adding that "more than half of the world's inventors are nuts." But Burns encouraged his members to embrace their eccentricities—to revel proudly in being "screwy," to "glory in being called nuts."[93] This was perhaps the greatest benefit of membership in the National Inventors Congress. Like the earlier associations of women inventors and African American inventors, the NIC was a welcoming, supportive community where idiosyncratic amateur inventors were championed and celebrated.

With open arms, the National Inventors Congress welcomed inventors of all ages, from eight to eighty. Eight-year-old David Suddeth of Enid, Oklahoma, was the star of the NIC's March 1937 convention in Chicago. He delivered a lecture to the assembled congress and presented two labor-saving inventions: a device that peeled and mashed boiled potatoes in one operation and a window washer that simultaneously cleaned both the inside and outside panes. Two months later, the *Washington Post* reported that Suddeth's potato peeler was scheduled to go into commercial production. On the opposite end of the spectrum, eighty-year-old R. R. Snowden presented a "method for super-inducing rain" in January 1935 to relieve the catastrophic Dust Bowl drought. Snowden proposed using an airplane to release ammonium nitrate particles into the atmosphere. These particles, Snowden believed, would attract moisture until a raindrop was heavy enough to fall.[94]

Despite its predominantly White male membership, the NIC welcomed men and women from all ethnic and racial backgrounds. News coverage of the 1933 Los Angeles convention described a young Filipino man who displayed a coconut grater, and an African American man who presented a self-cleaning cuspidor. When the NIC returned to Los Angles in 1940, the exhibitors included Mr. and Mrs. Quincy James of Pasadena, an African American couple who displayed a novel cream whipper and an arrangement for locking ropes in place.[95] "American inventors are a motley and cosmopolitan crew," observed a reporter at the NIC's 1934 fair in Omaha, Nebraska. "Neither age, race or previous condition of servitude, seems to

have any bearing on the makeup of the men and women who are inventing things. Whites, negroes, and Chinese, were included in those entering
gadgets at the congress."[96]

Burns was especially proud of women's growing participation and influence within the NIC. In 1933, the Los Angeles convention began with a pag-
eant to elect a "Queen Patent Pending" from the women exhibitors, but this
trivialization was soon abandoned in favor of a women's auxiliary and an
NIC resolution affirming women's creativity. In fact, the NIC's female members were some of its most successful inventors. At the NIC's 1937 expo in
Chicago, Anne Hebner claimed the top prize for her "glider garter," which
moved freely to prevent runs in women's hosiery. Likewise, at New York's
1937 fair, Albany schoolteacher Kathryn Coburn earned a prize—and a potential deal—for her automatic yarn-winding machine. A manufacturer offered
her $23,000 for the gadget, which promised to emancipate boys and girls
from standing "with arms outstretched while grandmother does her spinning."[97] According to Burns, women constituted approximately 7 percent of
the nation's inventors, but the NIC's female members far exceeded that figure
by contributing 42 percent of the inventions at the 1940 Los Angeles expo.[98]

As the nation navigated the economic difficulties of the Great Depression, Burns supplemented the NIC's commercialization mission with a turn
toward political activism. In 1933, shortly after Franklin D. Roosevelt's
inauguration, Burns called for a "New Deal" for American independent
inventors. According to Burns, massive unemployment had swelled the
NIC membership rolls as Americans "turned to inventiveness" out of sheer
desperation. Echoing causes once championed by the AAIM and the Inventors' Guild, the NIC called for faster processing of patent applications, a
special patent court staffed with technically trained judges, and new laws to
protect inventors from "racketeers, pirates, and parasites." In 1934, Burns
called for a one-year moratorium on the $30 patent application fee, since
inventors had "plenty of ideas, but not much cash." Burns also gathered
168,000 signatures on a petition to establish a federal "inventors' loan
fund" to finance the commercialization of new inventions. In the proposed
scheme, the Reconstruction Finance Corporation would lend money to
assist the inventors of meritorious inventions, using a lien on the patent
for collateral. Burns reasoned that Roosevelt's New Deal was already subsidizing writers, architects, and artists, so it was "only fair that the inventors
be included."[99] The NIC's political agenda apparently gained little traction.

Burns was not invited to testify before the Temporary National Economic Committee's patent reform hearings in 1938 and 1939 (chapter 5).

For reasons that are unclear, the National Inventors Congress changed its name to "Inventors of America" (IofA) in 1939. Meanwhile, IofA members mobilized for preparedness in the months before the United States entered World War II. Spurred by patriotism (and hopes for lucrative procurement contracts), IofA members exhibited a cluster bomb, a model air raid shelter, and an amphibious vehicle at the organization's July 1941 New York expo.[100]

However, the onset of war seemed to trigger the IofA's demise. In September 1941, Burns hosted a final IofA convention in Los Angeles. In December, the Japanese attacked Pearl Harbor and drew the United States into World War II. Despite the increased demand for military inventions, the IofA was forced to go on hiatus. In a 1943 interview, Burns reported that "our hands are tied by wartime priorities." Burns did not elaborate, but presumably most of the IofA's members headed to the factories, farms, and front lines. Any remaining inventors would have been hampered by wartime restrictions on raw materials. Many independent inventors probably also redirected their energies away from the IofA and toward the National Inventors Council, a federal clearinghouse for defense inventions established in 1940 that had a direct line to the War Department (chapter 6).[101]

Ever the optimist, Burns predicted the IofA's postwar return and "the gosh darndest rush of inventions, the likes of which the world has never seen." He also reaffirmed his belief that "most inventions still come from talented—and often mildly insane—amateurs, not from the boys in the laboratories of the big corporations" who "always think of why things can't be done." Burns was right about the postwar rush of inventions, but the IofA never returned after World War II. From its founding in 1924 to its final expo in 1941, the NIC/IofA was active for only seventeen years. Like the earlier communities of women inventors and African American inventors, the NIC/IofA could not sustain its support for eccentric independents. Albert G. Burns, "the nation's gadget chief," died in 1951.[102]

Conclusion

"It seems like we never will accomplish anything toward an organization of inventors," wrote R. A. Summers, a beleaguered independent inventor, who sent this pessimistic assessment to Orville Wright in 1918 as they discovered that the National Institute of Inventors was a scam. "There is the Inventors Exchange at Los Angeles," Summers wrote, but "it got into a gang of crooks and busted up. There was the Inventors League Chicago—gone like the rest of the protectives to Hell I guess. It's getting so the poor inventor cannot trust any one at all."[103] Summers's lament captured both the eagerness of independents to establish professional organizations and a profound disappointment in their inability to sustain them.

Despite their independence and autonomy, individual inventors developed a collective consciousness and established a series of informal communities and formal membership organizations to confront their professional challenges. These associations helped inventors commercialize their inventions, lobby for political reforms, and garner recognition and respect from a skeptical public that assumed they did not exist. Unfortunately, inventors' eagerness to find solutions in professional organizations made them vulnerable to exploitation by scammers. Moreover, their legitimate organizations were fragile, short-lived, and ultimately ineffective. With the exception of the Franklin Institute, each group collapsed within ten to twenty years of its founding. A few groups struggled financially, others dissolved when their leaders shifted to new endeavors, and some never recovered after the world wars. Ultimately, the same social, political, and financial precariousness that motivated individual inventors to seek strength in numbers contributed to each group's demise.

The ephemeral nature of inventors' organizations is especially striking when compared with the durable associations—such as the American Physical Society (founded 1889) and the American Institute of Chemical Engineers (founded 1908)—that scientists and engineers formed during the same era. Along with attorneys and physicians, scientists and engineers successfully established their disciplines as "professions," which were characterized by extensive training and expertise in a canon of specialized knowledge; a high degree of independence, autonomy, and prestige; and self-governing organizations that controlled admittance into the profession and upheld its ethical standards.[104]

Inventors founded numerous membership organizations, but they failed to achieve the other elements of professionalization. First, there were no educational requirements or qualifying exams to become an independent inventor, so *anyone* could join an inventors' organization. The welcoming, egalitarian nature of inventors' organizations was beneficial for women and minorities, but it prevented the exclusivity required of a profession. Second, unlike the tight-knit associations that scientists and engineers formed around academic subdisciplines, inventors' organizations welcomed members with every kind of expertise. Without a shared intellectual background to unify members, inventors' organizations struggled to maintain cohesion. Finally, inventors enjoyed the independence and autonomy of a profession, but they did not always command a high degree of prestige and intellectual authority. Successful inventors such as Elihu Thomson were well respected, but most "gadgeteers" and "crackpots" were precisely the kind of low-status amateurs that elite scientists and engineers sought to exclude from their ranks. In the end, inventors did not professionalize, because their occupation was too heterogeneous and not exclusive. Instead, independent inventors have maintained an ambiguous occupational status similar to medical technicians, computer programmers, and other "semi-professions."[105]

Without the benefit of stable professional organizations, inventors were less likely to access the technical training, professional connections, and marketing opportunities they needed to commercialize their inventions. They were also less unified in countering the disparaging rhetoric and coordinated PR campaigns of the industrial researchers who were encroaching on their turf. Finally, inventors were unable to achieve political reforms to correct the inequities in the patent system. Indeed, the patent system proved remarkably resistant to reform, even during the progressive fervor of President Franklin D. Roosevelt's New Deal.

5   The Elusiveness of Patent Reform

The patent system of the United States, more than any other in the world, offers hope, encouragement, opportunity and recompense to an individual or a company of small resources. It is as democratic as the Constitution which begot it.
—Conway Coe, Commissioner of the US Patent Office, January 1939

A large portion of the U.S. population is persuaded that any monopoly, even the temporary one created by the patent grant, is fundamentally evil; and the argument that patents help "the big get bigger" has a particularly convincing ring when it coincides with evidence that the lone inventor is being denied a fair opportunity.
—Edmund Van Deusen, "The Inventor in Eclipse," *Fortune*, December 1954

In April 1938, during the dog days of the Great Depression, President Franklin D. Roosevelt called on Congress to make "a thorough study of the concentration of economic power in American industry" and its impact on the economy. In his "Monopoly" message, Roosevelt warned that collectivism and anticompetitive practices among big businesses had perpetuated the Great Depression and if left unchecked could even lead to fascism, as witnessed by the "unhappy events abroad" in places such as Germany, Italy, and Spain.[1] The resulting Temporary National Economic Committee (TNEC) spent the next thirty months conducting a comprehensive investigation of the nation's economic ills. By the time the hearings concluded in March 1941, the committee had accumulated a mountain of evidence—over 20,000 pages of testimony from 552 witnesses and forty-three additional monographs commissioned on specific subjects—eighty-two printed volumes in all. The TNEC's thirty-one hearings examined antitrust policies, tax codes, and business practices in various industries, including insurance, oil, and banking.[2]

However, the TNEC did not begin by tackling these core economic issues; instead, it began with patents.[3] In December 1938, the TNEC hearings commenced with two hearings on the use and abuse of patents in the maintenance of corporate monopolies. The committee solicited testimony from witnesses such as Patent Commissioner Conway Coe, independent television inventor Philo Farnsworth, and Bell Labs' research director Frank B. Jewett. Meanwhile, the National Association of Manufacturers (NAM), a conservative, big-business lobbying group, developed a sophisticated public relations campaign designed to thwart some of the TNEC's suggested reforms.

Why all the fuss about *patents*? With so many thorny economic issues complicating the Great Depression, why did FDR's "Monopoly" committee devote so much attention to arcane questions of US patent policy? In short, the TNEC recognized that abusive patent practices were at the heart of a rigged economy that had collapsed during the Great Depression. Several of the largest technical firms—including General Motors (GM), General Electric (GE), and the Radio Corporation of America (RCA)—had used their massive patent portfolios and aggressive litigation to crowd out independent inventors, suppress new technologies, limit competition, fix prices, and generally exert tyrannical control over their industries. Many observers worried that these abuses would eventually bring all technologies under the control of the largest corporations, wipe out independent inventors and small businesses, and stifle the innovations that could lead to economic recovery. So patents became a key, if underappreciated, New Deal reform issue.[4]

Independent inventors had been seeking patent reforms for decades. In the 1890s, the American Association of Inventors and Manufacturers complained that an overcrowded, backlogged Patent Office, low examiner pay, and shoddy examinations resulted in weak patents and a rash of infringement lawsuits. During the 1910s, Leo Baekeland and the Inventors' Guild tried unsuccessfully to streamline the patent and judicial systems to prevent well-heeled corporations from using excessive litigation as a weapon against independent inventors. After those reform efforts stalled (chapter 4), corporate exploitation of the patent laws only intensified. As the economy slid into the Great Depression, a new generation of independents—including Farnsworth and Charles Eisler (bulb and tube-making machinery)—were still struggling against the aggressive practices of individual firms and patent-based cartels. After decades of futility, independents saw the TNEC hearings as another opportunity to reform a corrupted patent system, check the power of the big corporations, and level the playing field for individual inventors and small businesses.

During the hearings, progressive New Dealers reintroduced a series of familiar reform proposals, including a "twenty-year proposal" to discourage long-pending patent applications, a single court of patent appeals with nationwide jurisdiction, and limits on restrictive patent licenses. However, a combination of factors—including the NAM's conservative resistance, independent inventors' political disorganization, and the onset of World War II—resulted in little new legislation. Once again, independent inventors achieved few of their desired reforms, resulting in a patent system that, even today, remains more agreeable to corporations.

Aggressive Corporate Monopolies, Beleaguered Independent Inventors

The rise of corporate patenting had highlighted an inherent tension between the patent laws and the antitrust statutes. Article I, Section 8, of the US Constitution granted Congress the power to "promote the progress of science and the useful arts" by granting inventors the exclusive rights to their inventions for a limited period. The ensuing Patent Act of 1790 established a fundamental trade-off that balanced individual rights and the public good. When the US government issued a patent, it offered an inventor a fourteen-year (later seventeen-year) monopoly on the use of his or her invention. During this period, the inventor held the exclusive right to commercialize and profit from the invention with some legal recourse against piracy. In exchange for this government-sanctioned protection, the inventor was required to immediately publish a full description of the invention (the patent), which became freely available for public use after the term expired. The founders believed the patent's temporary period of exclusivity would economically incentivize inventors to create more new technologies, while the eventual free and widespread use of those technologies would benefit the public.[5]

Congress passed the Sherman Antitrust Act of 1890 to prevent monopolistic behavior by big firms and trusts, but the law inadvertently spurred the anticompetitive exploitation of corporate patents. The Sherman Act discouraged cartels and horizontal mergers, so firms such as GE and DuPont turned inward and invested in new R&D labs, since individual firms could legally exert tight market control over their own inventions.[6] By earning their own in-house patents, forging cross-licensing agreements with other firms, and buying up useful patents from independent inventors, corpora-tions could acquire massive patent portfolios that provided monopolistic control over the technologies in certain industries.[7]

By the 1930s, contemporary observers had become concerned that the big technology firms were using their patents in anticompetitive ways the founders had not intended. For example, AT&T sustained its telephone monopoly by aggressively defending Alexander Graham Bell's patent. An 1896 patent cross-licensing agreement helped rivals GE and Westinghouse exert monopolistic control over the electrical industry, including 93 percent of the US incandescent lamp market. In 1919, RCA pooled the key wire-less patents from AT&T, GE, Westinghouse, American Marconi, and United Fruit and then dominated the radio industry. Collectively, high-tech firms had come to recognize a strong patent portfolio as the most effective means of controlling competition in a given market, constituting a "monopoly of monopolies."[8]

The big corporations aggressively leveraged their patent portfolios to harass independent inventors who dared to compete in their industries. Károly "Charles" Eisler (1884–1973) was a Hungarian American inventor who found himself in the crosshairs of the incandescent lamp cartel. Eisler had worked as a journeyman engineer in Hungary, Germany, and around the United States before joining Westinghouse's equipment division in Bloomfield, New Jersey, in 1914. Drawing on his prior lamp-making experi-ence in Hungary, Eisler designed machines that automated the production of Westinghouse light bulbs. GE, Westinghouse, and their international cartel partners maintained their incandescent lamp monopoly by control-ling the patents on the lamps themselves, plus the machinery and process patents for manufacturing them.[9]

After five years at Westinghouse, Eisler was recruited to serve as the chief engineer of Save Electric, a Brooklyn-based lighting start-up. His Westing-house colleagues warned him not to take the position, since GE and the lamp cartel would no doubt put Save out of business through injunctions and infringement lawsuits. Save's president, Max Ettinger, was rebuffed when he approached GE to obtain lamp-making licenses, so he ordered Eisler to spend the next six months developing lamp-making machinery that he believed would circumvent the cartel's patents. With Eisler's new machines producing 25,000 lamps per day, Save quickly became the largest independent lamp maker operating outside the cartel. In 1920, GE filed a series of infringement lawsuits against Save Electric and eventually made a buyout offer. According to Eisler, GE paid Save Electric "one million dollars to quit the incandescent lamp manufacturing business." Eisler was suddenly unemployed, just as his Westinghouse coworkers had predicted.[10]

Then Eisler had an epiphany: he had redesigned Save's lamp-making machines and, within a year, the firm's owners had sold out to GE for a handsome profit, so why not do the same thing for himself? Eisler acquired a small machine shop in Newark once used by Thomas Edison, then again redesigned and patented improved lamp-making machines. Eisler believed his designs would avoid the cartel's patents by employing an angled exhaust tube in the bulb stem, which he later called the "million-dollar bend." His tungsten filament winders, bulb-frosting machines, and vacuum sealers operated more slowly and with less automation than the cartel's machines, but they were cheaper and, he believed, noninfringing. The Eisler Engineering Company opened for business in 1920, and orders poured in from independent lamp manufacturers who operated outside the cartel. Eisler also adapted his machinery to supply the soaring demand for radio tubes (figure 5.1), and his business boomed. By 1924, Eisler employed 450 workers at his expanded Newark plant.[11]

Predictably, GE initiated four separate infringement lawsuits against Eisler between 1924 and 1928. "Everything humanly possible was done to put me out of business," Eisler recalled, as the cartel sought to prevent him from supplying machinery to the independent lamp firms. Over the years, the lamp cartel had used similar pressure tactics to shut down or buy out about twenty-five small firms. The remaining holdouts were now in jeopardy of contributory infringement for using Eisler's bulb-making machines. In 1924, Eisler rallied his customers to form the Incandescent Lamp Manufacturers' Association (ILMA); the legal defense fund raised $100,000, including $45,000 from Eisler. At trial, Eisler could show prior art from his original Hungarian designs (1912) that predated his time at Westinghouse. He had also published a 1916 article in *Machinery* magazine describing his lamp-making techniques. In 1927, during the litigation, GE attempted to buy out Eisler, but he refused to capitulate. Finally, after nearly five years of expensive depositions, trials, and appeals, Eisler's patents were upheld, while several of the cartel's patents were invalidated. The New Jersey papers called Eisler a "trust buster," but he had won a Pyrrhic victory. Eisler ruefully noted that the cartel had harassed him and his customers to eliminate

Figure 5.1
Charles Eisler, from his advertisement, in 1929. Eisler invented machinery to auto-
mate the production of light bulbs and radio tubes. He faced harassment from the
GE-Westinghouse lamp cartel. Eisler Engineering Company Records, Archives Center,
National Museum of American History, Smithsonian Institution, Washington, DC.

a mere 2 percent of the incandescent lamp market. Over the next thirty
years, the cartel eventually sent $50 million in contracts to various machin-
ery firms but not a nickel of business to Eisler.[12]

Inventor Philo T. Farnsworth (figure 5.2) endured a similar struggle as
RCA used aggressive patent practices to extend its control of radio into television. Farnsworth (1906–1971) was a mechanically precocious Utah farm
boy who was born in a log cabin with no electricity. He grew up reading

Figure 5.2
Philo T. Farnsworth films dancers during his television demonstrations at the Franklin Institute in August 1934. Farnsworth and RCA spent a decade battling over their television patents. From the Collections of the Franklin Institute, Philadelphia, PA (cat. no. PR713).

about Edison, Bell, and the Wright brothers in *Science and Invention* magazine and closely followed the contemporary scientific advances of Albert Einstein. Farnsworth claimed that he conceived of electronic television at age fourteen while plowing parallel rows in a potato field. In a flash of inspiration, Farnsworth imagined a system in which an electron emitter acquired a picture by scanning a series of parallel lines, transmitted those signals wirelessly, and then reassembled the image on a faraway screen. At age nineteen, Farnsworth obtained start-up funding from some local Utah backers and eventually established a laboratory in San Francisco with additional financing from the Crocker family, who had made a fortune building the transcontinental railroad. By September 1928, at age twenty-two, Farnsworth had applied for several patents and had a working prototype. To attract additional backers and potential licensees, Farnsworth demonstrated his television system for reporters in a widely circulated story for the *San Francisco Chronicle*.[13]

As he read the *Chronicle* account, RCA's chief executive, David Sarnoff, could hardly believe he had been scooped by a twenty-two-year-old kid. Sarnoff intended to dominate television just as he had dominated radio—by controlling every important patent. After RCA was formed in 1919, Sarnoff pooled the wireless patents from the partnership's R&D labs and purchased all the important radio patents from independent inventors such as Edwin Armstrong. If a rival firm like Zenith or Philco hoped to market a radio set, it would have to license at least one of RCA's patents. With this leverage, Sarnoff employed the controversial practice of "package licensing," in which manufacturers were required to pay $5 million in royalties to license the entire package of RCA radio patents instead of paying lower royalties for the one or two patents they needed. Licensees complained that these extortionist payments were merely insurance against being sued by RCA, like paying protection money to the mob. Sarnoff routinely refused license requests, then brought infringement cases against any competitors who dared to operate outside the "Radio Trust."[14]

Sarnoff had developed a similar strategy for television, but it was not going so well. Among RCA's partner firms, AT&T's Herbert Ives and GE's Ernst Alexanderson were both pursuing mechanical television, a less-promising approach that used a spinning, perforated disk to scan images. Westinghouse's Vladimir Zworykin had filed a still-pending patent application on an electronic television system in 1923, but he had been unable to construct a working prototype. In 1929, Zworykin moved from Westinghouse to the new RCA laboratories in Camden, New Jersey, but progress on his television developments had stalled. In April 1930, Sarnoff sent Zworykin to San Francisco to meet Farnsworth and see a demonstration. Farnsworth's backers eagerly anticipated Zworykin's arrival; they were looking for a corporate licensee or buyer so they could recoup their investments as the Great Depression set in.[15]

Zworykin's visit, conducted under false pretenses, initiated a decade of harassment, as RCA sought to pry electronic television away from Farnsworth. After three days of demonstrations, Zworykin returned to RCA's labs and immediately tried to reverse engineer Farnsworth's system. In April 1931, Sarnoff himself visited Farnsworth's lab and offered $100,000 to purchase the start-up and its patents outright. Farnsworth and his backers refused the lowball offer, so Sarnoff turned up the heat. When Farnsworth signed a June 1931 consulting agreement to help Philco start a television business, RCA threatened to revoke Philco's radio licenses, which would have put the firm out of business.[16]

In May 1932, RCA initiated interference proceedings to overturn Farnsworth's patents. Interferences were administrative hearings, convened by Patent Office examiners, to determine the priority of competing applications that described similar technologies. RCA asserted that Zworykin's still-pending 1923 patent application had priority and should be issued immediately, thereby voiding Farnsworth's already issued patents from 1930. In July 1935—after three years of depositions and hearings—the Patent Office ruled in Farnsworth's favor, but RCA appealed and the case dragged on. The prolonged interference hearings consumed Farnsworth's time and money, diverted him from technical work, and allowed RCA's scientists to develop a competing television system, all while Farnsworth's seventeen-year patent clock ticked away.[17]

By the mid-1930s, monopolistic firms like GE and RCA were increasingly using aggressive patent practices to bully independent inventors like Eisler and Farnsworth. Meanwhile, consumers paid artificially high prices for light bulbs and radio sets and wondered whether legal machinations had delayed the introduction of a new technology like television. The Justice Department had launched multiple antitrust prosecutions against GE (in 1911, 1924, and 1930) and RCA (in 1930), but these actions did little to weaken the trusts and only highlighted the inherent tensions between the patent and antitrust laws.[18] As President Roosevelt searched for answers to the Great Depression, his administration eventually recognized the necessity of tackling corporate monopolies head-on.

Organizing the TNEC Hearings, 1937–1938

When Roosevelt assumed office in 1933, he faced a thorny dilemma. Economically, FDR needed the cooperation of big businesses to jumpstart the recovery, but politically he needed them to be scapegoats. Consequently, as Ellis Hawley has argued, Roosevelt's administration struggled to find a "coherent and logically consistent set of business policies." Initially, the New Deal turned to business-government cooperation, in which the National Recovery Administration (NRA) set prices, wages, and "codes of fair competition" in troubled industries. After the Supreme Court struck down the NRA in 1935, the administration helped trade associations self-regulate certain "sick industries" (e.g., coal, oil, and cotton) and "natural monopolies" (e.g., public utilities, railroads) that were essential to the economy.[19]

These planning efforts led to a tenuous recovery, and after his 1936 reelection, Roosevelt curbed stimulus spending to rebalance the federal budget, triggering the "Roosevelt Recession" of 1937. Antitrust hawks within the administration blamed big businesses for the downturn. They argued that several industries had enjoyed price gains via the endorsed collusion of the administration's planning efforts but had pocketed the resulting profits instead of expanding employment or increasing wages. Roosevelt vacillated among several possible responses before settling on a two-pronged approach combining renewed deficit spending and an invigorated antimonopoly campaign.[20] On April 14, 1938, Roosevelt sent his spending message to Congress and encouraged the renewal of work relief and public works programs. Two weeks later, on April 29, FDR sent Congress his antimonopoly message and called for the TNEC investigations.[21] The goal of the forthcoming hearings, a journalist observed, was to gather information and "lay the foundation for legislation out of which will emerge a 'modernized capitalism.'"[22]

During the spring and summer of 1938, the twelve members of the Temporary National Economic Committee were appointed from Roosevelt's administration and a mix of congressional Democrats and Republicans (figure 5.3). Senator Joseph C. O'Mahoney (D-WY) chaired the committee; some reporters jokingly referred to him as "Senator O'Monopoly." Notably, the committee featured FDR's antitrust czar, Thurman Arnold, and several administration officials who practiced "institutional economics," an emerging subfield that rejected laissez-faire principles in favor of government intervention in the economy.[23]

Reflecting the institutionalists' embrace of empirical data, the first hearing (Part 1), in November 1938, was an "Economic Prologue," in which TNEC staff members presented a bewildering array of charts and statistics on population, national income, prices, employment, wages, and other indicators. The next two substantive hearings investigated abuses of the patent system. In Part 2, held in December 1938, the Department of Justice's Antitrust Division examined the relationship between patents and the antitrust laws by studying their use in two sectors, the automotive and glass

Figure 5.3
Democratic members of the Temporary National Economic Committee congregate during preparatory meetings on July 1, 1938. *Left to right*: Chairman Joseph O'Mahoney, William O. Douglas, Edward C. Eicher, Hatton W. Sumners, Thurman Arnold, and William H. King. Photo by Harris & Ewing, image no. LC-DIG-hec-24801, Prints & Photographs Division, Library of Congress, Washington, DC.

container industries. In Part 3, held in January 1939, the Department of Commerce (with jurisdiction over the US Patent Office) organized hearings on proposals for changes in patent laws and procedures.[24]

Patents in the Automotive and Glass Container Industries

In the first patent hearing, held in December 1938, the TNEC examined patent practices in the automobile and glass container industries. Taking the lead was TNEC member Thurman Arnold, a former Yale law professor who now served as the assistant attorney general in charge of the Antitrust Division. Immediately following his appointment in March 1938, Arnold helped Roosevelt draft the speech calling for the TNEC hearings. He also began vigorously prosecuting the antitrust laws after years of sporadic enforcement. In justifying his choice of witnesses for the hearing, Arnold explained that the automobile industry represented a "nonaggressive use of the patent privilege," while the glass container industry represented an "aggressive" abuse of patents.[25]

Arnold began with the automotive industry and called witnesses from the Ford Motor Company, General Motors, the Packard Motor Car Company, and the Automobile Manufacturers Association. Collectively, the witnesses described the context for the industry's generally open exchange of patents. An automobile is a complex assemblage of hundreds of patented technologies—from oil filters to windshield wipers—so it is nearly impossible to build a vehicle without infringing on someone's patented component. This fact made the fledgling industry vulnerable to the abuses of inventor George Selden, one of the earliest "patent trolls."[26] In 1879, Selden filed a broad patent application covering an internal-combustion engine in combination with a four-wheeled chassis. Selden continuously amended the patent application and succeeded in stalling its issuance for sixteen years. When Selden's patent was finally issued in 1895, several automakers suddenly found themselves in jeopardy of infringement. Together with William C. Whitney, Selden formed the Association of Licensed Automobile Manufacturers (ALAM) and succeeded in extracting royalties from most automakers, even though Selden himself never manufactured an automobile from his own patent. Henry Ford and four other manufacturers contested the Selden patent, and after years of litigation, the patent was invalidated in 1911, one year before its expiration.[27]

After the Selden affair, the auto industry devised various strategies to manage its patents and avoid expensive lawsuits. Alfred Reeves, general manager of the Automobile Manufacturers Association (AMA), testified that thirty-four automakers—basically every key firm except Ford—belonged to his organization and shared their patents through the AMA patent pool. Members were granted nonexclusive, royalty-free licenses on any of the pool's 1,058 patents.[28] In contrast, Edsel Ford testified that the acrimonious Selden litigation had cemented his firm's resolve to avoid any industry associations or cross-licensing agreements. Instead, Ford took out patents to protect its intellectual property and to defend itself vigorously against charges of infringement. However, Ford generally did not sue infringers of its own patents but instead granted nonexclusive, royalty-free licenses to any petitioners. Finally, Packard vice president Milton Tibbetts testified that his firm had joined the AMA to benefit from its New York Auto Show, traffic research, and safety campaigns, but it was not part of AMA's patent pool. Packard routinely granted licenses on its patents, but unlike Ford, it charged a royalty to recover its development costs. In turn, Packard paid royalties to take licenses on the patents of other manufacturers. Over the past thirty years, Packard had collected $4,099,707 in royalties while paying $553,401 to license others' patents. However, given Packard's $25 million in annual revenues, royalty income and payments were relatively insignificant.[29]

Overall, there was a nearly free exchange of the patents produced by the R&D labs of the big automakers, yet patents remained a crucial bargaining chip for the independent inventors and small firms that supplied them. Joseph Farley, Ford's patent counsel, explained that most manufacturers were unwilling to acquire useful new devices from outside, independent inventors and small firms unless they could be assured of some patent protection. However, Farley suggested that these transactions were relatively rare, since most new automotive improvements emerged internally, from the "engineering departments of the corporations who are engaged in the actual manufacture of automobiles."[30]

This spurred a line of inquiry on the viability of independent inventors in the automotive industry, including some inconsistent testimony from General Motors' research director, Charles Kettering. On the one hand, Kettering confirmed the continuing contributions of independent inventors:

> **Mr. Patterson:** Mr. Kettering, there was some testimony yesterday to the effect that practically all of the valuable and worthwhile inventions came from the industrial laboratory and very few from the outside. Do you coincide with that?
>
> **Mr. Kettering:** Oh, no, no, no, no; there are a lot of very brilliant people outside of industry. We say we don't lock our laboratories up for the reason that we lock so much more out than we can in.[31]

However, at a different moment, Kettering insisted that a "one-man invention isn't very possible these days," since the complexity of automotive technology required that his research staff "work together as a group." Kettering—a former independent inventor who now directed one of the world's largest R&D labs—clarified that the United States was at "the transition point" between the era of "individual invention" and "group invention."[32] The committee would continue to ponder the sources of invention throughout the TNEC hearings.

The second set of witnesses provided more fireworks. Through testimony and various exhibits, the Department of Justice revealed how the Hartford-Empire Company and its affiliates had used patents and restrictive licensing provisions to gain monopoly control over the glass container industry.[33] Hartford-Empire had been founded in 1922 through the merger of small machine tool companies that Corning and General Electric had established to automate the manufacture of glassware and glass light bulbs. In 1924, Hartford-Empire negotiated a patent cross-licensing agreement with its primary competitor, Owens-Illinois. Through this alliance, the firms controlled nearly all key patents for the automated manufacturing of glass bulbs and containers. By 1938, the "Glass Trust" administered a pool of over 700 patents, such that 97 percent of the glass industry paid royalties and operated under licenses held by Hartford-Empire (67.4 percent) and Owens-Illinois (29.2 percent).[34]

Hartford-Empire utterly dominated the glassware industry by exploiting its impenetrable patent position. The firm did not have a plant. It did not manufacture or sell any bottles. It did not even fabricate its own patented glass-making machines, which were built and shipped to licensed glass producers by a contract manufacturer. Hartford-Empire leased its machines, so the firm derived no income from direct sales. However, the firm earned licensing fees and royalties on nearly every glass container and light bulb produced in the United States. These revenues came to over $6 million in 1937 and over $40 million between 1923 and 1937. Hartford-Empire employed dozens of industrial researchers to develop improvement patents and retained several lawyers to help it maintain its monopolistic grip on the industry. But exactly how did they do it? What were the defects in the patent system that allowed Hartford-Empire to control the glassware industry?[35]

The committee found that Hartford-Empire took advantage of the long pendency of certain patent applications to effectively extend the lives of their basic patents. By continuously amending their patent applications and purposefully employing dilatory tactics, Hartford-Empire might prolong the issuance of their patents for several years, thereby delaying the start of the seventeen-year expiration clock. Later, these "submarine" patents would be issued and then "surface" years after competitors were already using the process, which suddenly placed them in jeopardy of an infringement lawsuit. Then, under the threat of litigation, smaller competitors would typically capitulate and take out a license on the submarine patent on terms favorable to the cartel. This legal extortion was a tactic borrowed directly from George Selden's playbook.[36]

Hartford-Empire aggressively pursued infringement lawsuits against the two or three independent firms that tried to operate outside the Glass Trust. For example, the Obear-Nester Company was a small glassware maker that commanded 2 percent of the glass container market. The firm believed its patented "air feeder" method avoided infringement of the Hartford-Empire patents. However, in 1926, Hartford-Empire brought the first of three infringement lawsuits against Obear-Nester, resulting in twelve years of continuous and ongoing litigation. Hartford-Empire believed the prolonged, expensive cases would force Obear-Nester to take a license on the cartel's patents or bankrupt the smaller firm out of business. Obear-Nester held firm and eventually prevailed, but Hartford-Empire created a legal complication when it sued the other independent glassmakers in different jurisdictions. There was a "contrarity of opinion" among the appellate courts, as Hartford's patent claims were held valid in the eighth judicial circuit but invalid in the sixth.[37]

Another key to Hartford-Empire's monopoly was its highly restrictive patent licensing agreements. These licenses specified which firms could manufacture milk bottles, beer bottles, or cookware. They also restricted production (i.e., the number of bottles a licensee could manufacture) and even specified the types of customers and geographical territories where containers could be sold. Although none of the licenses contained illegal price restrictions, the effect on prices was implicit. If only one licensee could produce milk bottles, that manufacturer effectively had a price monopoly. The TNEC members questioned whether a firm's exclusive right to make, use, and sell its patented items should also extend so forcefully to control over its licensees.[38]

In addition, testimony and damning internal memos revealed that Hartford-Empire strategically used patents to "block" or "fence in" its rivals. As Hartford-Empire's engineers invented several ways to produce glassware, they took out "defensive" patents on these alternative methods. Hartford-Empire then suppressed these patents for seventeen years to block its rivals from using them while protecting its own favored processes. However, if a competitor succeeded in gaining a basic alternative patent, Hartford-Empire could "fence them in." By carefully studying the competing art and surrounding it with numerous improvement patents, it became nearly impossible for the owner to exploit the basic patent without infringing on Hartford-Empire's improvements. These tactics prevented rivals from inventing around Hartford's patent pool, practically forcing glassware manufacturers to take a license on the firm's preferred methods or face an infringement suit.[39]

Finally, if Hartford-Empire caught wind of a rival's patent application, it might hurriedly submit its own application on a similar aspect of the art. This would compel the Patent Office to declare an interference proceeding to determine which application held priority. Testimony revealed how Hartford-Empire had laid "a series of traps ... consisting of new applications" to "stage a delaying fight" with its competitor Whitall-Tatum. Interferences were bureaucratic proceedings, but as in infringement litigation, Hartford-Empire could use its superior legal and financial resources to wear down its competitors, subjecting them to time-consuming depositions, unanticipated legal fees, and time away from technical developments, all while forestalling the ultimate issuance of the competing patent.[40]

Overall, the first TNEC patent hearing revealed the nearly free exchange of intellectual property among automakers and the continuing importance of patents among the independent inventors who hoped to supply them with new devices. The hearings also demonstrated how the Glass Trust had used their patents to obstruct (rather than encourage) innovation, which subverted the original constitutional intent. Moreover, the patent system's systemic flaws worked disproportionately to the advantage of wealthy and powerful firms, a sobering reality for America's independent inventors and small businesses.

Proposed Reforms to the Patent Laws

In the second patent hearing, held in January 1939, the TNEC considered several legislative proposals to amend the patent laws. The first witness, Patent Commissioner Conway Coe, presented over a dozen charts and graphs outlining the historical utilization of the patent system. He showed that, in 1938, 157 large corporations (those with assets greater than $50 million) collectively earned 6,415 patents (17.2 percent), while individuals (42.9 percent), small corporations (34.5 percent), and a handful of foreign corporations (5.4 percent) accounted for the balance. In terms of ownership, Coe showed that between January 1, 1931, and June 30, 1938, foreign and domestic corporations of various sizes had purchased the rights to 9,548 patents, about 5.8 percent of the patents issued to individuals. However, there were only a few "hoarders"; that is, only twenty-two firms owned more than 1,000 patents. Coe believed these statistics affirmed that the patent system still offered "hope, encouragement, opportunity and recompense to an individual or a company of small resources. It is as democratic as the Constitution which begot it."[41]

Coe then offered several proposals to correct the structural defects in the patent system. First, several abuses stemmed from applicants' attempts to purposefully prolong the application process. As a remedy, Coe suggested the twenty-year proposal, in which a patent would expire seventeen years from the date of issuance or twenty years from the date of filing, whichever came first. Most inventors would still enjoy the usual seventeen-year protection; however, inventors or firms that delayed the prosecution of their applications for more than three years would only penalize themselves by diminishing the remaining patent term. To streamline interferences, Coe suggested abolishing the Patent Office's appeals process and having the examiner immediately issue the winning patent following the initial determination of priority. Any subsequent disputes over the validity of the issued patent could still be argued in the federal courts, but with the seventeen-year expiration clock already ticking toward free public use. Coe also proposed the creation of a single court of patent appeals with nationwide jurisdiction. This would discourage wealthy patent holders from suing their competitors in multiple federal districts and eliminate the ambiguity of contrary opinions across the various appellate circuits.[42] Coe's proposals echoed reforms first proposed by the Inventors' Guild in 1910 (chapter 4).

Coe also proposed several simple procedural changes designed to expedite the application process. The current law permitted an inventor to make public use of an invention for two years before filing an application; Coe suggested reducing the grace period to one year. Similarly, the law allowed an applicant two years after the issuance of another inventor's patent to assert his own priority claim to the invention; again, Coe proposed limiting the challenge period to one year. Finally, applicants were allowed six months to reply to official Patent Office correspondence; Coe suggested a thirty-day response period.[43]

Coe concluded his testimony (figure 5.4) by describing the patent system's intangible impact on the national psyche. Before making his policy

Figure 5.4
During his TNEC testimony on January 16, 1939, Patent Commissioner Conway Coe proposed several reforms while describing the patent system's "spiritual influence" on the national psyche. Photo by Harris & Ewing, image no. LC-DIG-hec-25844, Prints & Photographs Division, Library of Congress, Washington, DC.

recommendations, Coe had presented a series of sixteen famous patents to remind the TNEC of the nation's "indebtedness" to a patent system that had encouraged heroic inventors such as Eli Whitney, Cyrus McCormick, and Samuel Morse.[44] Now, in closing, Coe again suggested that, beyond the "evolvement of things purely mechanical," the patent system had instilled in America's citizen-inventors the virtues of "patience, resoluteness, sacrifice—suffering, too, if need be—in the pursuit of an ideal." Coe's romantic appeal to the heritage and traditions of the patent system—invoking its "spiritual influence in our national life and destinies"—would later be adopted by pro-business conservatives to oppose various reforms.[45]

Coe was a moderate, and his recommendations were essentially procedural in nature. While his proposals aimed to correct certain abuses, they did not change the fundamentals of the system.[46] However, the reforms suggested by Thurman Arnold, the Justice Department's antitrust czar, were more radical. Arnold recommended compulsory licensing statutes to mitigate defensive patenting and the suppression of technologies that firms never intended to use. If an allegedly suppressed patent had gone unutilized for three years, the statute would compel patent holders to grant licenses to petitioners at a royalty determined by the courts. Arnold also sought to ban any restrictions on output, price, use, or geography in patent licenses and proposed that all licensing agreements and contracts be registered with the Federal Trade Commission. To limit the use of excessive litigation as a "tool of business aggression," Arnold also recommended that infringement plaintiffs be required to first sue the original patent holder before pursuing any licensees for contributory infringement. Finally, Arnold proposed a rule requiring forfeiture of a patent to the public domain if the patent holder were convicted of engaging in anticompetitive activities. This penalty was designed to discourage abuses altogether.[47] Several of Arnold's proposed reforms—including compulsory licensing and limits on restrictive licensing—echoed provisions of the 1912 Oldfield patent bill that the Inventors' Guild had opposed.

Independent Inventors and R&D Executives Testify

During the remainder of the hearing, the committee heard testimony from ten diverse users of the patent system. The witnesses included independent inventors, small businessmen, corporate research directors, a scientific administrator, and an international patent attorney. Through their testimony, the committee learned about the growth of large-scale industrial research and its potential threat to independent inventors. The committee also solicited testimony on the various defects in the patent system and the merits of the potential remedies proposed by Coe and Arnold.[48]

Both independent inventors and small businessmen explained how the big firms had harassed them by manipulating the patent system. John Graham, president of Motor Improvements, Inc., described how General Motors abused the Patent Office's sloppy application and interference procedures to victimize his auto parts firm. According to Graham, independent inventor Ernest Sweetland had filed a patent application on an improved oil filter in 1920. Motor Improvements acquired the patent in 1923 and began manufacturing the "Purolator" oil filter in 1924 under the pending patent. Sweetland's patent was finally issued in 1926, but by then AC Spark Plug, a General Motors subsidiary, had begun making a similar filter, so Motor Improvements sued GM for infringement.

To circumvent the lawsuit and invalidate the Sweetland patent, GM paid an inventor named Cole $40,000 for rights to his long-pending 1918 application covering similar principles of filtration. Cole's claims had consistently been denied by the Patent Office, so after Sweetland's patent was issued in 1926, GM and Cole copied—verbatim—claims from Sweetland's issued patent and added them as an amendment to Cole's still-pending 1918 application. This forced an interference proceeding between the pending Cole application and the issued Sweetland patent. Incredibly, the Patent Office issued the Cole patent (with Sweetland's copied claims), so GM countersued Motor Improvements for infringement, since Cole's 1918 patent application antedated Sweetland's 1920 application. The various suits and countersuits raged on from 1926 to 1938 and cost Graham's firm $300,000. Eventually, the courts determined that the Cole patent was invalid, and GM settled with Motor Improvements for over $1 million.[49]

As he concluded his testimony, Graham agreed with Coe's suggestion that independent inventors and smaller firms were at "a decided disadvantage" in protracted disputes with "a large corporation or an adversary of considerable strength." Representative B. Carroll Reece (R-TN) struggled to reconcile Graham's mistreatment with earlier testimony describing cooperative cross-licensing in the auto industry. "I rather got the impression," Reece said sarcastically, that GM "never harmed anybody."[50]

The TNEC also heard testimony from independent inventor Philo T. Farnsworth. By January 1939, Farnsworth and his investors had organized the Farnsworth Television and Radio Corporation and he was racing with RCA to introduce the first commercial television sets. Many anticipated that Farnsworth would tell a harrowing tale of how David Sarnoff and RCA had used aggressive patent practices to sabotage his efforts. To preempt the anticipated negative publicity, RCA's chief patent lawyer, Otto Schairer, scheduled a speech before the National Industrial Conference Board to coincide with Farnsworth's testimony. Schairer hypocritically expressed support for Commissioner Coe's proposed patent reforms—including limits on long-pending patents and streamlined interferences—that would have curtailed the very abuses he and Sarnoff had perpetrated against Farnsworth.[51]

However, instead of attacking RCA, Farnsworth's TNEC testimony was muted and matter-of-fact. Without naming RCA, Farnsworth testified that he had been involved continuously in interference proceedings since 1927, some twenty to twenty-five altogether. The most serious case, Farnsworth said, had run for two years and cost $35,000 in legal expenses, which had diverted precious time and resources away from the commercialization of his invention. Chairman O'Mahoney asked directly whether "those who control radio" had tried to suppress the introduction of television. Farnsworth demurred and explained that the Federal Communications Commission and the Radio Manufacturers Association had merely been slow in setting technical standards for the emerging television industry. "It is my contention," Farnsworth said, "that the only thing holding back television is its own problems of getting it under way."[52]

Why had Farnsworth passed on a golden opportunity to expose RCA's abuses? During his testimony, Farnsworth conceded that it would be impossible to build a commercial television receiver without drawing on both Farnsworth and RCA patents. If Farnsworth lost the RCA interference case on appeal and his patents were invalidated, he was finished. However, if Farnsworth's patents were sustained, he and Sarnoff would still need to negotiate cross-licenses to use each other's technologies. Despite a decade of ongoing harassment, Farnsworth still needed RCA's cooperation to commercialize television, so he declined to publicly embarrass his nemesis.[53]

The TNEC also questioned corporate R&D managers to better understand how industrial research differed from the work of traditional individual inventors. In the earlier automotive hearings, GM's Charles Kettering had described how his team of researchers signed employment contracts stipulating that they assign ownership of any patents to the firm. GM's researchers received no bonuses or royalties for those patents, just their regular salaries. In terms of size and scope, GE's William Coolidge estimated that his lab employed about 300 scientists; Frank Jewett testified that Bell Labs spent $20 million annually on its R&D activities. Chairman O'Mahoney quickly surmised that only a firm with "a very large capital reservoir" could afford such sizable investments in organized research. Those R&D investments, the executives explained, helped protect their highly capitalized firms from being surprised by their competitors. Kettering, for example, likened his laboratory to "an insurance company" and its large annual budget as a "premium" for keeping the automaker "up to date technically."[54]

Had the massive scale of the corporate laboratories, the committee wondered, wiped out independent inventors? Were individual inventors still a meaningful source of inventions? On this point, the R&D executives were well coached and careful in their testimony. Vannevar Bush, president of the Carnegie Institution of Washington and former chairman of the National Research Council's Division of Engineering and Industrial Research, affirmed the persistence of independent inventors in an era of corporate R&D:

> **Mr. Dienner:** Is there any likelihood that henceforth all new ideas will be brought out through the research laboratories?
>
> **Dr. Bush:** No, I am quite sure that that is not the case. In the first place, there is no limit to the new ideas that can be produced . . . and while a great research laboratory is a very important factor in this country in advancing science and producing new industrial combinations, it cannot by any means fulfill the entire need. The independent, the small group, the individual who grasps a situation, by reason of his detachment is oftentimes an exceedingly important factor in bringing to a head things that otherwise might not appear for a long time.[55]

Although Bush had launched the Industrial Research Institute and obviously believed in the superiority of corporate R&D (chapter 2), it would have been politically perilous to suggest—to a monopoly committee, no less—that there were no longer opportunities for individual inventors.

Instead, Bush and Coolidge spoke in platitudes about how the "pioneering" discoveries of industrial research extended the "frontiers" of knowledge, which in turn opened "new vistas" for independents:

> **Mr. Chairman:** Is not the individual now . . . placed at a tremendous disadvantage to compete with your entire staff, the studies of which are so, what shall I say, stimulated by this tremendous organization that you maintain?
>
> **Dr. Coolidge:** I don't think so. The tree of knowledge is always growing, it is always putting on new branches, so that . . . the frontiers of scientific knowledge are always being extended. . . . [Our] work, you see, is published in detail. . . . It opens the field, then, doesn't it . . . to the individual inventor as well?[56]

Similarly, when the chairman pressed Bush, he quickly redirected to the frontier:

Mr. Chairman: Is the individual who operates outside of a collective group protected in our present system sufficiently?

Dr. Bush: In my opinion, he is not . . . [but] I think the day of the pioneer is not past, the day of the individual inventor is not past, for fine as these cooperative groups may be and necessary as they are to our general progress in this country, they do not cover the entire field.[57]

Bush and the R&D executives had probably adopted this "pioneering" rhetoric from the NRC's Maurice Holland, whose book *Industrial Explorers* (1929) had employed similar romantic characterizations of industrial researchers.[58] Soon after the hearings, the National Association of Manufacturers celebrated inventors as "Modern Pioneers" in their campaign to oppose certain TNEC-inspired reforms.

The independent inventors, of course, testified to their continuing viability in an era of corporate R&D. Farnsworth, for example, conceded that technological advances were increasingly becoming the "product of collective and cooperative effort," yet he maintained that the most "important inventions are made by individuals and almost invariably by individuals with very limited means." The witnesses testified that firms of all sizes—even corporations with their own R&D laboratories—continued to license and purchase patents from independent inventors. Clarence Carlton, vice president of the Motor Wheel Corporation, testified that the "unattached inventor" still enjoyed numerous opportunities given the constant "purchases of patents going on all of the time" in the automotive industry. Likewise, independent inventor Maurice Graham described how he had earned $113,000 in royalties by licensing his patented nonburning toaster to the McGraw Electric Company. "As an independent inventor," Graham testified, "it is my contention that if you have got something that has any merit to it you won't have any trouble finding plenty of people in the large organizations that are glad to listen to you and see what they can do."[59]

Further testimony highlighted some of the key differences between the independent and corporate modes of invention. AT&T's Jewett admitted that several of the most important contributions to the telephonic art had been produced by independent inventors. Beyond Bell's original invention, AT&T had purchased two patents from outside inventors that enabled long-distance transmissions: Michael Pupin's loading coil and Lee De Forest's audion amplifier. However, Bell Labs marshaled the combined efforts of dozens of scientists and engineers to solve the complex problems of integrating and scaling those fundamental inventions across a vast long-distance network. "I think it is inevitable," Jewett elaborated, "that the great bulk of what you might call run-of-the-mine patents in an industry like ours will inevitably come from your own people, from your running a research department. I think that it is equally the case that those few fundamental patents, the things which really mark big changes in the art, are more likely to come from the outside than from the inside."[60]

Essentially, Jewett explained that independent inventors (like Bell and Farnsworth) tended to generate the kinds of blockbuster ideas that spurred entire new industries in their infancy. However, as technologies matured, corporate R&D labs tended to make incremental improvements to their existing products and processes. In other words, independent inventors often made revolutionary discoveries, whereas corporate researchers typically made evolutionary improvements.[61]

Finally, the independent inventors and corporate research directors offered testimony on the relative merits of the reform proposals suggested by Patent Commissioner Conway Coe and Assistant Attorney General Thurman Arnold. Nearly every witness approved of the twenty-year proposal and Coe's other suggestions for streamlining the application process, except for patent attorneys, who stood to lose billable hours. Likewise, both independent inventors and corporate witnesses supported Coe's proposals to streamline interference procedures and create a single court of patent appeals, which together would reduce the expensive hearings and lawsuits that the bigger firms had used as weapons.[62]

However, there was nearly unanimous opposition to Arnold's more radical reforms. Both independent inventors and corporate executives objected to Arnold's compulsory licensing proposal, which would allow petitioners to license allegedly suppressed patents. Compulsory licenses, the witnesses argued, effectively foreshortened the patent monopoly from seventeen to three years. This shorter period of exclusivity would fundamentally weaken the economic incentive that spurred inventive activity and thereby nullify the founders' intent in establishing the patent system. Without an adequate period of exclusivity, inventors would have trouble attracting licensees or backers willing to invest in commercializing the invention. Corporations, in turn, would have far less time to recoup their R&D expenditures in the marketplace. Moreover, the chairman noted, compulsory licensing might inadvertently promote the further concentration of corporate power. Under a compulsory licensing statute, for example, an independent inventor like Farnsworth could be forced to license his television patents to a powerful rival, such as RCA. Overall, both independent inventors and corporate researchers opposed compulsory licensing, while advocating for strong, full-term patents.[63]

Independent inventors also joined the corporate executives in opposing Arnold's call for limitations on restrictive licenses. The proposed restrictions, they argued, undermined an inventor's exclusive constitutional right to dispose of his patent in any way he saw fit. Although the Glass Trust had admittedly built its monopoly by abusing highly restrictive licensing agreements, Commissioner Coe noted that independent inventors and small businesses still needed restrictive licenses for their own protection. In a letter to Chairman O'Mahoney, Coe described how a small-scale inventor of an improved internal-combustion engine might hypothetically license the engine to auto, motorcycle, and marine firms for use in transportation but restrict stationary applications so he could build his own power machinery business.[64] Overall, most witnesses were opposed to weakening patent protections as a remedy for anticompetitive business practices. International patent attorney Lawrence Langner testified that the solution "lies not in breaking down the patent monopoly, not in reducing the incentive to invention, but in strengthening your antitrust laws to prevent the unreasonable use of patents."[65]

In Arnold's mind, it was naive for Coe and other witnesses to oppose compulsory licensing, licensing restrictions, and other proposed reforms because of some perceived negative impact for independent inventors— their day had passed:

> In dealing with the patent system, it is no longer helpful to think in terms of the erratic and solitary genius. The problem no longer hinges on the activities of a Watt, a Stephenson, a Whitney, or an Edison. To think in those terms is to consider history and not the reality which we face today. The patent system and the scope of the patent privilege must now be considered against the background of such aggregates of capital as I. G. Farben, General Electric Company, The Radio Corporation of America, The American Telephone and Telegraph Company, The Hartford-Empire Company, and the patent pools and international cartel arrangements to which they are parties. This is not to say that patents are not important in the operations of smaller enterprises. But, even here, the emphasis has shifted from the individual to the company or the group.[66]

Arnold believed that independent inventors had been supplanted by corporate R&D labs, whose anticompetitive abuses warranted a radical overhaul of the US patent system. However, most of the witnesses tended to align with Coe's more moderate approach to reform. "I would be very much concerned," Jewett testified, "if anything was contemplated which struck at the roots, the fundamentals, of the system itself." Farnsworth likewise hoped that Congress would improve the patent system as much as possible "without changing it basically."[67]

Farnsworth may have regretted his moderate stance. Not long after his TNEC testimony, RCA delivered another crushing blow just as Farnsworth prepared to launch his commercial television business. In March 1939, the Farnsworth Television & Radio Corporation raised $3 million through an initial public stock offering. Farnsworth and his business partners acquired a vacant factory in Fort Wayne, Indiana, to manufacture television sets and dispatched salesmen across the country. On April 30, Farnsworth was in New York City to attend a board meeting for his new company. While walking back to his hotel, he noticed a crowd standing before a department store window. They stared in amazement at an RCA television set as David Sarnoff broadcast the opening ceremonies of the New York World's Fair. As Sarnoff yielded the podium to President Roosevelt and other dignitaries, the reporters assembled at the fairgrounds and the crowd gathered at the department store had no idea that RCA was knowingly and deliberately infringing Farnsworth's patents. Rather, Sarnoff's brazen publicity stunt created the impression that RCA alone was responsible for developing commercial television. In a poignant and bitter irony, Farnsworth watched in disbelief as Sarnoff used the new medium of television to erase his role in its invention.[68]

In its *Final Report*, the TNEC acknowledged that patents had become "a device to control whole industries, to suppress competition, to restrict output, to enhance prices, to suppress inventions, and to discourage inventiveness."[69] As the TNEC patent hearings concluded, America's independent inventors, industrial researchers, business leaders, and patent attorneys were anxious to see what Congress might do.

The National Association of Manufacturers and Its Modern
Pioneers Program

The National Association of Manufacturers (NAM), a conservative, pro-business trade association, took a particular interest in the TNEC's proposed reforms, since several of its members—including General Electric, Eastman Kodak, Westinghouse, and RCA—were high-technology firms that relied on patents.[70] The association had generally been a vocal critic of the Roosevelt administration, especially the National Industrial Recovery Act and its attempts to foster government-directed economic planning and unionism. The NAM had disputed the TNEC's overall body of work and even published a polemic titled *Fact and Fancy in the TNEC Monographs*.[71]

The NAM supported some of the TNEC's proposed patent reforms and vehemently opposed others. For example, the NAM endorsed Commissioner Coe's procedural recommendations to streamline applications, interference proceedings, and the appeals process. Despite its laissez-faire positions on many issues, the NAM recognized the government's crucial role (via the Patent Office and courts) in securing intellectual property rights. Coe's procedural recommendations promised to rationalize those aspects of the system. In fact, NAM executives developed a cooperative relationship with Coe after the hearings. For example, the NAM organized a series of town hall meetings in which Coe discussed his reform proposals and other aspects of the patent system. NAM officials subsequently accepted the commissioner's invitation to tour the Patent Office.[72]

However, the NAM vigorously opposed Thurman Arnold's more radical reform proposals. Like the independent inventors, the NAM believed that compulsory licensing statutes would effectively reduce the patent term from seventeen to three years and remove the incentive to commercialize new inventions. Likewise, the NAM viewed Arnold's proposal to eliminate restrictive licenses as unwarranted government meddling in a firm's absolute right to dictate the terms by which it managed its intellectual property.[73]

The NAM mobilized to thwart Arnold's proposed reforms. The leader of the NAM's counterattack was Robert Lund, executive vice president of Lambert Pharmaceuticals, former NAM president, and chair of the NAM's Patent Committee. Much of Lund's response to the TNEC hearings can be characterized as the standard nuts-and-bolts work of any trade association.

Lund probably coached at least four or five NAM members who testified as witnesses before the TNEC hearings.[74] He retained an ace corporate IP attorney, former AT&T patent counsel George E. Folk, to write a book and several pamphlets articulating the NAM's positions on various patent issues.[75] Lund also raised $25,000 from NAM's corporate members to commission a white paper, "Contribution of the American Patent System to the American Standard of Living."[76] The study, conducted by the National Industrial Conference Board (NICB) and the American Engineering Council (AEC), would draw on interviews with various business leaders (mostly NAM members) to reach the foregone conclusion that "the patent system deserves to be maintained in its present form" and not "radically reformed because of possible abuses in exceptional cases." This tendentious study would provide the "factual" basis for NAM's defense of the patent system and several quotable passages to share with the media.[77] Finally, Lund and the NAM's lobbyists conferred frequently with members of the House and Senate patent committees and presented their case for limited, not radical, reform.

Lund also mobilized the NAM's public relations expertise in the fight against patent reform. Lund had spent the early 1930s developing the association's sophisticated approach to public relations in order to combat various aspects of the New Deal. He knew that 1940 would mark the 150th anniversary of the original 1790 patent law, and he hoped to take advantage of that observance to defend the institution. Specifically, Lund proposed to honor "the Modern Pioneers on the Frontier of American Industry" by bestowing awards on America's most celebrated inventors and research scientists. These highly publicized awards would bring positive attention to the patent system, giving the NAM a platform for defending it against liberal attacks.[78]

In July 1939, Lund hammered out the essentials of the Modern Pioneers program. The NAM would solicit nominations of successful inventors and research scientists who might qualify as Modern Pioneers and then appoint a committee led by MIT president Karl T. Compton to evaluate them. Awardees would be honored at regional Modern Pioneers banquets to be held in thirteen industrial cities in early 1940. These regional banquets would culminate in a national ceremony held in New York City on February 27, designated as Modern Pioneers Day.[79]

By honoring America's Modern Pioneers, Lund drew on the powerful mythology of the American frontier to defend the patent system. In a press release announcing the campaign, Lund wrote, "The pioneer of the geo-graphical frontier of yesterday ventured forth into the wilderness and con-quered new territory. . . . The pioneer on the modern frontier of science and technology likewise ventures into the unknown and conquers it." In Lund's reasoning, homesteaders were spurred to settle the Old West by the promise of secure land claims. By analogy, the Modern Pioneers needed the incentive of secure intellectual property claims to advance the technologi-cal frontier. This was possible only with a strong and stable patent system, not one watered down by liberal reforms.[80] This language clearly echoed the appeals to "pioneering" and "the frontier" in the evasive TNEC testi-mony of Vannevar Bush and the corporate research directors. It is not clear whether Lund had already formulated this rhetorical strategy before coach-ing those witnesses or whether their "pioneering" testimony subsequently inspired the Modern Pioneers campaign.

Lund also marshaled the ideology of self-reliance and rugged individu-alism, which had been rhetorical hallmarks of the NAM's public relations campaigns during the 1930s. In his press release, Lund argued that the pat-ent system had "typified, perhaps better than any other American institu-tion, the American principle of reward for individual initiative." This line of discourse was somewhat ironic, considering that the NAM represented many large firms, whose corporate R&D labs epitomized the value of col-lective, team-based invention. The Modern Pioneers program honored a handful of industrial researchers, including William Coolidge (GE), Vladi-mir Zworykin (RCA), Charles Kettering (GM), and DuPont's eleven-member "nylon group." However, the majority of honorees were individual inde-pendent inventors such as Leo Baekeland (Bakelite), Edwin Armstrong (radio), Willis Carrier (air conditioning), and Edwin Land (Polaroid film).[81] The TNEC hearings had shown that corporations employed aggressive pat-ent practices to sabotage the efforts of individual inventors and small busi-nesses. However, the Modern Pioneers program undermined those critiques by celebrating the achievements of multiple independent inventors.

Here, Lund seemed to borrow from Patent Commissioner Conway Coe's playbook. Coe had opened his TNEC testimony by highlighting heroic individual inventors like Eli Whitney and Cyrus McCormick, while appeal-ing to the heritage, traditions, and "spiritual influence" of the American patent system. Coe's and Lund's rhetoric implied that there could hardly be anything wrong with a patent system that had inspired the work of these legendary inventors. Therefore, Congress should improve the system around the edges but leave it essentially unchanged. By associating the patent system with "the frontier" and "the individual"—two of America's most treasured cultural touchstones—Lund hoped to put the patent system on a pedestal and shield it against the most radical reforms.

In January and February 1940, the Modern Pioneers program came to fruition. From the 1,026 nominations it received, the selection committee honored 572 Modern Pioneers at regional banquets that collectively drew nearly 10,000 attendees.[82] On the evening before the final banquet, Lund delivered a speech over RCA's national radio network, the National Broadcasting Company (NBC). He marked the one hundred fiftieth anniversary of the patent laws and described the Modern Pioneers program. Then, using preliminary data from the NICB-AEC study, Lund explained that patented inventions had increased productivity and wages; he also dismissed allegations that valuable inventions had been suppressed. Finally, Lund addressed the TNEC reform proposals directly. He endorsed Commissioner Coe's procedural recommendations and vehemently rejected Thurman Arnold's proposed compulsory licensing statute and limits on restrictive licensing. If Arnold's ideas were enacted, Lund warned, they would "dangerously impair the patent system as an incentive to invention . . . discourage research and invention and hinder business enterprise." Lund concluded by assuring Americans that the NAM stood guard to "protect the institutions essential to the birth of ideas."[83]

The following evening—Modern Pioneers Day, February 27, 1940—more than 1,500 people attended the national awards banquet at New York's Waldorf-Astoria hotel. The program, again broadcast over NBC radio, featured actors' dramatizations of the honorees' invention stories. In addition, speeches by Coe, Lund, and Kettering echoed the NAM's pioneering rhetoric and its call for modest, not radical, reform.[84]

The NAM kept detailed publicity metrics on the Modern Pioneers program, which generated over one hundred hours of radio airtime and more than 5,000 press clippings. As the campaign concluded, Lund estimated that millions of Americans had become more interested in the patent system through the regional banquets and attendant national media coverage. At the conclusion of the Modern Pioneers program, Lund proudly reported that the NAM had responded to the "sniping at the United States Patent System" and diffused fears that patents had become "instruments of undesirable monopoly." Lund hoped that the NAM's lobbying efforts had prevented "drastic changes" that "would destroy the fundamentals of the patent system."[85]

Conclusion

Ultimately, the TNEC's reform proposals achieved a mixed legislative record. Within months of the hearings, legislators approved several procedural amendments to streamline operations at the Patent Office. However, Congress was unable to push through any major reforms: the twenty-year proposal to reduce the long pendency of applications; a single court of patent appeals to rationalize infringement lawsuits; limitations on restrictive licensing; or a compulsory licensing statute to prevent the suppression of patents.[86] In the end, why was meaningful patent reform so elusive?

With its extensive lobbying and public relations efforts, the NAM was adept at influencing the debate in ways that reflected the interests of its corporate members. In contrast, independent inventors lacked a comparable professional organization to counter the NAM and advocate for their concerns. Earlier groups—including the American Association of Inventors and Manufacturers (AAIM) in the 1890s and the Inventors' Guild in the 1910s—had lobbied for similar patent reforms. However, by 1939–1940, those organizations were long gone, and independent inventors were too politically disorganized to rally behind ideas they had been championing for decades.[87]

In addition, the nation's mobilization for World War II foreclosed the possibility for various domestic reforms. When the TNEC submitted its *Final Report* in March 1941, the United States was already engaged in preparedness efforts, and this probably stalled legislative action on its proposals.[88] However, mobilization efforts empowered the Justice Department to expand its antitrust crusade. Thurman Arnold and his successor, Wendell Berge, broke up international patent pools between US and German firms to prevent aid to the Nazis and used antitrust prosecutions to ensure adequate output and fair prices on key military materials such as aluminum, nitrates, and synthetic rubber. The Justice Department even included compulsory licensing provisions in several of its postwar consent decrees, as it forced R&D firms to issue nonexclusive patent licenses on aluminum (Alcoa), transistors (AT&T), and nylon (DuPont). In fact, the United States imposed its antitrust traditions on Europe and Japan after the war, to prevent business collectivism from sliding again into totalitarianism—the same menace that Roosevelt had referenced when he called for the TNEC hearings in April 1938. Mobilization may have hindered the passage of new patent laws, but it eventually broadened antitrust enforcement, especially where patents had been abused.[89]

The urgency of national defense also stimulated the activities of independent inventors. In May 1940—eighteen months before the Pearl Harbor attack drew the United States into World War II—patent attorney and TNEC witness Lawrence Langner drafted a letter to President Roosevelt with a proposal to mobilize "the inventive genius of the United States."[90] Langner reminded Roosevelt—the former assistant secretary of the navy (1913–1920)—that Thomas Edison had led a similar mobilization of inventors during World War I. Indeed, American independent inventors were crucial national assets as they served their country during both World War I and World War II.

6 Invent for Victory

One of the imperative needs of the Navy . . . is machinery and facilities for utilizing the natural inventive genius of Americans to meet the new conditions of warfare.
—Secretary of the Navy Josephus Daniels to Thomas Edison, July 7, 1915

Invent for Victory! All Americans who have an invention or an idea which might be useful to their country are urged to send it immediately to the National Inventors Council, Department of Commerce, Washington, DC.
—Recruiting poster, National Inventors Council, 1941

During the first half of the twentieth century, the United States fought and won two world wars. Both conflicts represented the intensification of industrialized warfare, in which superior technology—such as machine guns, airplanes, submarines, and atomic weapons—proved decisive for victory. Under these new conditions of warfare, inventors, engineers, and scientists became crucial national assets.[1]

Thousands of American inventors proudly served their country during World War I and World War II. Many were drafted into the armed forces to serve in technical roles or on the front lines. However, most civilian inventors mobilized by evaluating and developing new defense technologies for the military. The urgency of national defense provided immense opportunities for some independents; however, wartime restrictions created frustrations for others. The two wars also served as another backdrop in the ongoing tensions between independent inventors, corporate researchers, and academic scientists as they competed for funding, recognition, and status.

Inventors took the first steps toward American preparedness for World War I. In 1915, an aging Thomas Edison marshaled "the natural inventive genius" of everyday Americans to develop new military technologies for the US Navy. In an early form of government crowdsourcing, Edison's Naval Consulting Board (NCB) evaluated 110,000 ideas submitted by America's grassroots inventors. Unfortunately, the navy implemented only one of these ideas by the war's end, which undermined the nation's confidence in its independent inventors. Meanwhile, the scientific community thrived during the war, as academic physicists developed submarine detection techniques and industrial chemists synthesized a host of explosives during the "Chemists' War." Historians therefore have seen World War I as a turning point marking the rapid growth of corporate R&D and the decline of independent inventors.[2]

In considering the mobilization for World War II, historians have placed an even greater emphasis on the role of academic and industrial scientists while virtually ignoring independent inventors. In this standard interpretation, academic scientists (e.g., Robert Oppenheimer) and scientific administrators (e.g., Vannevar Bush) led the nation's technical mobilization by building key government institutions (e.g., Office of Scientific Research and Development) to coordinate the wartime activities of university, industrial, and government laboratories (e.g., MIT's Radiation Laboratory). Massive wartime efforts, such at the Manhattan Project, marked the apotheosis of "Big Science" and produced the atomic weapons, proximity fuses, radar, penicillin, and electronic computers that won the "Physicists' War."[3]

However, this interpretation overlooks the considerable contributions of individual inventors during World War II. In August 1940, the federal government again mobilized America's independent inventors by forming the National Inventors Council within the Department of Commerce. Led by Charles Kettering of General Motors—and drawing explicitly on the model of the earlier Naval Consulting Board—the council invited prominent independent inventors to work on specific military problems and evaluate the defense ideas submitted by America's civilian inventors. The independents rebounded from their poor showing during World War I and produced several key innovations, including land mine detectors and improved walkietalkie batteries. Many of these stories remained hidden for decades, since the National Inventors Council's records were not declassified until early in the new millennium.[4]

This chapter documents how independent inventors mobilized to serve their country during World War I and World War II. Their wartime contributions serve as further proof that independent inventors had not been supplanted by academic and industrial scientists. Rather, independent inventors were still making valuable—and patriotic—contributions midway through the twentieth century.

World War I: The Genesis of the Naval Consulting Board

In a May 1915 interview, Thomas Edison expressed "the deepest horror over the destruction of the *Lusitania*." On May 7, 1915, a German U-boat torpedoed the British passenger ship off the coast of Ireland, killing 1,198 of the 1,959 passengers aboard, including 128 Americans. Since 1914, President Woodrow Wilson had maintained a strict policy of American neutrality regarding the Great War. Edison, too, opposed military action against Germany, because he believed the United States was "not prepared to fight."[5]

The European combatants had spent the past few decades engaging in a deadly, high-tech arms race, and the fighting had achieved a new level of brutality. On land, soldiers employed machine guns, armored tanks, and poisonous chemical weapons in their trench warfare. In the air, zeppelins and airplanes dropped explosives on enemy positions. At sea, huge dreadnoughts, sleek submarines, and underwater torpedoes threatened both military and merchant vessels. American inventors, such as the Wright brothers, had pioneered some of these technologies, only to watch other nations perfect their military uses. According to Edison, the United States was ill prepared for the new conditions of technological warfare: "We haven't any troops, we haven't any ammunition, we are an unorganized mob."[6]

On May 30, 1915, Edison sat with *New York Times* reporter Edward Marshall and laid out his "Plan for Preparedness." He suggested that the United States could prepare for war without unduly raising taxes or creating a large standing army. Rather, technology was the answer, since modern industrialized warfare had become "more a matter of machines than of men." Edison believed the government should assemble a corps of technically trained men to build an enormous reserve fleet of airplanes, submarines, and naval vessels. Edison also proposed building "a great research laboratory" to develop new cutting-edge weapons. If the United States entered the war, Edison argued, the nation could recall the reservist technicians, mobilize the stored weapons, and employ the laboratory to develop new "instruments of warfare."[7] Edison, in short, proposed a scientific and technological approach to American mobilization.

Edison's ideas resonated with Wilson's secretary of the navy, Josephus Daniels. On July 7, 1915, Daniels sent a letter to Edison inviting him to lead a proposed naval advisory board that would harness the "natural inventive genius of Americans to meet the new conditions of warfare." Edison and his fellow board members would invent new weaponry and selectively implement the best ideas sent to the board by service members and America's civilian inventors. To facilitate this work, Daniels agreed that the navy needed a dedicated research laboratory, charged with building new weapons and perfecting "the crude ideas that are submitted to the Department, by our naturally inventive people." Edison immediately accepted Daniels's invitation; his "board of civilian inventors and engineers" was the first step in America's mobilization for World War I.[8]

In his memoir, Daniels implied that he serendipitously read Edison's interview in the *Times* and then extended his invitation. In fact, the advisory board was orchestrated by Miller Reese Hutchison, Edison's secretary and the chief engineer of the Edison Storage Battery Company. Hutchison devised the advisory board to facilitate sales of Edison storage batteries for navy submarines. In a much later letter to Daniels, Hutchison admitted that he had conceived the idea for the board, "drummed it into Mr. Edison's head," arranged the Marshall interview, and paid for Marshall's travel to Washington, DC, where he brought the interview to Daniels's attention. Hutchison then helped draft Daniels's July 7 invitation letter to Edison and Edison's affirmative reply.[9]

Daniels benefited politically from the arrangement. By adopting Edison's ideas on technological mobilization, Daniels could demonstrate action on preparedness to mollify Republican hawks, but in a measured, prudent way that satisfied both budget watchdogs and Democratic pacifists. Daniels also took advantage of Edison's popularity and public stature to rally members of Congress, naval officers, and the public to the board's cause. At sixty-eight years old, America's most heroic inventor was applying his unique talents during a national emergency, and that was something everyone could support.[10]

Daniels visited Edison in New Jersey on July 15 to discuss appointments for the board. At the suggestion of inventor Frank J. Sprague, Edison and Daniels asked the Inventors' Guild and other technical societies to nominate two members each.[11] There were some conspicuous exclusions. Edison purposefully omitted three organizations that represented academic and theoretical scientists—the American Physical Society, the American Association for the Advancement of Science, and the National Academy of Sciences (NAS). This was payback for Edison, as the NAS had barred him and other inventors from its exclusive membership. Edison, Hutchison explained, wanted the board to be "composed of *practical* men who are accustomed to *doing* things, and not *talking* about it." As historian Thomas Hughes has suggested, this snub precipitated a "simmering struggle between independent inventors, on one hand, and scientists, especially physicists, on the other" as each camp competed to develop new military technologies.[12]

By September 1915, the society nominations were in, and Daniels announced the members of the Naval Consulting Board (NCB; see table 6.1, figure 6.1). In addition to the delegates from the technical societies, Daniels

Table 6.1
Members of the Naval Consulting Board

Officers
Chairman: Thomas Edison
First Vice Chairman: William L. Saunders
Second Vice Chairman: Peter Cooper Hewitt
Secretary: Thomas Robins
Secretary of the Navy: Josephus Daniels, ex officio
Assistant Secretary of the Navy: Franklin D. Roosevelt, ex officio

Appointed by the Secretary of the Navy	American Society of Aeronautic Engineers
Thomas Edison	Elmer A. Sperry
Miller Reese Hutchison	Henry A. Wise Wood (resigned
David W. Brunton	December 24, 1915)
Rear Admiral William S. Smith	Bion J. Arnold (replaced Wood)
Inventors' Guild	**American Aeronautical Society**
Thomas Robins	Matthew B. Sellers
Peter Cooper Hewitt	Hudson Maxim
American Chemical Society	**American Society of Automotive Engineers**
Leo H. Baekeland	Howard E. Coffin
Willis R. Whitney	Andrew L. Riker
American Institute of Electrical Engineers	**American Institute of Mining Engineers**
Frank J. Sprague	William L. Saunders
Benjamin G. Lamme	Benjamin B. Thayer
American Mathematical Society	**American Electrochemical Society**
Robert S. Woodward	Lawrence Addicks
Arthur G. Webster	Joseph W. Richards
American Society of Civil Engineers	**American Society of Mechanical Engineers**
Andrew M. Hunt	William L. Emmet
Alfred Craven	Spencer Miller

Figure 6.1
The Naval Consulting Board, circa 1915–1916. *On the front step, from left to right:*
Frank J. Sprague, unknown (Benjamin G. Lamme?), Miller Reese Hutchison, Thomas
Edison, Josephus Daniels, and Franklin D. Roosevelt.
Source: Lloyd N. Scott, *Naval Consulting Board of the United States* (Washington, DC:
Government Printing Office, 1920), 3.

appointed Edison, Hutchison, mining engineer David W. Brunton, and Rear
Admiral William S. Smith as his personal representatives to the board. He
also appointed Assistant Secretary of the Navy Franklin Delano Roosevelt
as an ex officio member. The press seemed underwhelmed by a panel of
nominees they did not recognize. The *New York Times* had speculated that
world-famous inventors such as Orville Wright, Alexander Graham Bell,
and Nikola Tesla would serve alongside Edison. Instead, they got Leo Baekeland, Hudson Maxim, Frank Sprague, and Elmer Sperry. In a perfect description of their "post-heroic" status, the *Times* observed that "the twenty-three
members were chosen for fitness rather than notoriety."[13]

As civilians, the board's inventors and engineers served voluntarily and
without compensation, so most observers assumed they were financially
disinterested. Members even paid their own expenses until 1917's Naval

Appropriations Act earmarked $1.5 million for the board's activities.[14] However, as with Edison and his storage batteries, several board members saw their patriotic service as a commercial opportunity. For example, inventor Hudson Maxim had licensed his smokeless gunpowder patents to DuPont; both he and the firm stood to profit handsomely from America's entrance into the war. In September 1915, Maxim published *Defenseless America*, a book arguing for increased military preparedness. The book was underwritten with $35,000 from Maxim's patron, Pierre S. du Pont.[15] Both Elmer Sperry (Sperry Gyroscope Company) and Willis R. Whitney (General Electric) would also use their board positions to steer weapons contracts to their firms.

The Naval Consulting Board convened its first meeting on October 6–7, 1915, just five months after the *Lusitania* disaster. Over the next few months, the members elected officers, adopted rules of governance, and established sixteen (later twenty-one) standing committees organized around specific technical problems, with committee assignments determined by each inventor's expertise. Elmer Sperry, for example, chaired three committees: Aeronautics, Aids to Navigation, and Mines & Torpedoes. The board also outlined its four major tasks: (1) a survey of America's industrial readiness; (2) the evaluation and selective development of defense ideas submitted by America's inventors; (3) the development of new military inventions by board members; and (4) the establishment of a Naval Research Laboratory.[16]

The Naval Consulting Board at Work

The board's Industrial Preparedness Campaign set out to determine the nation's wartime production capacity while educating the public on the urgency of mobilization. The United States possessed an abundance of raw materials, factories, tools, and skilled workers. The campaign sought to determine how those assets could be converted from civilian to military production if the United States entered the war. Howard Coffin, vice president of the Hudson Motor Car Company, led the board and several technical societies in conducting industrial capacity surveys in each of the (then) forty-eight states. The board's propaganda campaign also helped convince isolationists of the need for preparedness. In May 1916, for example, the Naval Consulting Board joined with 135,000 other marchers in a Citizens' Preparedness Parade in New York City. Congress eventually established the

Council of National Defense; thereafter, the board ceded responsibility for coordinating transportation, industrial and farm production, and public morale.[17]

Meanwhile, the Naval Consulting Board served as a clearinghouse for evaluating the defense ideas submitted by America's independent inventors. When the United States entered the war on April 6, 1917, the NCB's New York City offices were deluged by a surge in submissions, with a peak of 600 letters per day. The board eventually recruited a mix of junior naval officers and seamen to aid the sifting operation, with a peak staff of twenty-two technical examiners, typists, and file clerks working alongside NCB secretary Thomas Robins. The NCB office staff served as a first line of evaluation. If an idea showed merit, the clerical staff passed it to one of the board's twenty-one technical committees for further evaluation and comment. If the committee members liked the idea, they presented it to the full board and asked Daniels for an appropriation to conduct further experiments. The board's liaison officer, Rear Admiral William Smith, coordinated with the relevant naval bureau.[18]

Most of the ideas inventors submitted to the board were naive and impractical. Inventors saw the problems clearly enough; they understood that the navy needed help with submarine detection, anti-torpedo defenses, and lifesaving devices. The board had even provided background information by circulating technical bulletins on topics such as "The Enemy Submarine" and the "Problems of Aeroplane Development." However, many inventors simply lacked the basic scientific knowledge or practical naval experience to properly address those challenges. For example, several amateur inventors suggested deploying mounted nets or buoys alongside ships to protect them from torpedoes; they failed to realize that the drag from the nets would slow the ships to a crawl. Others suggested hunting for enemy submarines by using glass-bottomed boats equipped with searchlights, without accounting for seawater's opacity.[19]

Evaluating the public's impractical ideas was frustrating for the board's professional inventors. Hudson Maxim angrily objected when the board spent $25,000 to test a device that sent "a repelling wave toward an oncoming torpedo," which predictably had no effect. Frank Sprague wrote of his disdain for serving as "a clerk to comment on other people's inventions," which left little time to work on his own defense ideas.[20] The amateur inventors became frustrated, too, and generated voluminous correspondence to protest their rejections. The board subsequently noted that it was "impossible to convince any inventor, however worthless his invention might be, that it was valueless." *Scientific American* praised the courtesy and persistence of the board's examiners and implored America's patriotic citizens to "stop sending in half-baked, impossible, impractical, visionary ideas."[21]

Not to be outdone, the US Army also solicited defense ideas from independent inventors. Late in the war, in May 1918, the US Army established its Inventions Section to secure the "prompt and thorough investigation of inventions submitted to the War Department." Like the NCB, the Inventions Section recruited an advisory board of technical experts to pass judgment on all ideas submitted. In the six months before the war concluded, the Inventions Section examined approximately 25,000 ideas. However, only about twenty-five suggestions (0.1 percent) merited further consideration, and it appears that none were implemented.[22]

In the end, the Naval Consulting Board sifted through nearly 110,000 suggestions from America's inventors. Only 110 ideas merited any serious investigation, and of these, only one device—a flight trainer, submitted by W. Guy Ruggles (figure 6.2)—was implemented before the armistice. In his official history of the NCB, Captain Lloyd Scott observed that the nation's inventors were ignorant of the special requirements of naval operations and unfamiliar with the current state of the military arts, since many of their suggestions described items already in use or long abandoned for something better. "The inventions of isolated inventors," Scott concluded, ultimately "had little or no value."[23]

The poor performance of America's independent inventors stood in sharp contrast to the strong résumé of work by the board's professional members, whom Daniels called the "keenest and most inventive minds that we can gather together." Hudson Maxim, for example, developed a new combustible material he called "motorite" for propelling torpedoes; he believed it would improve on the usual method of releasing compressed air. Maxim also developed a concept for a torpedo-proof ship. He suggested lining the hull with steel barrels of pulverized coal to simultaneously fuel the ship and protect its vital systems. The armistice arrived before either idea could be adequately tested.[24]

Chemical inventor Leo H. Baekeland was heartbroken about the war's devastating impact in his native Belgium. During his board service, Baekeland advanced two American industries traditionally dominated by

Figure 6.2
Of the 110,000 ideas submitted to the Naval Consulting Board by America's inventors, a flight trainer designed by W. Guy Ruggles was the only device implemented before the armistice.
Source: Lloyd N. Scott, *Naval Consulting Board of the United States* (Washington, DC: Government Printing Office, 1920), 148.

Germany: nitrates and optical glass. Through speeches and journal articles, Baekeland educated the public about the nation's dangerous dependence on Chilean bird guano for the nitrates used in explosives. Baekeland helped win a $20 million congressional appropriation to build the first nitric acid plant in the United States, which came online just as the war concluded in 1918. Baekeland also worked closely with the Bausch & Lomb Optical Company to develop improved binocular field glasses, submarine periscopes, and artillery range finders.[25]

In January–February 1916, former naval officer Frank J. Sprague returned to civilian duty with a highly publicized four-week inspection cruise aboard the USS *New York*. Sprague submitted a detailed report to Secretary Daniels with recommendations for improving shipboard communications, lighting, signaling, steering, and gun firing. In addition, Sprague invented a delayed-action fuse that allowed armor-piercing shells to detonate just after first contact to maximize damage. Sprague also developed a pressure-sensitive firing mechanism that enabled U-boat hunters to detonate their explosive charges at variable preset depths. Years later, Sprague's son Desmond recalled "the old man daily risk[ing] his life in testing out depth charges and shells" in the waters off Sandy Hook, New Jersey. Despite a favorable recommendation from the Special Board on Naval Ordnance, Sprague's fire control devices were not implemented during the war.[26]

In contrast, inventor Elmer Sperry leveraged several of his ideas into development contracts. He was probably the most effective member of the NCB, thanks to his prior navy experience. Between 1907 and 1914, Sperry had integrated his patented gyroscopes into several navy applications, including stability, guidance, range-finding, and feedback-control systems. While working with the Naval Consulting Board, Sperry incorporated his gyrocompass into a patented targeting and gunfire control system. By 1920, nineteen US dreadnoughts, eleven second-line battleships, and nine armored cruisers had installed Sperry's fire control system. Sperry also developed a long-range, high-intensity searchlight, controls for naval torpedoes, and a bombsight for use on airplanes.[27]

Sperry's most audacious wartime development was the "aerial torpedo." Earlier, in 1911 and 1912, Sperry had worked with aviation pioneer Glenn Curtis to develop a gyrostabilized "automatic pilot" to enhance the safety of early airplanes. Once the war began, Sperry believed he could modify this control system to create a pilotless flying bomb. The Naval Consulting Board approved the aerial torpedo project in April 1917 and awarded a $200,000 development contract to the Sperry Gyroscope Company. Sperry's work drew the attention of Charles Kettering, the inventor of the electric automobile starter and Orville Wright's partner in the Dayton Wright Airplane Company. With help from Sperry, Kettering initiated a parallel project to develop a flying bomb for the US Army. In 1917 and 1918, both inventors tested their projects, with mixed results. With the armistice in November 1918, further development work was suspended. However, Sperry's "aerial torpedo" and Kettering's "bug" anticipated Germany's V-1 and V-2 guided missiles by two decades.[28]

Tensions and Disappointments

Wartime service inspired both cooperation and competition among America's inventors, engineers, and scientists. After they were excluded from the Naval Consulting Board, American scientists made their own plans for national service. In April 1916, the National Academy of Sciences, led by astronomer George Ellery Hale, organized the National Research Council (NRC) within the NAS to mobilize American scientists. Hale wisely recruited both academics and corporate scientists, including research directors John J. Carty (AT&T) and Willis R. Whitney (General Electric), the latter of whom served concurrently on the NCB. Hale approached the NCB's vice chairman, William Saunders, to offer the scientific community's expertise in combating the submarine menace. Saunders rebuffed Hale, assuring him that the board's inventors and engineers did not need the scientists' help—another affront in a growing feud.[29]

When Germany resumed unrestricted submarine warfare in February 1917, the board redoubled its anti-submarine efforts. The results were mixed. Sperry's experimental system used anti-submarine nets to trigger wireless transmitter buoys, which signaled nearby patrol boats to drop depth charges. During a highly publicized test in June 1917, the nets snared a test submarine, but the signaling buoys—snarled in the nets—failed to transmit to the patrol boats. A second NCB team, led by Whitney, established an experimental station at Nahant, Massachusetts, to work on sound detection methods for locating enemy subs. Whitney assembled corporate researchers from GE, AT&T, and the Submarine Signal Company but excluded university physicists, because the navy feared their presence might complicate the patent situation.[30]

Here was yet another slight to the academic scientists. With the failings of Sperry's group in the news and the academics excluded from the board's Nahant group, the NRC initiated its own submarine detection unit in June 1917, based in New London, Connecticut. Robert Millikan, the NRC's vice chairman, led the council's anti-submarine group. He recruited academic physicists from Yale, Chicago, Rice, Cornell, Wisconsin, and Harvard.[31]

Along with Whitney, Leo Baekeland served concurrently on both the NCB and NRC. Privately, Baekeland worried about the "limitations and restricted views" of the NRC's "pure scientists." Echoing Edison's preference for "practical men," Baekeland had more confidence in the board's "engineers, manufacturers, and inventors," who had "won their spurs by constructive work."[32] Fortunately, petty rivalries did not interfere with patriotic duty. In mid-June, the board's Benjamin Lamme graciously wrote Millikan and suggested that "the subject of submarines is big enough for all of us." Whitney urged Daniels to help fund the NRC's New London group. "If a little of the energy which, I fear, may otherwise develop into hard feelings could be utilized in the submarine work," Whitney counseled, "time would be saved."[33]

Eventually, Whitney's Nahant group developed, and GE manufactured, a stethoscope-like sound detector called the "C-tube," which featured a pair of rotating, underwater "ears." Using the binaural principle, sub chasers could get a bearing on a target by rotating the C-tube until it found the loudest, clearest signal. Using a similar principle, Millikan's New London group developed an "M-detector" with multiple underwater listening tubes connected to an earphone. Listeners turned a circular dial to select the tube with the loudest signal, which revealed the submarine's heading.[34]

By July 1918, the navy had outfitted more than one hundred sub chasers with C-tubes and M-detectors. The ships located and harassed German U-boats in the English Channel but registered no confirmed kills. However, with help from some underwater nets, detector-equipped sub chasers managed to destroy a few Austro-Hungarian subs in the Adriatic. In the technological battle against the submarine menace, Sperry's system of nets and transmitter buoys had been a highly publicized failure. Meanwhile, the moderate success of the C-tubes and M-detectors fueled the argument that industrial and university scientists were superior to independent inventors.[35]

Overall, the Naval Consulting Board's results were disappointing. Its professional inventors had managed to implement only a handful of their development projects. With the exception of the Ruggles flight trainer, America's amateur inventors had failed to produce anything worthwhile. Collectively, this lackluster performance had sullied the reputations of America's independent inventors. This included Edison himself, who had grown increasingly discouraged in his role as board chairman. Edison complained that the navy had "pigeon-holed" all forty-five of his inventive ideas and generally resented "any interference by civilians."[36]

More seriously, Edison's board tenure was marked by scandal. His assistant, Miller Reese Hutchison, had conceived the board as a way to sell storage batteries to the navy, but his plan backfired in tragedy. On January 15, 1916, the experimental submarine E-2 exploded while docked at the Brooklyn Navy Yard, killing five sailors and injuring nine others. A public inquiry found that leaking hydrogen from Edison's batteries had caused the explosion. Despite the support of Secretary Daniels, Edison believed that certain navy officers (and his battery industry competitors) had used the tragedy to besmirch his reputation. Edison blamed the incident on the incompetence of the navy's technical staff, and this hardened his opinion that civilians—not navy officers—should control the proposed Naval Research Laboratory.[37]

During his interview with Edward Marshall in 1915, Edison had proposed "a great research laboratory, jointly under military and naval and civilian control." Secretary Daniels quickly adopted this idea for the navy; after all, dozens of successful inventions had flowed from Edison's own laboratories at Menlo Park and West Orange, New Jersey. However, Edison disagreed sharply with the navy and his fellow board members as they discussed the details of implementation. The navy wanted to locate the lab in Annapolis, Maryland (near the United States Naval Academy) or Washington, DC (near the War Department). Edison wanted the lab in Sandy Hook, New Jersey (near suppliers in New York City). The navy wanted naval officers to run the lab; Edison pushed for civilian control. The navy imagined a sophisticated scientific laboratory, focused on basic research, experimental testing, and quality control. Edison wanted "a constructing laboratory—more properly a universal machine shop" to build the storehouse of ships, guns, and weapons he had described to the *New York Times* in 1915.[38]

Secretary Daniels had leveraged Edison's reputation to gain public and congressional support for the Naval Consulting Board. Now Edison's popularity complicated Daniels's efforts to move forward on the proposed laboratory. Fearing negative publicity, Daniels was unwilling to override Edison, and allowed a stalemate to persist after the November 1918 armistice. In March 1919, the entire board threatened to resign over the disagreement, and Daniels was forced to act or potentially lose the lab altogether when President Wilson's term expired. In December 1920, the navy began constructing the Naval Research Laboratory in Washington, DC, as a scientific research laboratory under the control of naval officers—all contrary to Edison's vision. Edison resigned from the Naval Consulting Board in January 1921.[39]

The settlement of the laboratory debate ended the active work of the Naval Consulting Board. In later years, board members attended an annual Armistice Day reunion dinner—a purely social occasion. In contrast, the National Research Council secured a permanent charter from President Wilson and initiated its interwar campaign to promote industrial research.[40]

Although they never aired their views publicly, most of the inventors on the Naval Consulting Board came away disappointed by the experience. Edison was a shared source of frustration. Sprague, in particular, was jealous when the newspapers referred to the NCB as the "Edison Advisory Board." Others complained about Edison's imperious attitude and stubbornness regarding the proposed Naval Research Laboratory.[41] In addition, the board's efforts to harness the "natural inventive genius of Americans" ended in failure. Writing in 1920, the board's official historian, Captain Lloyd Scott, concluded that the NCB would have benefited from "a selective mobilization of those who are qualified to do the work, and not the mobilization of the untrained talent of the country."[42]

Yet even the board's professional inventors—including Sprague, Maxim, Baekeland, and Edison himself—grew frustrated when their proposed solutions were not adopted by the navy. Sprague suspected that the navy never took the board seriously, that it was a politically expedient tool for Secretary Daniels and President Wilson. However, the board's short operational window probably explains its meager results. In 1915 and 1916, the board focused on its preparedness campaign and overcoming the public's isolationism. The board shifted toward technical work in 1917, but the war ended just eighteen months later. With the November 1918 armistice, several promising projects—including the aerial torpedoes—were abandoned before they could be perfected.[43]

Historians have seen World War I as a tipping point marking the decline of independent inventors and the ascendance of science-based industrial research. They contrast the poor showing of America's grassroots inventors, Edison's cut-and-try experimentalism, the naval laboratory stalemate, and the breakup of the Naval Consulting Board with the scientists' theoretically informed research, successful anti-submarine efforts, and the permanence of the National Research Council.[44] Yet these interpretations overlook the Naval Consulting Board's greatest legacy. Twenty years after the board's dissolution, independent inventors adopted its organizational structure to mobilize for World War II, but with far greater impact.

World War II: The Genesis of the National Inventors Council

In May 1940, international patent attorney Lawrence Langner read the newspapers with growing alarm. Germany had ignited World War II by invading Poland in September 1939, which followed Italy's invasion of Ethiopia (1935) and Japan's attack on China (1937). Congress had passed a series of Neutrality Acts, but Langner assumed the United States would eventually be drawn into these conflicts. Like Thomas Edison in 1915, Langner feared that the United States was technologically unprepared for the coming war. Using letterhead from his position on the National Advisory Council of the House Committee on Patents, Langner drafted a letter to President Franklin D. Roosevelt with a proposal "to establish an 'Inventions Council for National Defense'" and once again mobilize "the inventive genius of the United States."[45]

Langner was a colorful character who had pursued "a life in two fields, theatre and invention." He was born in Wales and raised in London. As a teenager, he took clerical jobs at a West End theater and a patent law firm before earning his license as a chartered patent agent. In 1910, Langner moved to New York City to open the US branch of his London-based law firm and then founded his own practice in 1913. In New York, he rediscovered his early love of the theater, and in December 1918 he cofounded the Theatre Guild, which produced several of the twentieth century's landmark musicals, including *Porgy and Bess* and *Oklahoma!* Although Langner was better known for his avocation, he was a brilliant and influential international patent attorney.[46]

In a letter dated May 14, 1940, Langner reminded Roosevelt—the former assistant secretary of the navy under Josephus Daniels—that "a Council of this type was formed during the last war." He granted that very few suggestions by "crackpot inventors" had been of any value but reminded Roosevelt that the Naval Consulting Board's professional inventors had been developing several promising technologies when the war concluded. Langner suggested that the proposed new council should use the earlier bureau as a model but improve on its shortcomings by serving as a liaison for both the army and navy.[47]

The following day, Langner sent a similar letter to the commissioner of patents, Conway Coe. He suggested that Coe should lead the proposed Inventors Council, and he offered to come to Washington, DC, to discuss the proposal in person. Coe agreed to a meeting and invited Vannevar Bush, president of the Carnegie Institution of Washington and chairman of the National Advisory Committee for Aeronautics (NACA), an organization already identified with national defense. Langner traveled to Washington, DC, accompanied by his friend and former client Thomas Midgley, Jr., the inventor of the tetraethyl-lead gasoline additive that prevented knocking in automobile and airplane engines. The meeting at the Patent Office—featuring Langner, Midgley, Coe, and Bush—resulted in three outcomes.

First, Coe agreed to take Langner's proposal to the secretary of war (Harry Hines Woodring), the secretary of the navy (Edison's son, Charles Edison), and his own boss, the secretary of commerce (Harry Hopkins), who was FDR's influential confidant. Second, the group decided to call their proposed organization the National Inventors Council (NIC). Third, the meeting stirred Vannevar Bush into action.[48]

For months, Bush and other like-minded members of the National Research Council—including James B. Conant (Harvard), Karl T. Compton (MIT), and Frank Jewett (AT&T/Bell Labs)—had contemplated a plan for mobilizing American scientists for defense preparedness. Scooped by Langner, Bush quickly pulled together his own proposal for the National Defense Research Council (NDRC), which he presented personally to Hopkins. Hopkins, in turn, tried to persuade Bush to lead the National Inventors Council. Bush declined and eventually convinced Hopkins and FDR of the wisdom of mobilizing scientists alongside inventors. Roosevelt established the NDRC on June 15, 1940, and to ensure proper coordination with the NIC, he appointed Conway Coe to serve both groups. Bush's NDRC, and its successor, the Office of Scientific Research and Development (OSRD), eventually contributed several innovations crucial to winning World War II, including proximity fuses, radar, and the atomic bomb. However, Langner's proposal for the National Inventors Council was the first to urge "the mobilization of scientists and inventors in the National Defense."[49]

The nascent council needed a leader. With Bush unavailable, the small planning group nominated their close friend Charles Kettering, president of the General Motors Research Corporation and one of America's most prominent inventors and business leaders. Kettering had studied electrical engineering at Ohio State University and then invented the electric cash register while working for the National Cash Register Company (NCR) in Dayton, Ohio. While moonlighting from NCR, Kettering developed the first electric automobile starter, founded the Dayton Engineering Laboratories Company (DELCO) to commercialize it, and retained Langner in 1913 to pursue the European patents. Kettering and Midgley had worked together for years, first at NCR, then at DELCO, and finally at General Motors after GM acquired DELCO in 1918. More recently, Kettering had worked alongside Coe on a committee charged with celebrating the sesquicentennial of the 1790 patent law. On July 11, 1940, Hopkins appointed Kettering as chairman of the National Inventors Council.[50]

Kettering was an ideal choice. He personally embodied the potential of both the independent inventor and the industrial researcher, giving him credibility in both camps. Kettering had worked with the military during World War I, manufacturing aircraft, testing fuels, and developing the aerial torpedo alongside Elmer Sperry. His firm had plenty of experience evaluating the thousands of ideas submitted each year to GM's New Devices Section. Most importantly, Kettering was famous. He was the engineer journalists sought when they needed an informed opinion on invention and technology, and he explained scientific concepts on the radio in a plainspoken, accessible manner. Kettering's popularity would inspire America's inventors to produce a steady stream of new ideas. His impeccable credentials would help legitimize the NIC amid lingering skepticism over inventors' lackluster performance during World War I.[51]

In his appointment letter, Hopkins charged Kettering with assembling a council comprised of "outstanding American inventors and business men" to aid in technological preparedness. Kettering, Midgley, Langner, and Coe drew on their professional networks to staff the NIC.[52] They avoided the animus generated by Edison's exclusions and nominated university-based scientists and engineers; a science journalist; technical officers from both the army and navy; and corporate research directors from General Electric, Chrysler, DuPont, and Bakelite. Finally, Kettering appointed his Dayton colleague Orville Wright. Curiously, the sixty-nine-year-old airplane pioneer was the only independent inventor on a council assembled to mobilize them. The military officers and Coe served the council in their official capacities, while the civilian members served as government consultants for $1 per year. The council existed without the benefit of an executive order or legislative authority; it was an advisory committee to the Department of Commerce.[53]

The National Inventors Council at Work

Kettering convened the first meeting of the National Inventors Council on August 6, 1940, at the Department of Commerce in Washington, DC (figure 6.3). From the beginning, the prior example of the Naval Consulting Board directly influenced the mission and operations of the National Inventors Council.

In his opening remarks, Kettering outlined the NIC's two primary functions: (1) the "organization of the capable inventors of the country" to work

Figure 6.3
The first meeting of the National Inventors Council, August 6, 1940, at the Department of Commerce, Washington, DC. *Standing left to right:* Fin Sparre, Webster N. Jones, Frederick Feiker, Lawrence Langner, Watson Davis. *Seated left to right:* Frederick Zeder, Conway P. Coe, Chairman Charles Kettering, Thomas Midgley, William Coolidge. *Not present:* George Baekeland, Admiral Harold Bowen, General Joseph O. Mauborgne, General William H. Tschappet, Orville Wright. Photo by Harris & Ewing, image no. SIA2008-4787, Science Service Records, Smithsonian Institution Archives, Washington, DC.

on specific army and navy problems; and (2) the evaluation of "unsolicited inventions sent by the general public." Langner then introduced an organizational plan that quoted liberally from *The Naval Consulting Board of the United States* (1920), Captain Lloyd Scott's official history.[54] Following the NCB's model, the NIC organized twelve technical committees, covering all military requirements "from toothpicks to tanks." Individual NIC members chaired the committees according to their expertise, and they could nominate additional $1-a-year consultants to aid the committee's work.[55] For example, General Electric's William Coolidge chaired two NIC committees—Instruments, and Signals and Communication—and he subsequently appointed forty-eight additional consultants, mostly colleagues drawn from the GE Research Laboratories.[56]

The council members and their committee consultants served the NIC in an advisory capacity, working around their jobs in industry or academia. The NIC also employed a full-time staff of salaried examiners and clerks at the Department of Commerce who handled the council's day-to-day business (figure 6.4). Langner suggested hiring twenty-three staff members, the

Figure 6.4
In this photo dated March 25, 1942, clerks Everett E. Smith (*left*) and Mildred Steinmetz (*right*) sort through the deluge of inventive suggestions that arrived at the National Inventors Council in the months following the Pearl Harbor attack. General Subject Files, National Inventors Council, Records of the National Institute of Standards and Technology, Record Group 167, National Archives at College Park, College Park, MD.

identical staffing level employed by the Naval Consulting Board during World War I. The NIC's volume of suggestions and staffing levels fluctuated with the prosecution of the war, peaking at fifty-five salaried staff immediately after the Pearl Harbor attack.[57]

Initially, the council and its staff operated under a $115,000 appropriation from the President's Emergency Fund. In subsequent years, the NIC's average annual appropriation ($125,000) became a standard line item under the Department of Commerce's budget. The council operated under budget in every year from 1941 to 1946 inclusive.[58]

Like the NCB, the National Inventors Council distributed a series of *Information Bulletins* to provide submission guidelines and manage the expectations of America's inventors. Inventors were instructed to submit their ideas in writing to the Department of Commerce and include any drawings or relevant details about the concept's patent status. The NIC could not pass judgment on the patentability of an idea or provide an endorsement for those seeking financial backers. It could only evaluate the military potential of an idea and recommend it to the armed services for further development. To preserve secrecy regarding the technologies the military might (or might not) be pursuing, the NIC could not divulge the specific reasons for an idea's rejection. It could not discuss compensation; rather, all specifics were to be negotiated between the inventor and the relevant military authorities. However, the NIC hoped that inventors might donate any adopted inventions to the US government "in a spirit of patriotism."[59]

The NIC's evaluation process was nearly identical to the NCB's procedure, but it now served all military branches. It received inventions and ideas directly from citizen-inventors and service members, as well as ideas forwarded from the army, navy, and other government agencies. In this sense, it was a true clearinghouse. The NIC clerical staff acknowledged receipt of each suggestion with a quick reply letter, then classified and routed the file to the appropriate examiner, who was aligned to one or more of the NIC's twelve technical committees. The NIC's examiners were trained in engineering, chemistry, and physics, but their greatest asset was "sufficient imagination and vision" to recognize the military potential of incoming ideas, regardless of how unorthodox they might appear. The frontline examiners provisionally rated each idea as meritorious or impractical. Chief Engineer Leonard Lent then performed a second review of all cases, ensuring that at least two engineers reviewed every suggestion. Impractical ideas were sent a polite, but brief, rejection letter. Meritorious ideas were forwarded to the chairman of the relevant technical committee for a deeper evaluation.[60] This is where the process often bogged down, as the chairman and his appointed consultants typically spent several weeks considering the merits of dozens of ideas. Finally, the chairman presented his committee's report to the full council. If the council affirmed the chairman's positive recommendation, the suggestion was forwarded to the relevant branch of the military.[61]

The NIC's recommendation was no guarantee for the inventor, however, as the army or navy could still reject the idea. If the military adopted the proposal, the inventor would develop and test the idea, working with the military, academic, and industrial labs coordinated by the National Defense Research Committee (NDRC). The separation of the NIC's evaluation function from the NDRC's development and testing function was a procedural departure from the Naval Consulting Board, which had coordinated all cases from start to finish. Finally, the council asked each service branch to report on the final disposition of each submission so the NIC could track statistics on the number of suggestions implemented.[62]

In a second key mission, the NIC recruited "the professional inventive genius of the country" to work on specific problems provided by the military. During World War I, the Naval Consulting Board encouraged Edison, Sprague, and the other board members to do the actual inventing. During World War II, the NIC played matchmaker: it identified the nation's top professional inventors and invited them to address the military's specific problems. To assemble a roster of the leading inventors, Coe surveyed his corps of patent examiners, while Langner reached out to his fellow patent attorneys. Langner also compiled a list of needs from the army and navy. By December 1940, the NIC had a comprehensive list of 2,000 prominent inventors and a military wish list with fifty items, including an improved metal service helmet, a lightweight gasoline generator, and an improved wire for field telephony.[63]

Langner took charge of these procurement requests and developed a set of procedures. After he gathered the technical requirements for each military need, the NIC technical staff searched the patent record and its own files of inventors' suggestions to determine whether a potential solution already existed. Langner then shared the file with the chairman of the relevant technical committee, and together they selected five names with the appropriate expertise from the master list of 2,000 inventors. After passing a background check, each of the five inventors was appointed as a $1-a-year NIC consultant, given a copy of the problem specifications, and invited to develop a solution, working in collaboration with the chairman and his technical committee. If no solutions were forthcoming, the NIC might send the problem to the NDRC or abandon it. Inventors with potential solutions were referred to the relevant military branch for further development.[64]

The National Inventors Council's similarity to the Naval Consulting Board immediately generated skepticism and critical press coverage. Calling the council a "Crackpots' Haven," *Time* magazine observed that "one of the gadflies that made official Washington miserable in World War I was the crackpot inventor, buzzing with mosquitoey ideas for winning the war—schemes for crashproof airplanes, inescapable torpedoes, [and] mole-burrowing bombs." *Business Week* reported a rumor that the council might be a decoy organization, designed to keep amateur inventors away from the army, navy, and NDRC.[65]

To counter these criticisms and recruit the nation's inventors, NIC member Watson Davis mounted a public relations campaign. Davis, director of the Science Service and editor of the *Science News Letter*, secured a *Reader's Digest* feature on the NIC in January 1941. Author Arthur Chase reported that the National Inventors Council was keeping and forwarding about 4 percent of the 150 submissions it received each day, which doubled the 2 percent rate of the earlier Naval Consulting Board. President Roosevelt had urged the United States to become the "arsenal of democracy," and Chase assured individual inventors that they played a crucial role in the nation's defense, alongside university and corporate scientists. The article concluded with instructions for submitting ideas to the NIC and a rallying cry: "Inventors of America: your country is calling you."[66] After Chase's article, submissions to the NIC temporarily doubled from 150 to 300 per day and ultimately generated about 5,000 to 7,500 new ideas overall. Davis eventually placed similar articles in *Collier's*, *Popular Science*, *Radio News*, and other magazines.[67]

The NIC complemented Davis's print campaign with a series of national radio interviews, recruitment speeches, and a promotional poster. On consecutive Saturdays in February 1942, Davis interviewed the NIC's William Coolidge and Lawrence Langner on his weekly CBS radio program *Adventures in Science*. Charles Kettering also spoke occasionally of the National Inventors Council during his Sunday NBC radio talks on science and invention during the GM-sponsored program *Symphony of the Air*.[68] Meanwhile, Chief Engineer Leonard B. Lent barnstormed the nation, promoting the NIC and recruiting inventors in his frequent speeches to various technical societies.[69] Finally, the NIC developed a promotional poster (figure 6.5) encouraging the nation to "Invent for Victory." Commerce Secretary Jesse Jones introduced the poster in *Life* magazine, and the NIC ultimately placed thousands of posters in post offices, factories, and college and high school science clubs.[70]

Collectively, these public relations efforts mobilized thousands of independent inventors to work with the National Inventors Council. Their experiences varied widely. Working with the NIC provided great opportunities for some independents but frustration and disillusionment for others.

Figure 6.5
To mobilize independent inventors, the National Inventors Council placed thousands of posters in post offices, industrial factories, and science clubs. General Subject Files, National Inventors Council, Records of the National Institute of Standards and Technology, Record Group 167, National Archives at College Park, College Park, MD.

The NIC's open-minded attitude toward Wade Stiles resulted in one of its greatest success stories. In November 1940, Stiles brought some strange equipment to his interview with NIC examiner Leonard Hardland in Washington, DC. Stiles was a professional treasure hunter, and he had benefited from a device invented by his Miami neighbor, electrician Charles A. Hedden. Hedden's patented "metal locator" consisted of a round electrified plate with two balanced metallic coils. When the plate came near a piece of metal, it altered the inductive balance of the coils' electromagnetic field, generating a signal that chirped in the operator's headphones. Stiles had used the device to locate a sunken ship sixty feet underwater and reasoned that it might be used to locate enemy submarines. Hardland worried that the device's limited range made it unsuitable for spotting submarines, but he immediately saw its potential for detecting land mines.[71]

Hardland arranged for Hedden to visit the NIC in Washington, DC, and conduct additional tests nearby at Fort Belvoir, Virginia. A committee of top military brass watched anxiously as Hedden combed a field with his device; they cheered with excitement when he detected and dug up a defused mine. With further development by the Army Corps of Engineers, the device detected metal objects buried thirty inches underground, surpassing the six- or seven-inch range of the best British or German detectors. Once in operation, Hedden's mine detector (figure 6.6) saved countless Allied lives and proved crucial in clearing beaches for amphibious operations across North Africa, Italy, France, and the Pacific. Moreover, the feel-good story of Wade Stiles and Charles Hedden delivered on the promise and potential of the National Inventors Council. As Lawrence Langner wrote in a 1943 report, the mine detector was a "tribute to the intelligence of the staff of the National Inventors Council," which immediately saw a military application in a seemingly crackpot invention.[72]

The NIC was also proud of the mercury dry battery, developed by electrochemical inventor Samuel Ruben. On July 11, 1941, Ruben received a letter inviting him to submit his defense ideas to the NIC. He spent the next few months developing a fascinating array of devices, including an incendiary bomb, a poison gas detector, and batteries that could operate at −40° F. In 1942, Ruben traveled to Washington, DC, to present his ideas to the NIC's staff director, Thomas R. Taylor. As Ruben presented his low-temperature batteries, Taylor described the army's struggles with poor battery performance

Figure 6.6
US Army soldiers clear a French roadside on July 13, 1944, using mine detectors developed by inventor Charles Hedden and the National Inventors Council. Photograph no. 531193 (photographer Boyd Norton), Records of the Office of the Chief Signal Officer, Record Group 111, National Archives at College Park, College Park, MD.

at high temperatures and high humidity. The standard zinc-carbon batteries in use at the beginning of the war were unsealed and thus vulnerable to ambient conditions. Consequently, the batteries sent to the Pacific theater for use in "walkie-talkie" radios were plagued by corrosion at the terminals, spontaneous discharges during transit and storage, and a short five-hour service life.[73]

Ruben returned to his laboratory in New Rochelle, New York, abandoned his low-temperature experiments, and immediately commenced investigations into a "tropical" battery. Working closely with engineers from the Signal Corps Laboratories in Fort Monmouth, New Jersey, Ruben invented a miniature zinc-mercury cell with a sealed, airtight casing. Initial tests confirmed that Ruben's prototype resisted ambient effects, stayed fresh on the shelf during storage, and operated for twenty-five hours. Ruben's individual cells were tiny, so the Signal Corps linked seventy-two cells together in a battery pack sized to fit its walkie-talkies. Given the battery's strategic importance, Ruben was not permitted to apply for a patent until July 10, 1945, just before the end of the war.[74]

Ruben's longtime licensee, the P. R. Mallory Company, had no prior experience with batteries, but at the Signal Corps' insistence, the firm agreed to manufacture the cells beginning in 1943. After overcoming exploding batteries and other production difficulties, Mallory eventually manufactured over one million mercury cells per day and even sublicensed further production to Ray-O-Vac, Sprague, Magnavox, and British Eveready. Although he was owed approximately $2 million per year in royalties, Ruben felt it would be unconscionable to receive such large payments during wartime. Instead, the Signal Corps paid him $150,000 per year through the end of the war.[75]

Because of the mercury cell's superior performance and durability, the army realized a net savings of $4.50 on each battery pack. More importantly, soldiers carrying walkie-talkies enjoyed greater transmission range and a fivefold improvement in service life. After the war, Ruben and Mallory adapted the miniature mercury battery for use in hearing aids, pacemakers, and electric watches. They also introduced a modified line of alkaline batteries with a "copper top" to prevent corrosion. In short, Ruben's tropical battery emerged from the National Inventors Council and helped Mallory evolve into the brand we know today as Duracell.[76]

In contrast, Everett Bickley's wartime service was a source of both pride and frustration. Bickley was a Philadelphia-based inventor of optical sorting machines that separated usable beans, seeds, and ball bearings from substandard rejects.[77] Bickley served as a $1-a-year NIC consultant and evaluated ideas in metallurgy, machinery, and tools for Committee #10. He also presented thirty-two of his own ideas to the NIC. For example, his "Cat's Eye" employed low-intensity fluorescent lights and special detectors for signaling during blackouts. Bickley's "Talking Fish" was a sonic decoy that ships and submarines towed behind them to mask their true position from enemy sonar.[78]

Bickley felt disillusioned after the war. The army and navy had rejected all the ideas he had submitted through the NIC. Moreover, he believed his service and other wartime restrictions had prevented him from fully exploiting his own commercial inventions. In a March 1945 letter to new commerce secretary Henry A. Wallace, Bickley listed his grievances: "My trained employees have been taken away by the draft. . . . Materials have been cut off. Prices have increased in spite of the ceilings. . . . Increased income tax rates have taken what might have been used to expand. . . . As a result, I have been stopped and prevented from developing my patents in the way contemplated when the term of 17 years was determined for the life of a patent." Bickley asked Wallace, who directed the Patent Office, to "extend the life of all unexpired patents at the beginning of the war, for an additional period, equal to the duration." Bickley and his daughter Audrey even traveled to Washington, DC, to lobby several congressmen for passage of an extension bill, but it never passed. Although he had proudly "invented for Victory" and freely donated his "time, money, and energy to the cause," Bickley ended up frustrated by his experience with the National Inventors Council.[79]

The NIC's most famous inventor was Hollywood actress Hedy Lamarr. As the *New York Times* reported in October 1941, Lamarr sent the NIC a "red hot" idea concerning remote control for weaponry, but her experience also ended in frustration. Hedwig Eva Marie Kiesler was born in Vienna, Austria, in 1914, and she initially pursued her stage and film career in Europe. In 1934, she married Viennese arms manufacturer Friedrich "Fritz" Mandl and observed many conversations about the latest weaponry as she accompanied him to dinner parties with potential clients such us Benito Mussolini and Adolf Hitler. As she grew wary of her husband's controlling nature and fascist politics, Kiesler left Mandl in 1937, ending the first of her six marriages. A chance encounter with MGM's Louis B. Mayer punched her ticket to Hollywood; he renamed her Hedy Lamarr and christened her "the most beautiful woman in the world." As war broke out across Europe, Lamarr starred in a number of films, including *Algiers* (1938), *Boom Town* (1940), and *Tortilla Flat* (1942).[80]

At a Hollywood party in August 1940, Lamarr met a kindred spirit in avant-garde composer George Antheil. The Trenton, New Jersey, native made a living by writing film scores, but he had earned a reputation as the self-proclaimed "bad boy of music" for his modernist compositions. For example, his *Ballet Mécanique* (1926) featured several synchronized player pianos, along with airplane propellers, bells, and sirens. With the war escalating in the summer of 1940, Lamarr confided to Antheil that "she knew a good deal about munitions and various secret weapons" from her first husband and was considering "quitting M.G.M. and going to Washington, D.C. to offer her services to the newly established Inventors' Council." Instead, she and Antheil stayed in Hollywood and developed their "frequency hopping" concept to guide radio-controlled torpedoes.[81]

In 1940, the US Navy's anti-submarine torpedoes were technically unsophisticated and often missed their marks. As a remedy, the Lamarr-Antheil concept combined both a radio guidance system and a defense against possible countermeasures. A radio transmitter onboard a surface ship or airplane would remotely control the torpedo's rudder but rapidly switch through different control frequencies at split-second intervals. A receiver on the torpedo was synchronized with the transmitter and would receive the ever-changing signals. The rapid "hopping" between control frequencies would prevent enemy eavesdroppers from acquiring and jamming the signal. Just as he had coordinated the player pianos in *Ballet Mécanique*, Antheil proposed synchronizing the transmitter and receiver with rotating, slotted paper rolls that switched through eighty-eight frequencies, the same number as the keys on a piano. Lamarr and Antheil submitted their idea to the National Inventors Council in December 1940; they later received US Patent 2,292,387 for their "Secret Communication System" in August 1942.[82]

The NIC duly passed along the "red hot" idea; however, the navy rejected the proposed mechanism as too bulky. Antheil believed the device could be built at small scale, but he suspected that the "brass-headed gentlemen in Washington" simply could not imagine putting "a player piano in a torpedo." After the navy's rejection, Antheil returned to composing while Lamarr leveraged her celebrity to sell $25 million in war bonds. With the development of smaller transistors and microchips, frequency hopping eventually became the basis for most military anti-jamming devices and other forms of "spread spectrum technology," including mobile phones, Wi-Fi, Bluetooth, and GPS. Thanks to Lamarr and Antheil, today's wireless devices can communicate securely and without interference.[83] Like Sperry and Kettering's World War I aerial torpedo, the frequency-hopping concept was simply ahead of its time.

An Evolving Role for the National Inventors Council

Despite the inevitable rejections and frustrations, American inventors could take pride in their improved performance. In September 1942, the National

Inventors Council paused to celebrate its second anniversary with a luncheon at the capital's Hotel Washington. In two years, it had conducted over 10,000 interviews and processed 104,431 inventive suggestions. From these suggestions, the council had forwarded 4,367 meritorious ideas (4.2 percent) to the military for a "quality" metric much higher than that of the Naval Consulting Board. Noting Hedden's mine detector and other successes, the *New York Times* admitted in 1943 that "the record of the National Inventors Council is far better than anything achieved in the First World War"—not because "this generation is any more inventive" than its predecessors but because the NIC represented "a better system of clearing and appraising inventions."[84]

However, the NIC still had plenty of critics. Alfred Toombs of *American Magazine* seemed to relish the bizarre behavior and wacky ideas (figure 6.7) of America's citizen-inventors: "There is the inventor, for instance, who suggests that this country send a few hundred bombers over Germany on some rainy day to spray the entire country with cement and turn the Nazi fighters into concrete statues. Another has developed a bomb loaded with *eau de skunk*, which he wants to drop on hostile ships, contending that the odor would cause the crew to jump overboard, leaving the ship intact for us to use." Toombs also described the man who appeared at the NIC offices for an interview "wearing a green turban, scarlet cloak, purple sandals, and a full beard." When he failed to convince the staff of his plan for "extracting energy from the air," he exited in a huff and exclaimed, "I couldn't expect ya'll Occidentals to understand." Still, the NIC's staff remained optimistic. As he looked at a basket of unopened mail, staff director Thomas R. Taylor freely admitted that "there will be junk here, plenty of it," yet "there may be an idea that will win the war!"[85]

Like the earlier Naval Consulting Board, the National Inventors Council's inventors were criticized for being woefully ignorant of the state of the military arts. A *New York Times* op-ed chided "clerks in the Middle West, who have never seen the ocean" but were convinced they could defend Allied ships against enemy submarines. In response, the NIC again turned to public relations to cut down on duplicates and useless ideas. A September 1942 press release begged inventors to stop suggesting ship-borne torpedo nets, which were proven ineffective during World War I. Similarly, an October 1942 *Popular Science* article urged inventors to refrain from suggesting miniature submarines, death-ray guns, floating air mines, explosives

Figure 6.7
Journalists and cartoonists enjoyed poking fun at the bizarre behavior and wacky ideas of America's citizen-inventors.
Source: Baltimore Sun, May 20, 1943, a clipping in General Subject Files, National Inventors Council, Records of the National Institute of Standards and Technology, Record Group 167, National Archives at College Park, College Park, MD.

dropped via parachute, or any "superman ideas" involving special suits or body armor. Finally, the NIC advised prospective inventors to consult a five-page bibliography of "Instructive Literature" in the NIC's updated *Information Bulletin* to determine whether their bright ideas had already been implemented or rejected.[86]

In addition, two internal government audits questioned the efficacy of the NIC's procurement function, which charged five professional inventors with solving a specific military problem. In January 1941, the Patent Office's C. C. Henry reported that any inventions before the NIC that required experimentation or research should be referred to the National

Defense Research Committee or else risk "usurping . . . or duplicating" the NDRC's functions. Only the NDRC—not the National Inventors Council— had the congressional authority and funds to sponsor experimental research and development by entering into contracts with university and corporate laboratories. A second report, in August 1942, determined that the NDRC was already working on sixty of the NIC's 144 open procurement problems, confirming fears of duplicate effort. The report urged the NIC to cede development responsibility to the NDRC and confine itself to evaluating and referring the public's inventive suggestions.[87]

Given these objections, the NIC modified its approach. Instead of identifying five inventors to solve a particular problem, the NIC attempted to publicize the military's growing list of requests and focus all the nation's independents on those urgent needs. The army and navy initially balked at widely publicizing their technical problems, fearing that such publicity might reveal too much to the Germans and Japanese. Yet, given the volume of unhelpful and duplicate ideas deluging the NIC, the armed forces were beginning to see the wisdom of providing inventors with some specific targets on which to focus their energies. E. L. Gustus, research director for the Army Service Forces, stated that "the individual inventor . . . is in a position similar to a blindfolded hunter. He can fire away imaginatively at random, but his bag of successful ideas would be much greater if he could see his quarry—our problems—and aim accordingly."[88]

Citing concerns about secrecy, the navy initially declined to publicize its problems. However, in July 1943, the NIC convinced the army to release a list of twenty declassified requests, such as a method for detecting nonmetallic mines and a substitute material for rubber wire insulation. The NIC shared the vetted list with two dozen engineering and technical societies, who reprinted the list in their journals. After receiving about 250 promising solutions, the NIC began contemplating an even broader dissemination.[89] In May 1944, the NIC released a longer list of declassified army problems to the engineering societies and major news outlets such as the *New York Times*. The navy eventually relented and released the first of several problem lists in January 1945.[90]

Meanwhile, Congress began considering a series of bills to establish the National Inventors Council on a permanent legal basis. In January 1942, the Senate held hearings on S. 2078, a bill that would endow the NIC with statutory authority and appropriations, with accompanying congressional oversight. In October 1942, the Senate considered a related bill (S. 2721) to permanently establish and coordinate the work of the NIC, the NDRC, and other wartime technical agencies. The ensuing debates revealed additional criticisms of the NIC as the United States crept toward a state of permanent mobilization.[91]

In his testimony regarding S. 2721, *New York Times* science editor Waldemar Kaempffert acknowledged the public demand for the National Inventors Council but argued that the nation could no longer rely on "the heroic inventor of the Morse or the Bell or the Edison type." Independents could still produce "fountain pens, vacuum cleaners, and contrivances of that kind," said Kaempffert, but for serious military problems in "metallurgy or illumination or synthetic rubber, your lone garret inventor is simply, hopelessly lost." Chicago inventor Hiram Sheridan testified that his fellow independents were still contributing valuable ideas but characterized the NIC bureaucracy as aloof and unresponsive. Calling the NIC a "black hole," Sheridan complained that it provided little information on the status of either adopted or rejected ideas, leaving inventors disillusioned.[92]

Senator Edwin C. Johnson (D-CO), the sponsor of S. 2078, worried that the NIC lacked the statutory authority or funding to conduct its own experiments. He also feared that it had been unduly influenced by corporate and military prejudice against independent inventors. In a letter to the *New York Times*, inventor W. Houston Kenyon likewise complained that "the council has no power and no money to build or test models," so "if the inventor on paper fails to interest the particular officer of the Army or Navy who reads it, the idea ends right there." With S. 2078, Kenyon hoped for an unbiased council "free of military preconceptions" and "equipped with sufficient funds to translate enthusiasm into practical demonstration."[93]

The National Inventors Council actually opposed both bills. In a statement regarding S. 2078, council members argued that they were already doing their job well, had adequate appropriations under Commerce's budget, and believed that additional congressional oversight would impede their work. In his testimony, Langner flatly rejected the accusation that the NIC had been co-opted by military or corporate interests. Moreover, he countered that additional congressional appropriations for prototyping were unnecessary, since the council was due to receive an additional $150,000 for a new Model Building Fund.[94]

Established in March 1942, the Model Building Fund helped the NIC's independent inventors construct prototypes when the scale of the project did not warrant a full-blown NDRC development contract. The models were built by the National Bureau of Standards (NBS), a sister agency within the Department of Commerce, or under contract by an NBS vendor. The NBS also tested the prototypes in coordination with the army or navy, especially if certain facilities (e.g., a firing range) or conditions (e.g., an underwater test) were required. A typical small-scale project was the $600 disbursed to inventor William Allwood to build ten improved mortar fuses at $60 apiece. In a somewhat bigger project, the fund allocated $10,000 to inventor E. L. White of the American Isteg Steel Corporation, who spent only $7,000 to test an improved steel bar for reinforcing concrete barriers. Between March 1942 and November 1945, the NBS allocated $54,432 for the construction and testing of forty-three models before returning $88,000 in unused funds to the Treasury.[95]

Both S. 2078 and S. 2721 stalled in committee, and the NIC remained content to operate on an emergency, advisory basis during the war. However, in May 1945, the council held internal deliberations about its postwar future. A report circulated by Langner affirmed the NIC's desire to "continue after the war in its present functions." With a reduced staff and less frequent meetings, the proposed peacetime council hoped to continue publishing lists of the military's declassified problems and evaluate the public's many valuable suggestions for postwar needs, such as artificial limbs and improved wheelchairs.[96]

Congress continued to search for a coherent federal science and technology policy after Japan's surrender. In December 1945, Senator William Fulbright (D-AR) introduced a bill (S. 1248) to consolidate the NIC and several other wartime advisory groups within the Department of Commerce as an "applied science" complement to Vannevar Bush's "pure science" proposal for the National Science Foundation.[97] When the bill passed in 1946, the peacetime National Inventors Council was established on a firm statutory basis within the Invention and Engineering Division (IED) of Commerce's newly established Office of Technical Services (OTS). As planned, the NIC continued to evaluate the public's defense-related ideas and publish the military's problem lists. More broadly, the IED and its "Inventors Service" answered general inquiries from individual inventors, offered tentative appraisals of new ideas, and provided guidance on patenting and commercialization. As the government's "open door" for America's individual inventors, the IED and peacetime NIC provided a crucial advisory service for independents—a service they could not sustain through their own, short-lived professional groups (chapter 4).[98]

In 1946, Lawrence Langner and Watson Davis assembled the nineteen-page *Administrative History of the National Inventors Council*, which assessed the council's accomplishments and shortcomings. They argued that the NIC had succeeded in its role as an efficient clearinghouse. Rather than making multiple pitches across the different service branches, inventors could send their ideas to a single bureau. Likewise, the army and navy could present their technical problems to a single agency and leave the sifting of public suggestions to the NIC. The NIC's major weakness was its lack of statutory authority. As a mere advisory body, the council had limited jurisdiction and funds to assist inventors in building prototypes, and it could not always secure status reports on the inventions it had referred to the armed forces. As a result, many inventors were kept in the dark about their NIC submissions, and morale suffered. Given secrecy concerns and incomplete reporting from the army and navy, Langner and Davis predicted that "the full extent of the Council's effectiveness will never be known."[99]

Nevertheless, the National Inventors Council and America's inventors had performed well, especially when compared with the NCB's dismal record in World War I. As of June 30, 1946, the NIC had received 208,975 suggestions and forwarded 8,615 meritorious ideas (4.1 percent), resulting in 106 confirmed inventions in use by the armed forces. Certainly, the NIC had received thousands of outlandish ideas from "crackpots" and uninformed amateur inventors. However, it had successfully advanced several good ideas that helped win the war, such as Samuel Ruben's tropical battery and Charles Hedden's land mine detector.[100]

The National Inventors Council's strong record served as a vindication for America's independent inventors, who had endured the stigma associated with World War I's Naval Consulting Board. The success of the NIC had demonstrated that individual inventors could still make valuable contributions in an era increasingly marked by university and corporate research labs. The *Administrative History* duly praised the wartime R&D projects that had produced radar, penicillin, and the atomic bomb. However, "it would be calamitous," the council suggested, "if such institutionalization resulted in the stifling exclusion and discouragement of the individual, free-lance inventor."[101]

Conclusion

American independent inventors patriotically served their country by mobilizing for World War I and World War II. In each conflict, independent inventors—not scientists—took the first steps toward preparedness when they established the Naval Consulting Board (NCB) and the National Inventors Council (NIC) to coordinate their activities. After inventors seized the initiative, American scientists established their own wartime institutions, and thanks to greater military and government support, they were more successful at sustaining them. The National Research Council (NRC) is still active, and the legacy of the National Defense Research Council (NDRC) lives on through federal agencies such as the National Science Foundation and the Atomic Energy Commission. However, the NCB disbanded around 1920, and the NIC eventually dissolved in 1973 (chapter 7).

The mobilization for World War I inflamed a growing rivalry between inventors and scientists. This professional animus motivated the NRC to promote industrial research and disparage the independents after the war (chapter 2). In contrast, inventors and scientists were more cooperative during World War II. The NIC's corps of $1-a-year consultants included a mix of independent inventors, corporate researchers, and university scientists, and they coordinated with the NDRC to develop the NIC's most promising inventions. Journalists occasionally revived these tensions when they praised the sophisticated corporate laboratories and maligned the NIC's inventors as "crackpots."

Despite the NCB's poor record, the NIC borrowed—and improved on—its organizational model. During World War I, the NCB and Army Inventions Section ran parallel but separate operations. In contrast, the National Inventors Council was a true clearinghouse—a one-stop destination for independent inventors and all branches of the armed forces. The NIC was also more respectful toward America's citizen-inventors, so it was more successful at harnessing their talents. The NIC's longer, six-year operational window (1940–1946) also enabled the implementation of more ideas. After three years (1915–1918), several promising NCB inventions—such as the aerial torpedo—were abandoned following the armistice.

Independent inventors enjoyed more opportunities than usual during times of war, when well-funded government patrons urgently sought unconventional ideas that might confer a military advantage.[102] Although inventors served without compensation, their wartime work sometimes led to lucrative defense contracts and commercial applications for their military inventions. However, inventors also endured numerous frustrations during both wars, especially the military version of the "not invented here" attitude. In addition, materials shortages, high taxes, and the time allotted to wartime service prevented inventors from developing their other, nonmilitary inventions.

Overall, inventors' wartime mobilizations offer further proof of their ongoing contributions during the twentieth century. In addition, the Naval Consulting Board and the National Inventors Council established an important precedent for open innovation. At the turn of the twenty-first century, several federal agencies and R&D firms revived crowdsourcing strategies first deployed during World War I.[103]

7 Postwar Eclipse, Twenty-First-Century Resurgence

The typical lone inventor of the eighteenth and nineteenth century has all but disappeared. In his place in the mid-twentieth century came the industrial research laboratory.
—Harvard University president James B. Conant, 1951

Corporate America is slashing in-house R&D and turning to the country's basements, garages, and small companies for innovation.
—Ian Mount, "The Return of the Lone Inventor," *Fortune Small Business*, 2005

Although independent inventors had served admirably during World War II, most postwar observers believed that they were obsolete. "There is no more pleasant fiction," observed economist John Kenneth Galbraith, than the belief that "technical change is the product of the matchless ingenuity of the small man." Rather, Galbraith argued in *American Capitalism* (1952), "technical development has long since become the preserve of the scientist and the engineer" working in large firms. However, conditions shifted considerably over the next half century. By 2005, *Fortune Small Business* celebrated "the return of the lone inventor," while *The Economist* (2007) reported that technology firms had "left the big corporate R&D laboratory behind."[1] How did independent inventors achieve such a dramatic turnaround? What happened to American independent inventors and corporate R&D labs after World War II?

During the period from roughly 1950 to 2015, independent inventors navigated a series of opportunities and challenges that were often familiar but sometimes very different from those encountered by their predecessors. Postwar independents continued to compete (and sometimes partner) with corporate labs while pursuing a mix of commercialization strategies. Likewise, independent inventors remained both energized and frustrated in their efforts to supply inventions for the US military and the federal government. Meanwhile, postwar independents established—and generally failed to sustain—professional organizations to support their commercialization efforts and advance their political agenda. In the policy arena, inventors benefited from a pro-patent judiciary and aggressive antitrust enforcement, but they were unable to block fundamental changes to the patent laws that favored the big corporations. Nevertheless, independent inventors enjoyed a resurgence at the end of the twentieth century and managed to rehabilitate their public image. Individual inventors—who were once derided as crackpots or assumed to be extinct—became celebrated as entrepreneurial mavericks who presided over the high-tech industries of the twenty-first century.

Before exploring the postwar experiences of independent inventors, it will be helpful to review the parallel fortunes of the corporate R&D labs. Historians have described how Cold War military spending and aggressive antitrust enforcement spurred the explosive growth and intensification of corporate R&D while generally assuming that independent inventors were a negligible factor in the postwar innovation economy.[2] However, those same postwar policies and macroeconomic trends had profoundly different impacts for independent inventors. The postwar experiences of independent inventors stand in sharper relief when compared against the well-documented record of their corporate counterparts.

The Rise, Fall, and Restructuring of Postwar R&D

During the postwar period, the number of industrial research personnel grew rapidly, while the number of independent inventors appeared to remain stable. A 1965 study by the Patent, Trademark, and Copyright Foundation estimated that there were approximately 225,000 "independent" and 235,000 "employee" patentees in the US population. Over the next several decades, however, corporate research personnel ballooned to an estimated 960,000 in 2014.[3] There was no comparable employment data for independent inventors, but the patent data suggested a similar divergence from near parity at midcentury (figure 7.1). In 1950, US individuals earned 18,960 patents (46.5 percent), while US corporations earned 21,782 patents (53.5 percent). Over the next sixty years, however, the number of

Figure 7.1

After 1950, corporate patenting grew substantially, while individuals' annual patent totals remained constant at about 10,000 to 15,000 patents per year.

Sources: For 1950–2000 inclusive, historic patent data from the United States Patent and Trademark Office (USPTO) have been conveniently compiled in *Historical Statistics of the United States: Earliest Times to the Present*, millennial ed., ed. Susan Carter et al. (New York: Cambridge University Press, 2006), 3:425–429, table Cg27–37. For 2001–2015 inclusive, see USPTO, "Patenting by Organizations," table A1-1b, https://www.uspto.gov/web/offices/ac/ido/oeip/taf/reports_topo.htm.

corporate patents grew continuously, while independents' annual patent totals essentially remained constant at about 10,000 to 15,000 patents per year. In 2015, for example, individual inventors earned 13,643 patents (9.3 percent) as usual, but this represented a much smaller percentage when compared with the 133,434 patents (90.7 percent) earned by corporations. So independent inventors remained a steady and appreciable source of new inventions after World War II, but they represented a *much smaller relative share* of the innovation economy because of the explosive postwar growth of corporate R&D.[4]

That growth was fueled initially by federal military spending. University and corporate scientists organized by Vannevar Bush's Office of Scientific Research and Development (OSRD) had helped win World War II, and those high levels of federal R&D spending continued throughout the permanent mobilization of the Cold War. Federal R&D investments accounted for more than 50 percent of total national R&D spending from 1953 to 1978, with defense projects accounting for 80 percent of federally funded R&D throughout the 1950s and rarely dropping below 50 percent through 2005.[5] After

World War II, new federal agencies such as the Atomic Energy Commission (AEC), the Advanced Research Projects Agency (ARPA), and the National Aeronautics and Space Administration (NASA) contracted with established firms such as General Electric, DuPont, Boeing, RCA, and IBM to build nuclear reactors, jet engines, rockets, advanced communications equipment, and electronic computers. Given the large scale of these "Big Science" projects and the proven success of the wartime contract model, government funders favored the big corporations and their R&D labs over individual inventors. Many of these military procurement contracts included secrecy provisions, so R&D firms increasingly severed their external alliances with inventors and small firms to pursue a closed, proprietary approach to innovation.[6]

Furthermore, an aggressive antitrust policy encouraged R&D firms to turn inward. During the postwar period, the Department of Justice (DoJ) brought antitrust suits against General Electric, Westinghouse, DuPont, Alcoa, Corning, AT&T, RCA, IBM, and many other firms. Given the government's antimonopoly posture, the big firms were hesitant to join industry cartels, forge alliances with independent inventors, or acquire complementary technologies through horizontal mergers and acquisitions. Instead, high-tech firms intensified their internal R&D activities to legally develop more new products in-house. DuPont, for example, invested heavily in fundamental research to develop new synthetic fibers such as Lycra, Mylar, and Kevlar. Moreover, corporations preferred to reinvest their substantial profits into R&D, advertising, and marketing rather than pay the 50 percent corporate tax rates still in place after World War II.[7]

Antitrust penalties also forced the big firms to share their hard-earned research products. Following the TNEC hearings, Congress decided not to build compulsory licensing into the patent statutes, but the DoJ and Federal Trade Commission (FTC) frequently deployed the practice to penalize monopolistic behavior and encourage competition.[8] Between 1945 and 1975, nearly one hundred firms were forced to offer licenses on their patented technologies, including AT&T, Chrysler, Corning, DuPont, Eastman Kodak, Exxon, Ford, General Motors, IBM, and Westinghouse. For example, a 1956 consent decree forced AT&T to offer licenses on the transistor and 8,600 other patents to dozens of foreign and domestic firms.[9]

R&D firms did not experience the long-term negative impacts of this antitrust climate until the 1970s. In the meantime, the immediate postwar decades were a golden age for corporate R&D. With lucrative Cold War procurement contracts, near monopolies in their markets, robust profits, and only a few emerging competitors, the big R&D firms could easily afford the significant investments in personnel and facilities required to pursue industrial research on a massive scale. In 1950, AT&T's Bell Labs employed 6,000 people, one-third of whom were professional scientists and engineers. This workforce represented approximately 1 percent of the entire science and engineering workforce in the United States and included four future Nobel laureates in physics and five future winners of the National Medal of Science. Those scientists developed several remarkable innovations, including solar cells, lasers, global satellite telephony, miniature microphones, the C++ software language, and the UNIX computer operating system.[10]

By the 1970s, industrial research labs began to struggle with a host of challenges. Decades of vigorous antitrust enforcement gradually eroded the monopolistic positions of the R&D giants and created new foreign and domestic competitors. AT&T remained dominant in telephony, but it ceded the electronics sector after being forced to license its semiconductor patents to Texas Instruments, Intel, and Motorola. Similarly, in 1973 the FTC forced Xerox to license its photocopier patents with the expectation that Kodak, IBM, and other domestic firms would enter the market. Instead, Japan's government-subsidized imaging cartel—including Canon, Toshiba, Sharp, Panasonic, Konica, and Minolta—licensed the patents and began flooding the US market with cheaper copiers. By 1979, Xerox's 100 percent market share had plummeted to only 19 percent.[11] With smaller market shares and slimmer profits, the big high-tech firms had fewer resources available to invest in research. Moreover, decades of self-imposed isolation and reliance on military procurement contracts made the corporate R&D labs insular and less responsive to consumer markets.

As a result, American high-tech corporations struggled with poor research productivity and endured some embarrassing commercial failures. In 1973, Xerox's Palo Alto Research Center (PARC) introduced the experimental Alto, arguably the first personal computer. However, Xerox could persuade only one customer (Siemens) to buy the Star, the Alto's commercial successor. Moreover, an exodus of Xerox researchers went on to found 3Com, Silicon Graphics, and Adobe by commercializing technologies originally invented at PARC.[12] RCA introduced its Selectavision VideoDisc player in 1981, but it was outsold by videocassette recorders (VCRs) from Sony and Matsushita (Panasonic). This was a self-inflicted wound since Matsushita had developed its VCR by licensing RCA's patents. After losing the instant camera market to Polaroid, Eastman Kodak invented the first digital camera in 1975, shelved it to protect its film business, and then watched as Sony, Nikon, Casio, and Minolta dominated the sector. In 1996, after decades of antitrust pressure, AT&T spun off Bell Labs and its manufacturing operations into a new company named Lucent Technologies. Lucent became overextended during the dot.com rush to install fiber-optic infrastructure and began liquidating assets. In 2006, Lucent vacated Bell Labs' Holmdel, New Jersey, research campus (figure 7.2) and put it up for sale. The salad days of corporate R&D were over.[13]

These difficulties forced many high-tech firms to downsize their research operations and reopen their closed, proprietary approach to innovation. In a revival of the hybrid innovation strategies from the early twentieth century, several firms began to acquire external patents and partner with independent inventors to complement the inventions they developed internally. After experiencing a postwar golden age and a humbling decline, corporate R&D labs pivoted to "open innovation" in the new millennium.[14]

Figure 7.2
In 2006, Lucent Technologies shuttered the Bell Labs campus in Holmdel, New Jersey, designed by architect Eero Saarinen. Photo by Rob Dobi.

The rise, fall, and restructuring of postwar R&D has been well documented, but we know comparatively little about the parallel fortunes of independent inventors. Their postwar (1950–2015) experiences often mirrored—but sometimes diverged—from those of their predecessors.

Commercialization: The More Things Change . . .

After World War II, independent inventors survived much as they had during the first half of the twentieth century. In short, independent inventors could found a start-up to manufacture and market an invention as an inventor-entrepreneur (make), assign the patent to an established firm for a onetime fee (sell), or license the invention and earn royalties and consulting fees by partnering with a manufacturer (ally). To maximize their chances for success, inventors pursued several of these strategies simultaneously in a mixed, hybrid approach to innovation while alternating between periods of employment and independence.

The independents struggled commercially during the 1950s and 1960s as the corporate R&D labs dominated the overall context of innovation. As the demands of military secrecy and aggressive antitrust policies encouraged corporate labs to turn inward and intensify their research operations, there was generally less demand for the inventions developed by independent inventors.

Nevertheless, numerous postwar independents persisted in relative obscurity alongside the more heralded industrial labs. College student Robert W. Gore conducted experiments in his family's basement using samples of polytetrafluoroethylene (PTFE, or Teflon) provided by his father, Wilbert, a DuPont chemist. In 1958, Wilbert left DuPont and founded W. L. Gore & Associates to commercialize his son's idea for PTFE-insulated computer cables. In 1969, Robert developed PTFE fibers into Gore-Tex, a waterproof but breathable fabric used in outdoor apparel.[15] Stanford Ovshinsky was a working-class machinist and self-taught inventor with a high school education. He became an expert in condensed matter physics and developed several key inventions, including rewriteable CDs and DVDs and the rechargeable nickel metal hydride batteries that power digital cameras and hybrid vehicles.[16] Marion O'Brien Donovan, a college-educated Connecticut homemaker, used a sewing machine and a plastic shower curtain to invent the "Boater," a reusable diaper cover that was leak-proof (like a boat). After failing to find a licensee, she founded Donovan Enterprises to commercialize the Boater and eventually sold her firm for $1 million in 1951. She continued to market numerous household inventions through the 1990s, including the "Zippity-Do," an elastic zipper pull for women's dresses, and the "Big Hang-Up," a closet organizer (figure 7.3).[17]

In 1975, Donovan appeared alongside fellow inventor Jacob Rabinow during an episode of Barbara Walters's daytime talk show. After completing bachelor's and master's degrees in engineering at the City College of New York, Rabinow began his career at the National Bureau of Standards (NBS) in 1938. He spent fifteen years evaluating inventors' ideas and building prototypes for the National Inventors Council before leaving the NBS in 1954 to become an independent inventor. Rabinow eventually received 230 US patents on a variety of mechanical, optical, and electrical devices, including a magnetic clutch used in tape and disk drives (1956), an automated

Figure 7.3
Inventor Marion O'Brien Donovan with the "Big Hang-Up" in the late 1970s. Marion O'Brien Donovan Papers, Archives Center, National Museum of American History, Smithsonian Institution, Washington, DC.

mail-sorting machine (1957), an improved phonograph arm (1959), and a reading machine that pioneered optical character recognition (1960).

Rabinow pursued a hybrid mix of commercialization strategies as he alternated between employment and independence. He licensed his automated mail sorter to Burroughs and collected royalties and then founded Rabinow Engineering to manufacture, install, and service his reading machines for multiple clients. After selling his firm to Control Data Corporation (CDC), Rabinow led research efforts at CDC's Advanced Development Laboratory. While still employed at CDC, Rabinow founded the RABCO Company in 1968 to commercialize his phonograph arm; Harmon-Kardon acquired the firm in 1972. Rabinow eventually left CDC and returned to the National Bureau of Standards as its chief engineer. After retiring in 1975, Rabinow continued to work sporadically as an industrial consultant until his death in 1999.[18]

By the 1970s and 1980s, the same antitrust policies that frustrated industrial researchers were creating new entrepreneurial opportunities for independent inventors and small start-ups. In 1969, the DoJ brought an antitrust case against International Business Machines (IBM), the dominant player in the computing industry. IBM had long required clients to lease its entire bundle of hardware, software, and technical support services for a single negotiated price; DoJ alleged that this practice stifled competition. IBM moved immediately to unbundle and individually price its hardware, software, peripherals, and engineering services.[19]

That same year, IBM engineer Alan Shugart left IBM's R&D laboratory in San Jose, California, to join Memorex in nearby Santa Clara. During his eighteen years at IBM, Shugart and his colleagues had developed a series of disk drives and 8.5-inch floppy storage disks to replace traditional punched card systems. With the antitrust lawsuit creating space for new competitors, Memorex recruited Shugart to develop IBM-compatible disk storage systems, which they sold to Wang, Digital Equipment Company, and other firms in the emerging minicomputer sector. Shugart eventually founded his own firms, Shugart Associates (1972) and Seagate Technologies (1979), which patented, commercialized, and supplied 5.25-inch floppy disks and internal hard drives to manufacturers across the industry. Shugart had learned his trade while working for IBM and Memorex, but he was able to advance the technology and tap new markets as an independent inventor-entrepreneur. As a result, Seagate, not IBM, came to dominate the disk storage sector.[20]

IBM's attention to its long-running antitrust case distracted it from the emergence of the personal computer (PC) market. In the Bay Area, hobbyists from the Homebrew Computer Club assembled computer kits from the cheap, third-party microprocessors, disk drives, and peripherals that were more widely available in the wake of IBM's unbundling. As their designs matured, some entrepreneurial hobbyists sensed the emergence of a market for low-cost computers for use in homes and small businesses, a niche much different from the expensive mainframes IBM sold to universities, government agencies, and big businesses. More than twenty start-up firms originated from the Homebrew Computer Club, including North Star Computers, Cromemco, Processor Technology Corporation, and, most famously, Apple Computer.[21]

Apple cofounders Steve Jobs and Steve Wozniak became inventor-entrepreneurs while moonlighting from their day jobs as engineers at Atari and Hewlett-Packard (HP), respectively. In 1976, Jobs and Wozniak demonstrated their Apple I motherboard at the Homebrew club to wide acclaim. Club member Paul Terrell, owner of the Byte Shop computer store in Mountain View, agreed to buy fifty Apple I kits at $500 apiece, due in thirty days. Jobs sold his Volkswagen van, Wozniak sold his HP calculator, and they cobbled together $1,300 in start-up capital. Wozniak and Jobs assembled the kits in the Jobs family garage and delivered them to the Byte Shop in twenty-nine days. After paying their parts supplier with the $25,000 from Terrell, they kept a modest profit. Jobs was twenty-one years old; Wozniak was twenty-five.[22]

IBM rushed to enter the PC market, but the firm did not possess the expertise to quickly develop all the necessary components and software within its R&D labs. Instead, IBM sourced the microprocessors from Intel and the operating system software from an Albuquerque start-up called Microsoft. Harvard dropout Bill Gates had founded the firm in 1975 with his childhood friend Paul Allen to supply a BASIC programming language compiler for the burgeoning community of kit companies and microcomputer hobbyists. IBM contracted with Microsoft to supply the disk operating system for the IBM PC (PC DOS), but the nonexclusive licensing deal allowed Microsoft to sell its own version, MS-DOS, for non-IBM computers. When the IBM PC shipped in 1981, competitors took note of its open, nonproprietary architecture. Compaq, Tandy/Radio Shack, and Dell soon introduced cheap IBM "clones" built with Intel processors and running

Microsoft's MS-DOS. Soon, both Intel and Microsoft boasted market capitalizations greater than IBM.[23]

In 1982, after thirteen years, the DoJ withdrew its antitrust case against IBM. With dozens of new competitors now making peripherals (e.g., Shugart's disk drives), software (e.g., Microsoft's MS-DOS), and personal computers (e.g., Apple, Compaq), the DoJ acknowledged that the case no longer had merit. Without a judgment or consent decree, the antitrust case had simultaneously persuaded IBM to unbundle its products and services while creating new opportunities for inventor-entrepreneurs in the computer industry. Moreover, the commercialization patterns from the 1970s and 1980s were beginning to echo the hybrid mix of strategies inventors had pursued during the early twentieth century. Shugart and the moonlighting Jobs and Wozniak each benefited from prior employment at established firms before starting their own businesses in the same sector. In addition, Gates and Allen forged a nonexclusive licensing deal to develop PC DOS for IBM while simultaneously making and selling copies of BASIC and other programs to hobbyists.

By the turn of the twenty-first century, IBM and many other firms were forced to reassess the scale, scope, and closed, proprietary nature of their R&D investments in the wake of increased competition, slimmer profits, and low research productivity. Specifically, corporations proactively began to adopt an "open innovation" strategy, in which their R&D labs forged alliances and licensed patents from independent inventors to complement the technologies they developed internally.[24] In this model, the research director coordinated the activities of the in-house R&D lab while simultaneously cultivating external sources of invention. Of course, this was not a new strategy. Rather, the embrace of open innovation merely revived the hybrid mix of inventor-firm alliances that were common during the first half of the twentieth century.[25]

In 2000, Procter & Gamble (P&G) restructured its R&D operations by actively seeking external collaborators. CEO A. G. Lafley set an ambitious goal: henceforth 50 percent of P&G products would originate from *within* the firm's R&D labs, while the other half would be acquired externally and developed *through* them. The new "Connect & Develop" strategy also demanded a shift in attitude, from resistance to technologies "not invented here" to enthusiasm for those "proudly found elsewhere." To complement its 7,500 in-house research employees, P&G deployed a team of seventy external technology evaluators to seek out promising ideas from a global network of independent inventors, design firms, and university professors. Through the Connect & Develop initiative, P&G acquired and commercialized several successful products, including its Swiffer Duster, Olay Regenerist skin care creams, and the Mr. Clean Magic Eraser. As P&G evolved "from R&D to C&D," it doubled its innovation success rate while reducing its research budget from 4.8 percent to 3.4 percent of sales.[26]

The opening of corporate innovation created greater demand for the services of independent inventors such as John Nottingham and John Spirk, who *Forbes* called "the most successful inventors you've never heard of." After completing their industrial design degrees at the Cleveland Institute of Art in 1972, Nottingham and Spirk declined job offers from the General Motors Research Center and Huffy Bicycles and founded their eponymous Cleveland design firm. Much like Wadsworth Mount or Samuel Ruben from an earlier era, Nottingham and Spirk thrived as inventor-consultants by inventing to order for a host of corporate clients. Through P&G's Connect & Develop initiative, Nottingham and Spirk developed the Swiffer Sweeper-Vac and the Crest SpinBrush, an electric toothbrush with a circular rotating head. They also earned and licensed the patents underlying Dirt Devil's line of handheld vacuum cleaners, Sherwin Williams's Twist & Pour paint can lids, and Pepto-Bismol's combined cap and dosage cup, a now ubiquitous design on liquid medicine bottles. Altogether, Nottingham and Spirk's inventions have generated $50 billion in sales for their clients.[27]

A new emphasis on innovation crowdsourcing also sparked the emergence of internet-based intermediaries that connected individual inventors with corporate patrons. In 2001, pharmaceutical giant Eli Lilly built its e.Lilly web portal to solicit external solutions to biochemical problems that had frustrated its in-house researchers. In 2005, Lilly spun off the portal as InnoCentive, an independent "innovation matchmaker" for thousands of corporate "seekers" and individual "solvers." On the InnoCentive website, corporations, nonprofits, and government agencies paid a fee to post a challenge, which was accompanied by a cash award ranging from $2,000 to $1,000,000. Registered solvers uploaded their potential solutions to the InnoCentive portal; if the seeker found a suitable solution, it paid the cash award to the winning solver and a 40 percent commission to InnoCentive.[28]

Corporations embraced InnoCentive and other matchmakers because their challenges tended to multiply and diversify the amount of talent and capital directed at a given problem. However, corporations paid only for the single solution that best achieved the prize criteria. Critics worried that crowdsourcing had shifted the risks of innovation from wealthy seekers to vulnerable solvers. However, independent inventors had always assumed these risks, and they responded in droves to the increased demand for their services.[29] By 2019, InnoCentive had enrolled more than 390,000 solvers from 190 countries, 60 percent of whom held advanced degrees. Those solvers had posted more than 162,000 potential solutions to 2,000 challenges. With a 30 percent success rate, InnoCentive's solvers had earned more than $20 million in awards.[30]

John Davis, a chemist from Bloomington, Illinois, was typical of InnoCentive's solvers. Davis knew that concrete would not set if the builder kept it vibrating. He claimed a $20,000 award from the Oil Spill Recovery Institute when he suggested the same technique to prevent crude oil from freezing in Alaskan storage tanks. Larry Kearns, a Bank of America customer service representative from Jacksonville, Florida, was one of InnoCentive's top solvers. In 2016, Kearns solved three challenges: developing digital credentials to match online job posters and job seekers ($2,000, posted by Credly); preventing rodent burrows in earthen embankments to stop dam and levee breaches ($2,500, Bureau of Reclamation); and developing improved desalination methods for freshwater production ($25,000, Bureau of Reclamation).[31] Kearns's work for the Bureau of Reclamation reflected the federal government's recruitment of independent inventors after a long hiatus.

Government Crowdsourcing and Pro-innovation Policies

During both World War I and World War II, the federal government established clearinghouses to solicit and evaluate the military inventions submitted by America's grassroots inventors. After World War II, the federal government mostly ignored independent inventors, while awarding huge procurement contracts to the largest high-tech firms. However, as the United States suffered through the economic malaise of the 1970s, the federal government enacted policies beneficial to independent inventors and small firms to stimulate growth. By the turn of the twenty-first century, the federal government had joined corporations in embracing open innovation and had revived crowdsourcing strategies first deployed during World War I.

During the Korean Conflict (1950–1953), the National Inventors Council (NIC) continued to evaluate military technologies submitted by independents, while dispensing general advice as the government's "open door" for American inventors. In the process, it attempted to rectify some of the frustrations independent inventors experienced when they invented for the US military. Because of secrecy concerns, inventors were sometimes prevented from patenting their defense inventions. In addition, many ideas submitted through the NIC were reduced to practice by the National Bureau of Standards, and those resulting in patents became the property of the US government, not the inventor. Finally, inventors were justifiably angry when they patriotically shared their promising military technologies but received no compensation when the Pentagon awarded the procurement contracts to big manufacturing firms. "To put it flatly," the NIC's John C. Green told Congress, "the armed services do not pay inventors when they use their ideas." Congressman Emmanuel Cellar (D-NY) proposed legislation to create a NIC-administered Invention Awards Board that would evaluate disgruntled inventors' petitions and compel the armed services to compensate them up to $50,000 for their adopted inventions. Between 1952 and 1957, Cellar sponsored the awards legislation in three consecutive sessions of Congress, but it never became law.[32]

Meanwhile, the National Inventors Council assumed an expanded role in the development of federal innovation policy. During the 1960s, economists such as Robert Solow, Richard Nelson, and Kenneth Arrow began to appreciate the degree to which technological innovation drove economic growth.[33] Inspired by these insights, the Department of Commerce hired some of its first self-styled innovation experts, such as former GE research director J. Herbert Holloman, former Arthur D. Little consultant Donald Schön, and former Bell Labs patent attorney Daniel DeSimone. Working across the Kennedy, Johnson, and Nixon administrations, these men helped normalize the idea that stimulating technological innovation was the government's responsibility and the key to ensuring military superiority and continuous economic growth. In 1966, DeSimone invited Samuel Ruben, Chester Carlson, and other inventors to the White House to meet with Vice President Hubert Humphrey to kick off the NIC's expanded mission: "to encourage the finest national climate for the inventive process."[34]

Things intensified in March 1972 when President Nixon delivered his "Special Message to the Congress on Science and Technology." Nixon announced several initiatives designed to "marshal science and technology in the work of strengthening our economy." Nixon proposed increased support for the Small Business Investment Companies, a Small Business Administration program that steered federal venture capital to high-risk technology start-ups. He also called for expanded licensing of patented government technologies for private sector commercialization.[35] Finally, Nixon announced the new Presidential Prize for Innovation, a series of $50,000 cash awards designed to recognize the achievements of American inventors and entrepreneurs. However, as the Watergate scandal enveloped Nixon's administration in 1972 and 1973, the prizes were never awarded. Then, in 1973—for reasons that are unclear—the National Inventors Council was spun out of the Department of Commerce and transferred to the private sector under the auspices of the University of New Hampshire's Franklin Pierce School of Law. Despite renewed attention to innovation policy, independent inventors lost the one federal bureau specifically dedicated to addressing their concerns.[36]

Nevertheless, Nixon's congressional message had been prescient. During the 1970s, the United States experienced a decade of economic volatility marked by wage and price controls, Rust Belt deindustrialization, the OPEC oil embargo, and "stagflation," a brutal combination of high unemployment, high inflation, and slow growth. Throughout the decade, America's manufacturing firms shuttered factories, slashed jobs, and moved overseas in search of cheaper labor markets.[37] However, economists increasingly observed that small businesses—especially high-tech start-ups—were driving new employment, wealth creation, and economic growth.[38] To stimulate the economy, politicians began to enact a series of pro-innovation policies that were especially beneficial for independent inventors and entrepreneurs.

In 1977, the National Science Foundation (NSF) piloted the Small Business Innovation Research (SBIR) program, which reserved 2–3 percent of the agency's research grants for firms with fewer than 500 employees. Most postwar federal research grants and procurement contracts had traditionally been awarded to a select group of top-tier research universities and corporate R&D laboratories. In contrast, the SBIR program steered research grants to smaller firms and encouraged the commercialization of their federally funded research, an intentional divergence from NSF's usual focus on "pure" science. After the successful NSF pilot, the Small Business Innovation Development Act of 1982 formally established SBIR programs in the eleven federal agencies (e.g., NASA, Department of Defense, and the Department of Energy) that annually spent more than $100 million on R&D. The SBIR grants, ranging between $150,000 and $2 million, provided crucial funding for high-tech start-ups such as Qualcomm, Symantec, Genzyme, and Da Vinci Surgical.[39]

New government policies also spurred the growth and institutionalization of the private sector venture capital (VC) industry. In 1978, the National Venture Capital Association (NVCA) successfully lobbied Congress to lower the tax rate on capital gains from 49.5 percent to 28 percent. Entrepreneurs and VCs enjoyed the huge equity gains that followed a start-up's acquisition or initial public stock offering, but they hated giving nearly half to the Internal Revenue Service. The new, lower tax rate attracted more investment capital for high-risk start-ups. Then, in 1979, the NVCA persuaded the Labor Department to relax the "prudent man" rule that required pension managers to invest workers' savings in low-risk, low-return vehicles such as corporate and municipal bonds. The revised policy permitted pension fund managers to invest up to 10 percent of their capital in high-risk venture funds. Pension fund managers, seeking better returns in a high-inflation era, poured billions of new dollars into venture capital. The VCs, flush with new cash, invested that money in independent inventors, entrepreneurs, and small start-ups in the biotech and computer industries.[40]

New legislation also facilitated the commercialization of government-funded research, which created opportunities for independent inventors and small start-ups. As noted, the US government retained all patents and royalties resulting from federal research grants, which diminished the incentives for commercialization. In 1980, the landmark Bayh-Dole Act permitted and encouraged the recipients of government-sponsored research to keep and commercialize any resulting patents. That same year, the Stevenson-Wydler Innovation Act required government agencies—such as the Oak Ridge, Brookhaven, and Lawrence Livermore national laboratories—to establish "technology transfer" offices to facilitate the licensing and commercialization of government-owned patents. A decade later, the High-Performance Computing Act of 1991, sponsored by Senator Albert Gore, Jr. (D-TN), permitted widespread commercial use of the government-funded ARPANET for computer-to-computer communications. Similar commercialization provisions accompanied the government-sponsored Human Genome Project (1990–2005).[41]

By the mid-1980s and early 1990s, the government's pro-innovation policies began to stimulate the efforts of independent inventors, scientists, venture capitalists, and their entrepreneurial firms. University professors Herbert Boyer (University of California—San Francisco) and Stanley Cohen (Stanford) used federal funding from the National Institutes of Health and the National Science Foundation to develop and patent their recombinant DNA technique, a gene splicing and cloning method that could produce new medicines. Encouraged by the Bayh-Dole Act, Boyer and venture capitalist Robert Swanson founded Genentech in 1980 and launched the biotech industry.[42] After technology transfer provisions made the military's global positioning system (GPS) publicly available, venture capitalist Ed Tuck hired Norm Hunt and a small team of engineers to build a commercial receiver. In May 1989, Magellan Systems Corporation introduced the NAV 1000, a handheld GPS receiver that was readily adopted by delivery drivers, boaters, and hikers.[43]

Similarly, the opening of the internet for commercial development created enormous opportunities for independent inventor-entrepreneurs. In 1994, Jeff Bezos founded Amazon.com in his Seattle garage to sell books via the emerging World Wide Web. Bezos and his colleagues earned dozens of e-commerce patents for authorizing online credit card payments, recommending products, and making "one-click" purchases, and eventually transformed Amazon into the world's leading retailer.[44] In 1998, Bezos personally invested $250,000 in Stanford graduate students Larry Page and Sergey Brin, who had turned their NSF-funded dissertation projects into an improved internet search engine. Google combined Page's patented PageRank algorithms, Brin's automated web indexer, and a clever method for selling advertising linked to the user's search terms. Like Bezos, Page and Brin started out in a garage, but they eventually built Google into one of the world's largest companies.[45]

The federal government also joined the private sector in embracing open innovation, as various agencies revived crowdsourcing practices first established by the Naval Consulting Board in 1915. NASA, for example, took note of the Ansari X Prize, a $10 million award established in 1996 by the Ansari family to stimulate private sector space travel. In 2004, aerospace engineer Burt Rutan and his financier, Microsoft cofounder Paul Allen, claimed the prize by twice launching their SpaceShipOne to an altitude of one hundred kilometers within two weeks. That same year, NASA announced its Centennial Challenges, a series of seven competitions that offered cash prizes to stimulate space-related innovations. Between 2007 and 2009, independent inventor Peter Homer won two different challenges and $450,000 for developing a more dexterous spacesuit glove, which he assembled in his dining room with a sewing machine (figure 7.4). Homer, a former engineer for Lockheed Martin and Northrup Aerospace, patented the technology and started his own firm, Flagsuit LLC, to commercialize it for use by NASA and the burgeoning space tourism industry.[46]

In December 2010, Congress passed the America COMPETES Reauthorization Act, which granted all government agencies broad authority to enact incentive prizes and crowdsourcing initiatives. The federal government established Challenge.gov, a web-based clearinghouse where citizen-inventors could submit solutions to government challenges that ranged from healthier school lunches (Department of Agriculture) to the development of new data tools for an emerging solar energy infrastructure (Department

Figure 7.4
In 2007, inventor Peter Homer developed a more dexterous spacesuit glove in the dining room of his home in Southwest Harbor, Maine. Photo by Jeff Riedel.

of Energy). In the latter solicitation, Elena Lucas and Daniel Roesler won DoE's $30,000 SunShot Catalyst competition in May 2015, filed for a patent, and founded UtilityAPI, a software start-up based in Oakland, California. Federal agencies were also encouraged to advertise their challenges with private sector intermediaries, such as the two Bureau of Reclamation challenges Larry Kearns found and won through InnoCentive. As of June 2017, Challenge.gov had hosted more than 750 competitions for one hundred federal agencies, with prizes cumulatively valued at $250 million.[47]

Overall, the government's pro-innovation policies and return to crowdsourcing generated increased demand for the ideas of independent inventors. To capitalize on these new opportunities, independents continued to seek support from communities of like-minded inventors.

Fragile Support Groups

Between 1890 and 1950, independent inventors formed a series of supportive but fragile professional organizations. Independent inventors continued to establish support groups during the postwar decades but tended to eschew national organizations in favor of dozens of local clubs. Like their predecessors from the early twentieth century, postwar inventors struggled to sustain their twenty-first-century communities.

The onset of World War II spurred the dissolution of the National Inventors Congress (a.k.a. Inventors of America), as its members headed off to serve in factories, on farms, and on the front lines. However, the United Inventors and Scientists of America (UISA) emerged in 1942 by specifically recruiting returning service members. Founder David Resnick was a Russian-born chemical inventor who had retired at age fifty-six after earning substantial royalties from a chemical stain remover. In 1945, he told the *Los Angeles Times* that James Watt had conceived the steam engine while home recovering from an injury, and he believed that "servicemen in hospitals would get well faster if they would put their time into working out inventions instead of worrying."[48]

Like the National Inventors Congress, the UISA was a commercial organization that connected its soldier-inventors and other independents with manufacturers and a curious public. For $15 in annual dues, the group's 1,400 members could display their inventions at UISA-sponsored expositions held throughout Southern California. Resnick also kept a permanent display of inventions—including a water-cushioned toilet seat, a folding inflatable surfboard, and an ear and nose hair trimmer—at UISA's downtown office. "All of the good inventions are snapped up right away by the manufacturers," Resnick reported, while "some of the impractical ideas gather dust." Resnick attempted to merge the UISA with other inventors' organizations to form a national group, but it always remained a one-man show and a local affair. Resnick was still leading the UISA at age ninety when the *Los Angeles Times* ran another profile in 1974, but the group presumably dissolved after his death.[49]

The UISA, based in Los Angeles, reflected a trend toward the fragmentation and localization of postwar inventors' groups. Before World War II, independents had been burned by the dissolution of a few high-profile national groups and the scam perpetrated by the National Institute of Inventors. In response, postwar inventors formed dozens of smaller, local organizations. Inventors undoubtedly found it easier to trust a local organization—or sniff out a fraud—while meeting with other members face-to-face. In 1991, sociologist Peter Whalley found fifty-six inventors' clubs operating across the United States; in 1992, microcomputer inventor Don Lancaster published a more comprehensive list of 238 inventors' associations. Lancaster reported that Anchorage, Alaska (population 260,283), supported four inventors' groups, a perfect reflection of the postwar trend toward fragmentation.[50] In 1990, the United Inventors Association (UIA) emerged as "a national umbrella organization" to coordinate the activities of local organizations that collectively represented 10,000 independent inventors.[51]

Local inventors' organizations generally operated on a small scale and provided commercialization support for their members. In Washington, DC, the Inventors Network of the Capital Area (INCA) held monthly meetings featuring guest lecturers who offered advice on obtaining angel funding and navigating the patent process. INCA also posted descriptions of members' patented inventions on the group's website to attract licensees. Similarly, the Chicago Inventors' Council hosted traditional inventors' fairs and occasionally circulated a "request for inventions" from local corporations. A smaller subset of groups, such as American Innovators for Patent Reform, advocated for legislation favorable to independent inventors.[52]

The emergence of crowdfunding, cheap tools, and internet sales platforms also made it easier for inventors to commercialize their new inventions while connecting with local and online communities. Crowdfunding websites such as Kickstarter, Indiegogo, and GoFundMe first appeared around 2008 and democratized the process of acquiring start-up capital. Aspiring inventors posted their funding goals for their projects and then invited friends and complete strangers to make small credit card pledges in exchange for a gift or premium. For example, Dallas inventor Josh Malone hired Eventys Partners, an invention marketing firm, to draw social media attention to his Kickstarter campaign for Bunch O Balloons, a patented garden hose attachment that could fill and tie one hundred water balloons in sixty seconds. Malone hoped to raise $10,000 by sending Bunch O Balloons to any donor who made a $15 pledge. Instead, Malone was amazed to receive $929,160 from 21,455 online backers.[53] Since Kickstarter's launch in April 2009, more than 16.8 million backers had pledged over $4.5 billion to support nearly 170,000 creative projects.[54]

Twenty-first-century inventors often required far less capital and material infrastructure than the twentieth-century inventors who worked in chemistry or electricity. As Facebook's Mark Zuckerberg proved, an aspiring digital inventor could purchase a $1,500 computer, download a free, open source software developer's kit, and build a new web-based business, smartphone app, or video game with relatively minimal investment. Furthermore, PayPal, Square, Amazon, Google Play, and the Apple App Store provided subscription-based platforms that helped inventors advertise, sell, and distribute both traditional and digital products. Granted, these digital platforms were only possible thanks to decades of massive private sector and government investments in internet and telecommunications infrastructure, such as fiber-optic cables and mobile telephony. Moreover, Zuckerberg needed huge venture capital investments to scale facebook.com from a dorm room invention into a multinational firm with fifteen data centers, 39,000 employees, and 2.41 billion monthly users. Yet, in 2019, there were 2.47 million apps available for download at Google Play, which proved that thousands of small-scale developers could invent and distribute new software alongside Nintendo and Disney.[55]

Inventors also enjoyed greater access to affordable tools by forming cooperative communities. In 2006, Jim Newton and Mark Hatch founded TechShop, a chain of ten "maker spaces" located in high-tech hubs such as Silicon Valley, Pittsburgh, and Detroit. For a membership fee of $150 per month, 9,000 members nationwide gained access to over $1 million worth of drill presses, welding stations, 3-D printers, and design software at each location. Each TechShop also offered courses in drafting, sewing, laser cutting, and marketing so aspiring inventors could handle various jobs themselves instead of hiring contractors. Like the Franklin Institute a century earlier, TechShop members could ask one another: Is this invention any good? Do you think it will sell? And how did you mill that part out of carbon fiber? Much like the inventive culture prevalent within nineteenth-century machine shops, each TechShop location functioned as an informal incubator for start-ups. In 2008, inventor-entrepreneur Phil Hughes built a patented liquid cooling system for computer servers at the TechShop in Menlo Park, California. When he licensed the technology to a manufacturer, Hughes hired three associates and essentially ran his firm, Clustered Systems Company, out of the Menlo Park location.[56]

Quirky, an online inventor community, offered a variation on Tech-Shop's do-it-yourself model.[57] Ben Kaufman founded the start-up in 2009, and its website quickly attracted one million registered members. Quirky solicited ideas from America's armchair inventors, used its members' collective wisdom to select and refine the best concepts, and then applied the firm's design expertise to turn simple napkin sketches into commercial products. "We started the company to make invention accessible," Kaufman said in 2013, "People come to our site, submit their ideas, and the best new ideas make their way all the way to retail shelves and we do all the heavy lifting in between."[58]

In January 2013, Maryland inventor Garthen Leslie caught Kaufman's appearance on *The Tonight Show with Jay Leno*, where the founder described how Quirky helped "ordinary people become successful inventors." While moonlighting from his day job as an IT consultant, Leslie had conceived of an internet-enabled window air conditioner that could be remotely controlled by a smartphone app. Leslie reasoned that a user could start the air conditioner during the evening commute and come home to a precooled apartment. That spring, Leslie joined the Quirky community and submitted his idea, where it sailed through the evaluation process. In March 2014, Kaufman invited Leslie to Quirky's New York City headquarters to check on the development process. Leslie was amazed to find Kaufman and 200 cheering Quirky employees gathered to witness the unveiling of "Aros" (figure 7.5), a working prototype of Leslie's smart air conditioner. Quirky's team of patent attorneys, product designers, software engineers, and marketers had partnered with General Electric to develop Leslie's idea from sketch to

Figure 7.5
Inventor Garthen Leslie appeared on the packaging for his Aros smart air conditioner, which Quirky codeveloped with General Electric in 2014. Photo by Quirky.

prototype in just four months. Some reviewers complained that the Aros unit was loud and inefficient and that the mobile app was buggy. Nevertheless, Aros sold $5 million in preorders on Amazon.com and then retailed at Walmart, Home Depot, Best Buy, and Target. In 2014, at age sixty-three, Leslie received the first of five $100,000 royalty checks.[59]

Both Quirky and TechShop struggled to stay afloat. Quirky declared bankruptcy in September 2015. Its complex operations—managing a community of one million inventors, transforming raw ideas into real products, and orchestrating manufacturing and distribution—proved too costly and broke down at scale. In December 2015, a Dutch private investment group named Q Holdings LLC paid $4.7 million at the bankruptcy auction to acquire Quirky's platform, inventory, and intellectual property. The firm relaunched in March 2017 with an open innovation business model. Instead of manufacturing its own products, Quirky licensed its portfolio of crowdsourced ideas to established firms for manufacture and sale under the Quirky name or their licensees' corporate brands.[60] The business press reserved judgment, but inventors were ecstatic. "Makers rejoice!," wrote Quirky member Taron Foxworth; another member, with the handle "Eagledancing," simply thanked Quirky for "continuing this dream for us inventors!"[61]

Just as Quirky 2.0 came back online, TechShop declared bankruptcy in November 2017. Only a few of the TechShop locations were able to maintain sufficient income through membership dues, and the company fell into debt. TechShop was unable to attract a white knight buyer, so the firm liquidated its assets. As he arrived at the shuttered Pittsburgh location, member Devin Montgomery hoped the closure would not "quell the spirit of making."[62] The Quirky and TechShop bankruptcies were stark reminders that twenty-first-century inventors struggled to sustain their professional communities, much like their twentieth-century predecessors.

Furthermore, inventors continued to face exploitation by unscrupulous scammers. In 1994, a congressional hearing convened by Senator Joseph Lieberman (D-CT) revealed that 25,000 independent inventors had received few commercial benefits for the $200 million they had collectively paid to patent promoters. Shady promotion firms routinely charged naive inventors up to $10,000 to evaluate, patent, and market their inventions and then pocketed the money without delivering the services. Lieberman's bill required patent promotion firms to register with the United States Patent and Trademark Office (USPTO), outlined standards for invention promotion contracts, and created a tracking system for inventors' complaints.[63]

The American Inventors Protection Act (AIPA) eventually became law in 1999, but it did not eliminate scams. In 2017, the Federal Trade Commission shut down World Patent Marketing (WPM), a Miami promotion firm that bilked nearly $26 million from 1,504 inventors. The scam received heightened attention when journalists discovered that Matthew Whitaker, President Donald Trump's acting attorney general, had served on WPM's advisory board. Whitaker had allegedly threatened legal action against inventors who complained about the firm's suspicious practices.[64]

Despite its anti-scam provisions, independent inventors did not support passage of the AIPA. Between 1995 and 1999, lawmakers inserted several patent law revisions into Lieberman's bill that were detrimental to independent inventors and beneficial for R&D firms. For decades, the patent system had been resistant to reform. However, the AIPA and subsequent statutes transformed the patent system at the turn of the twenty-first century.

The Perils of Patent Reform

During the early twentieth century, independent inventors failed to reform a broken patent system. That trend continued until the 1980s, when a pro-patent shift in the judiciary helped independent inventors gain new leverage. In response, corporate lobbyists secured fundamental changes to the patent laws that favored R&D labs at the expense of independent inventors.

The immediate postwar decades witnessed a few key changes to the patent laws. The Patent Act of 1952, for example, clarified some uncertainty about the standards of patentability. In a 1941 patent case, *Cuno Corp. v. Automatic Devices Corp.*, Supreme Court justice (and former TNEC member) William O. Douglas ruled that a "new device, however useful, must reveal the flash of creative genius, not merely the skill of the calling" in order to merit a patent. To many legal observers, this suggested that the "Eureka!" moments of individual inventors were patentable, while the prolonged, incremental efforts of industrial research teams were not. The business community complained that the higher standard threatened to undermine the entire business model of corporate R&D and called on Congress to clarify the threshold for patentability. The 1952 law established that a patent was not to be denied based on "the manner in which the invention was made." It was immaterial, a revision note explained, whether a patentable idea resulted from "long toil and experimentation or from a flash of genius." The 1952 law did not harm independent inventors, but it unequivocally affirmed industrial research as a viable mode of invention.[65]

Meanwhile, a series of judicial developments expanded the boundaries of patentability. In 1951, Bell Labs secured the first software patent by arguing that the numerical "Error Detecting and Correcting System" in its automated telephone switches was "embodied" in the physical hardware. Software patenting expanded dramatically over the next several decades, as thousands of programmers coded mainframe batch routines, desktop spreadsheets, video games, and e-commerce algorithms.[66] Likewise, the Supreme Court's 1980 decision in *Diamond v. Chakrabarty* permitted the patenting of genetically modified organisms (GMOs). General Electric researcher Ananda

Chakrabarty had used radiation to modify a bacterium's genome, which improved its ability to ingest crude oil. GE hoped to commercialize the bacteria to clean up oil spills, and the firm filed a lawsuit after the Patent Office rejected Chakrabarty's patent application. The Supreme Court decision opened a new frontier for big agricultural and pharmaceutical companies (e.g., Monsanto, Merck) and small start-ups (e.g., Genentech, Amgen) in the emerging biotech industry.[67]

In 1982, Congress established the Court of Appeals for the Federal Circuit (CAFC) to hear appeals on all patent cases. Since the Inventors' Guild in 1910, reformers had sought a specialized national appellate court, staffed with technically trained judges, to eliminate inconsistencies in patent decisions across the twelve regional courts of appeal. The single court of patent appeals ended the ambiguity of contrary appellate opinions when plaintiffs sued alleged infringers over the same patents in multiple circuits. President Ronald Reagan appointed intellectual property experts as the first judges on the CAFC bench, and by the mid-1980s, legal observers noticed a decidedly pro-patent shift in appellate decisions. Prior to the CAFC, infringement plaintiffs prevailed in only 35 percent of appeals, meaning that 65 percent of their patents were ruled invalid on appeal. After the CAFC, the tables turned completely, as plaintiffs prevailed in approximately 67 percent of appeals, with their patents upheld as valid and infringed.[68]

This legal shift gave new hope to beleaguered independents, such as Robert Kearns, who relied on strong patents as their only bargaining chips in negotiations with powerful corporations. In the late 1960s, Kearns demonstrated his patented intermittent windshield wiper for engineers at the Ford Motor Company, but they declined to take a license. In 1978, with the wipers in widespread use, Kearns sued Ford for infringement. Ford tried to bankrupt Kearns by inundating him with twelve years of pretrial filings, motions, and depositions, but he refused to capitulate. In 1990, a district court jury ruled in Kearns's favor. Given the pro-patent shift in the CAFC, Ford elected to drop its pending appeal and settle with Kearns for $10.2 million in royalties, interest, and damages.[69] With the courts upholding more of their patents, independent inventors gained new leverage in their licensing negotiations and infringement lawsuits.

Like Kearns, Jerome Lemelson became a hero to independent inventors—and a bête noire to corporate attorneys—as his inventions earned significant royalties and infringement damages.[70] Lemelson grew up on Staten

Island and earned a bachelor's and two master's degrees in engineering from New York University. After World War II, Lemelson moonlighted as an independent inventor while working at the Office of Naval Research and later at Republic Aviation. He earned his first patent in 1953 for a variation on a propeller beanie, but he struggled to build a business producing and selling toys. "In the beginning, I wanted to manufacture certain ideas I had in the toy and hobby field," Lemelson recalled, but after several disappointments, he broadened his interests and shifted his approach.[71]

Lemelson studied technical trade journals and followed his curiosity into numerous subfields of engineering and electronics. In lieu of entrepreneurship, he specialized in earning patents and seeking licenses from established firms as a prolific serial inventor. Lemelson worked in the attic of his New Jersey home and sketched new concepts during his frequent train trips into Manhattan. He even kept a bedside notebook and sometimes would wake his wife, Dorothy, to witness his late-night sketches. Lemelson developed a wide-ranging expertise, with inventions covering cordless telephones, fax machines, cassette players, information systems, industrial robots, integrated circuits, vehicle anti-collision systems, medical devices, and all kinds of children's toys. To save on attorneys' fees, Lemelson wrote his own patent applications and averaged one submission per month. With 606 issued patents, Lemelson became one of the most prolific and versatile inventors in American history.[72]

Like many postwar independents, Lemelson faced the "not invented here" attitude and suffered many rejections in his attempts to find corporate licensees. Income from his wife's interior design business sustained the family until he earned his first significant royalties. In 1974, Sony paid Lemelson $2 million to license his audiocassette drive system for use in its pathbreaking Walkman. In 1981, Lemelson declined an offer to join IBM's R&D labs; he later earned $6 million when the firm took a perpetual, nonexclusive license on his entire patent portfolio.[73]

However, defending his patents consumed much of Lemelson's time and royalty income. Lemelson lost his first infringement trial in 1968, when he failed to convince a judge that Kellogg had infringed his patent for a cutout face mask printed on a cereal box. In 1973, Lemelson initiated a decade-long suit against alleged infringers of his automated warehousing patents. That case was in progress in 1979 when the US Senate invited Lemelson to testify regarding proposed patent reforms—including a single court of patent appeals—that might encourage innovation and stimulate the stagnant economy. Lemelson asserted that corporations routinely infringed inventors' patents instead of taking licenses because the "antipatent philosophy" of the federal courts provided no deterrent to piracy. To illustrate, Lemelson described how a district court judge had granted a summary judgment and invalidated his Velcro dartboard patent without allowing him to testify. Nevertheless, in the midst of these lawsuits, Lemelson continued to generate a steady flow of new inventions; he earned 178 patents between 1970 and 1979.[74]

Lemelson's fortunes changed dramatically when he sought licensees for his "machine vision" inventions. In 1954, Lemelson had submitted a 150-page patent application for a universal manufacturing robot that could scan a component and then perform different welding, riveting, or inspection operations, depending on its specific label. Because the filing was so elaborate, the USPTO divided Lemelson's claims into several continuing applications. Lemelson resisted the ruling—he preferred a single patent and a single filing fee—but he prosecuted the first set of claims, and his first machine vision patents were issued in 1963 and 1966.[75] Following standard industry practice—and as allowed by USPTO rules—Lemelson waited until just before the 1963 patent was issued before prosecuting the remaining divided applications. Lemelson's original 1954 disclosure was ahead of its time, so it made sense to slow-walk the prosecution of his subsequent applications while waiting for potential machine vision licensees to emerge. The examination backlog at the USPTO and further subdivisions of his applications added more delays. When Lemelson's next long-pending patent was finally issued in 1982, machine vision technologies were in widespread use across several industries. Hundreds of manufacturing firms were by then potential licensees of the issued patent.[76]

Lemelson retained Chicago patent attorney Gerald Hosier, who pioneered the practice of representing patentees on a contingent-fee basis. Hosier earned one-third of any royalties or damages he secured for inventors but nothing if he was unsuccessful. This speculative practice was risky, but it also motivated Hosier to advocate tenaciously for his clients. In November 1989, Hosier informed every major firm in the electronics, semiconductor, and automobile industries that they were infringing Lemelson's machine vision patents. Hosier urged the firms to take licenses and pay Lemelson retroactive royalties or face an infringement lawsuit in the now pro-patent courts. After their negotiations ended in a stalemate, Lemelson and Hosier sued Toyota, Honda, Mazda, and Nissan in July 1992. The Japanese firms eventually paid Lemelson a $103 million settlement. The case set a precedent, and over the next several months, Hosier negotiated multimillion-dollar licensing deals with several consumer electronics firms.[77]

Meanwhile, Hosier continued to amend and prosecute the remaining divisional applications that were still pending from Lemelson's original 1954 disclosure. In August 1992, Ford, General Motors, Chrysler, and Motorola filed countersuits to invalidate Lemelson's patents, alleging that he had misled the Patent Office and deliberately delayed his applications. Critics accused Lemelson of exploiting loopholes in the patent laws. "They're doing their damnedest to defame me," Lemelson said in 1994.[78]

The lawsuits to invalidate Lemelson's patents continued for years, even as he fell ill and died from cancer in October 1997. However, by 1998, Hosier had convinced Motorola, the major US automakers, and their key suppliers to settle, earning another $850 million for the Lemelson estate. Altogether, Lemelson's machine vision inventions and other patents earned a remarkable $1.4 billion in licensing royalties and interest. The press offered conflicting characterizations. Critics denounced Lemelson for manipulating the patent system "to squeeze money out of big companies," said Denis Prager, a former adviser to Lemelson's philanthropic foundation, whereas "others romanticize him as a talented, lone-wolf inventor who stands up for the rights of the individual."[79]

With independents winning big settlements in the federal courts, corporations tried to shift perceptions in the court of public opinion. Nearly contemporaneously with Lemelson, Gilbert Hyatt was a frequent target of anti-inventor rhetoric. In 1970, Hyatt filed a patent application for a "Single Chip Integrated Circuit Computer Architecture," an early microprocessor. He founded a venture-backed start-up, Microcomputer, Inc., and arranged for Intel to manufacture the chips. However, after a few months, Hyatt's investors decided to wind down the start-up, and he spent the next several years consulting for aerospace and electronics firms in Southern California. After several continuations, Hyatt's long-pending microprocessor patent was finally issued in 1990 and sent shock waves through the semiconductor industry. In his frustration with Hyatt, Intel attorney Peter Detkin coined the term "patent troll" to denigrate independent inventors who sought royalties on "paper patents" they had never manufactured.

Hyatt eventually earned $150 million through licensing deals with Philips and other electronics firms.[80]

By the turn of the twenty-first century, numerous observers were lamenting the sad state of the US patent system. The USPTO was chronically underfunded and its examiners were overburdened. A tidal wave of software and internet-based inventions had generated a huge backlog of unexamined applications. Critics complained about long-pending applications, tedious priority disputes, lax examinations, and expensive infringement litigation. Moreover, many US patent provisions were out of sync with the European and Japanese patent systems, which complicated patent management for multinational firms. In short, the patent system still suffered from issues that inventors had first raised in the 1890s.[81]

For decades, high-tech corporations had resisted patent reforms because they could use their superior legal resources to exploit the ambiguities and inefficiencies in the system. However, the tables had turned, and independent inventors were now securing significant licensing revenues from high-tech corporations. In response, industry lobbyists embarked on a prolonged twenty-year campaign to reform the patent system. Ironically, the big R&D firms championed several reforms they had opposed during the early twentieth century to curb the same approaches they had once utilized to exploit individual inventors.

The first steps toward patent reform emerged in the context of international trade negotiations. During the mid-1990s, American pharmaceutical firms lobbied forcefully for stricter IP enforcement in developing nations where their medicines were exempt from patentability and regularly pirated. In exchange for these provisions, the United States agreed to update its patent policies to conform more closely to the European and Japanese systems. To comply with the 1995 Trade-Related Aspects of Intellectual Property Rights (TRIPS) agreement, Congress added several provisions to Senator Joseph Lieberman's pending anti-scam bill.[82]

Ultimately, the omnibus American Inventors Protection Act of 1999 (AIPA) enacted safeguards against unscrupulous patent promoters and introduced new transparency into the application process. As required by TRIPS, Congress changed the US patent term to twenty years from the date of filing (versus seventeen years from the date of issue). The AIPA also directed the USPTO to publish all pending patent applications, as in other international systems. These long-debated reforms discouraged long-pending patents, harmonized US and foreign patent laws, and leveled the international playing field. Supporters believed these reforms would cut down on dilatory applicants, duplicate patents, long priority disputes, and excessive litigation. However, the Alliance for American Innovation, a coalition of independent inventor groups and small businesses, countered that publishing unpatented applications would merely invite big corporations to steal their ideas and beat them to market. "They're ripping up our patent system and giving away our future," said Beverly Selby, the Alliance's executive director.[83]

In addition, the AIPA introduced a "prior commercial use defense" to reduce excessive litigation over the flood of broadly scoped dot.com patents. For example, nearly every e-commerce website used some variant of Stephen Messer's 1999 patent covering the tracking of commissions generated by web clicks on retail sites. The provision provided an infringement defense if these and other methods were already in commercial use before the relevant patents were filed. While the measure aimed to reduce litigation, Messer complained that it enabled "pirates" to steal good ideas from "pioneers."[84]

Finally, the AIPA introduced *inter partes* reexamination (IPR), an administrative process that expanded a petitioner's ability to challenge the validity of an issued patent. Reformers designed IPR to weed out poorly examined and improperly issued "mistake" patents and thereby reduce litigation. However, independent inventors feared that corporate attorneys would set up shop at the USPTO and systematically use IPR to invalidate the patents of their rivals.[85]

Despite the 1999 reforms, the US patent system continued to struggle with a backlog of 750,000 applications, an average examination period of thirty-four months, and excessive litigation. In his *Strategy for American Innovation* (2009), President Barack Obama called for additional patent reforms to facilitate the nation's recovery from the 2008 recession. An efficient, streamlined patent system, Obama argued, would help entrepreneurs protect their ideas, attract venture capital, start businesses, and create new jobs. In April 2010, a Department of Commerce report echoed Obama's arguments and warned that continuing Patent Office delays would "cost the U.S. economy billions of dollars annually in 'foregone innovation.'" Congress agreed, and in 2011, Senator Patrick Leahy (D-VT) joined with Representative Lamar Smith (R-TX) to enact the America Invents Act (AIA).[86]

The AIA allowed the USPTO to keep and manage the application fees
it collected rather than transferring those revenues to the Treasury. This
reform, first suggested in the 1890s, provided the USPTO with more financial resources to work through its massive backlog. The AIA also introduced
expanded post-grant review (PGR) procedures to complement the existing
inter partes reexaminations. These hearings, administered by the USPTO,
gave third parties another option for challenging the validity of issued patents instead of going to court. Finally, under the AIA, the USPTO switched
from a "first-to-invent" to a "first-to-file" test for application priority as in
other international systems. This sea change made the determination of
priority straightforward and eliminated interference proceedings entirely.[87]

The AIA's passage was a triumph for big corporations and another loss for
independent inventors. According to the *Washington Post*, the AIA marked
the culmination of a "long-running advocacy campaign" that sent "100 lobbying shops to Capitol Hill" in 2011 alone. Indeed, several research-intensive
companies, including Johnson & Johnson, Dow Chemical, General Electric,
and Procter & Gamble, had formed the Coalition for 21st Century Patent
Reform in 2005. In 2011, the group spent $1.4 million to ensure passage of
the AIA.[88]

Independent inventors, entrepreneurs, and small businesses vehemently
opposed the bill, but they had fewer resources at their disposal. For example, the National Small Business Association spent only $250,000 to oppose
the AIA. They argued that the first-to-file provisions strongly favored "large
incumbent corporations" and their teams of patent attorneys. Angel investor Valerie Gaydos suggested that the first-to-file provisions would create
a "race to the patent office," thereby exacerbating—instead of reducing—
the flood of poorly written and hastily examined patents. Alexander Poltorak, president of American Innovators for Patent Reform, also worried
that expanded post-grant reviews favored well-resourced corporations that
could systematically challenge dozens of rival patents in their industries.
With PGRs, Poltorak feared that the big firms had simply found a cheaper
and faster way to make small-scale inventors "abandon their inventions."[89]

These fears were borne out. Between September 2012 and May 2017,
the USPTO's Patent Trial and Appeal Board (PTAB) resolved 1,601 petitions
for *inter partes* reexaminations and post-grant reviews. In 286 cases (18 percent), all claims were confirmed as valid. However, at least one claim was
overturned in 271 cases (17 percent), and in 1,044 cases (65 percent) *all*

claims were canceled, effectively invalidating the patent. The PTAB had weeded out some bad patents, but its propensity to invalidate claims had a chilling effect on independent inventors. Fearing systematic post-grant challenges, some independents stopped applying for patents altogether.[90] Even pro-business conservatives recognized that the AIA's effects were one-sided. In 2015, *National Review* proclaimed that the "Bipartisan 'Stick It to Individual Inventors' Act" had hung "garage innovators and backyard tin-kerers out to dry."[91] New legislation had transformed the patent laws, but one thing remained the same: the inequities in the system continued to benefit wealthy corporations at the expense of independent inventors.

Changing Perceptions

During the twentieth century, corporations convinced many observers that their sophisticated R&D labs had supplanted individual inventors. Although the independents had not disappeared, they remained overshad-owed by the R&D labs in the decades following World War II. However, inventors managed to rehabilitate their public image. By the turn of the twenty-first century, they were celebrated as high-tech mavericks and celeb-rity entrepreneurs.

During the immediate postwar decades, independent inventors gener-ally remained invisible. The records of the National Inventors Council were sealed and classified, which prevented the public from fully appreciating inventors' wartime accomplishments. Those publications that reported on the independents tended to characterize them as eccentric crackpots, an image reinforced in comedic films such as Jerry Lewis's depiction of *The Nutty Professor* (1963).[92] *The Big Idea*, a weekly Philadelphia television show hosted by Donn Bennett, invited four independent inventors to showcase their inventions before a panel of experts. Approximately 1,600 inventors appeared on *The Big Idea* during the 1950s, but the show's local audience limited its reach.[93]

In contrast, industrial researchers enjoyed significant attention during R&D's postwar golden age. Popular books—such as James Phinney Baxter's Pulitzer Prize–winning *Scientists against Time* and General Leslie Groves's Manhattan Project memoir *Now It Can Be Told*—lionized the wartime service of university and corporate scientists. The National Association of Manu-facturers celebrated corporate R&D in 428 *Industry on Parade* newsreels that aired between 1950 and 1960. In 1956, Bell Labs researchers William Shockley, John Bardeen, and Walter Brattain garnered tremendous acclaim when they shared the Nobel Prize in Physics for developing the transistor.[94] R&D stalwarts such as AT&T, Chrysler, DuPont, Ford, General Motors, General Electric, IBM, and Kodak flaunted their research achievements by building extravagant pavilions for New York's 1964 World's Fair. Famed architect Eero Saarinen designed three new R&D centers—the General Motors Technical Center (Warren, Michigan, opened 1956), IBM's Thomas Watson Research Center (Yorktown Heights, New York, 1961), and AT&T's Bell Laboratories (Holmdel, New Jersey, 1962)—to showcase each firm's technical prowess. Some critics, such as William Whyte, questioned the corporate conformity of R&D's "organization men," but during the 1950s and 1960s, industrial researchers generally enjoyed a better reputation than independent inventors.[95]

Beginning in the 1970s, public perceptions of independent inventors and corporate R&D labs began to shift. As noted, the big corporations endured several commercial failures and missed opportunities, such as RCA's VideoDisc and Kodak's digital camera. Persistent postwar antitrust pressure revealed that AT&T, IBM, and several other firms had gouged consumers by engaging in monopolistic practices. Moreover, as a generation of baby boomers and hippies came of age, they tended not to trust corporate scientists and engineers, who were negatively associated with the Vietnam War, the military-industrial complex, environmental degradation, and domestic job losses resulting from automation and globalization. By the 1980s, Japanese firms outpaced America's blue-chip companies, as consumers increasingly turned to Sony, Nikon, and Toyota instead of RCA, Kodak, and GM.[96] Big corporate R&D labs, once the paragons of innovation, were now regarded as slow, outdated, conservative, and ineffective.

In contrast, independent inventor-entrepreneurs were increasingly celebrated as the new embodiment of American innovation—agile, cutting-edge, creative, and driving the economy. Steve Jobs, Steve Wozniak, Bill Gates, Paul Allen, Jeff Bezos, Mark Zuckerberg, Larry Page, and Sergey Brin began their careers as garage and dorm-room inventors; each earned multiple patents. With the help of venture capitalists, a commitment to sustained innovation, and the widespread adoption of their products, these inventor-entrepreneurs grew their start-ups into Apple, Microsoft, Amazon, Facebook, and Google, five of the biggest companies in the world. Much like

Colt, Singer, or Edison from an earlier era, these inventors became high-tech celebrities thanks to their success as entrepreneurs.[97] There were criticisms, of course. Gates was branded as a monopolist when the DoJ sued Microsoft for antitrust violations. Zuckerberg and Facebook were condemned for eroding privacy and spreading fake news. Jobs was denounced as a petulant tyrant, yet when he died in 2011, there was a period of national mourning similar to the memorials following Edison's passing in 1931.[98]

Granted, most independents remained unknown to the public; Jacob Rabinow, Garthen Leslie, and Josh Malone were not household names. Yet even the more obscure independents benefited from changing perceptions of big firms and the American tendency to embrace the underdog. By the start of the new millennium, corporate scientists and organization men were no longer considered the best sources for innovation. Instead, the nation revered hackers, hobbyists, college dropouts, brash mavericks, and quirky nerds—the kinds of people once denigrated as crackpots. White shirts, black ties, and lab coats were out; jeans, t-shirts, and sneakers were in. Many viewed garages, basements, and dorm rooms—not pristine laboratories on isolated corporate campuses—as the best places for invention. By the turn of the twenty-first century, most observers regarded independent inventors and celebrity entrepreneurs, not corporate R&D labs, as the wellspring of American innovation.[99]

Unfortunately, most women and minority inventors remained invisible amid the accolades for their White male contemporaries. *The Demographics of Innovation in the United States*, a 2016 study of independent, corporate, and university patentees, found that the median innovator was a White man in his late forties with an advanced degree. In contrast, women (12 percent), US-born minorities (8 percent), and African Americans (0.5 percent) were woefully underrepresented in the innovation economy.[100] Overall, women and African Americans were less likely to earn an advanced science or engineering degree, receive a patent, or commercialize their patents than their White male counterparts. Women and minorities also continued to experience unconscious bias and outright discrimination in a high-tech sector rife with racism and misogyny.[101] Given these entrenched historic inequities, women and minority inventors remained unfamiliar to many Americans. For example, in a provocative 2016 Microsoft television ad, several girls were asked "Can you name any women inventors?," but they were unable to provide any examples.[102]

Those young women might have named Lori Greiner, an independent inventor and self-made millionaire who became famous by selling her patented jewelry organizers on the QVC shopping channel. Greiner gained even more notoriety after becoming a judge on *Shark Tank*, an Emmy-winning reality TV show that debuted in 2009. Much like *The Big Idea* from the 1950s, independent inventors and entrepreneurs pitched their ideas to the "sharks," a panel featuring Greiner and five other investors. Between 2009 and 2019, more than 600 inventors and entrepreneurs made their pitches in the tank, resulting in over $100 million in development deals. Greiner financed and mentored dozens of independent inventors, including former car detailer Aaron Krause, who pitched an improved sponge in October 2012. Krause's Scrub Daddy employed a temperature-sensitive foam that was firm in cold water for tough scrubbing but soft in warm water for light cleaning. After Krause's appearance on *Shark Tank* (figure 7.6), the Scrub Daddy product line grew from $120,000 in retail sales to $50 million by 2015. In 2019, ABC renewed *Shark Tank* for an eleventh season, which reflected America's continuing fascination with plucky, independent inventors.[103]

Figure 7.6
In October 2012, Lori Greiner (*left*) invested in inventor-entrepreneur Aaron Krause and his Scrub Daddy sponge during season 4 of *Shark Tank*. Photo by Craig Sjodin © ABC/Getty Images.

Individual inventors also appeared on the big screen. Greg Kinnear starred as intermittent windshield wiper inventor Robert Kearns in *Flash of Genius* (2008). Ashton Kutcher and Michael Fassbender portrayed Apple's cofounder in *Jobs* (2013) and *Steve Jobs* (2015), respectively. Likewise, *Joy* (2015) starred Oscar winner Jennifer Lawrence as Joy Mangano, the QVC and Home Shopping Network star who invented the self-wringing Miracle Mop and dozens of other household products. Not surprisingly, these movies portrayed the independents as mercurial but scrappy underdogs who prevailed against skepticism and the corporations that tried to steal their ideas. Granted, fictional roles—such as Christopher Lloyd's Dr. Emmett Brown in *Back to the Future* (1985)—still poked fun at crackpots, while HBO's *Silicon Valley* (2014–2019) satirized self-absorbed "brogrammers." However, by the turn of the twenty-first century, independent inventor-entrepreneurs generally held more cultural cachet than any corporate scientist. As the real-life Jobs once observed in 1983, "It's better to be a pirate than join the navy."[104]

Conclusion

After World War II, independent inventors navigated a series of challenges and opportunities that often resembled those encountered by their predecessors. Between 1950 and 1975, a small cadre of independent inventors persisted in relative obscurity, while industrial researchers enjoyed a postwar golden age. Fueled by a booming economy and abundant Cold War military spending, the corporate R&D labs enjoyed big budgets, huge staffs, idyllic corporate campuses, prestigious Nobel prizes, and numerous commercial successes. Individual inventors were generally assumed to be crackpots, if not extinct.

The innovation landscape shifted considerably between 1975 and 2000. Corporate R&D labs struggled with a volatile economy, vigorous antitrust enforcement, new foreign and domestic competitors, and a shifting intellectual property environment, which resulted in lower research productivity and some embarrassing commercial failures. In contrast, inventors benefited from federal antitrust and innovation policies that favored entrepreneurial small businesses. Independent inventors began to reassert themselves in the economy, especially in the computing and IT sectors.

Between 2000 and 2015, independent inventors experienced a renaissance, while the R&D labs restructured their operations. Inventors enjoyed increased demand for their inventions as firms embraced open innovation and cultivated ideas from external sources. In a revival of earlier practices, corporations increasingly acquired and licensed patents from outside, independent inventors to complement the inventions they developed internally. Similarly, the federal government began hosting high-stakes invention challenges and crowdsourcing for inventions—strategies it first pursued during World War I. In addition, inventors benefited from greater access to the government grants, private venture capital, inexpensive tools, and digital platforms that helped them commercialize their inventions.

However, independent inventors still struggled to avoid predatory scams and sustain the maker spaces and online communities that enabled their work. A pro-patent shift in the courts helped independents win more royalties and infringement settlements, but they were unable to prevent patent reforms—such as first-to-file application priority and post-grant reviews—that disproportionately favored powerful corporations. Nevertheless, the independents managed to rehabilitate their public image. In the new millennium, hobbyists, hackers, and garage inventors—not corporate scientists—were hailed as the most promising sources of innovation.

8 Conclusion

Despite the great mass organizations which compete with him, the individual, genius type of inventor still holds the field as the most important single source of new ideas. . . . He has a hard fight, but holds his own pretty well against the regimented scientists of industry.
—"Lo, the Poor Inventor!" *Business Week*, December 21, 1929

Mr. Gates, what's your biggest worry? What's your business nightmare? . . . Gates said, "I'll tell you what I worry about. I worry about some guy in a garage inventing something, a new technology I have never thought about."
—Journalist Ken Auletta, interviewing Bill Gates, 1998

"Individual inventors matter."

Author John Seabrook offered this opinion in *Flash of Genius*, his biography of Robert Kearns, inventor of the intermittent windshield wiper.[1] This has not always been a mainstream view. In the early decades of the twentieth century, many contemporary observers—and later many historians—came to believe that corporate R&D labs had displaced independent inventors as the wellspring of American innovation. However, the independents were not supplanted. Rather, they endured alongside the corporate R&D labs as an important, if less visible, source of inventions. Between 1890 and 1950, independent inventors adopted a variety of flexible commercialization strategies, formed a series of fragile professional groups, lobbied for fairer patent laws, and mobilized for two world wars. After World War II, the independents were temporarily overshadowed during R&D's postwar golden age. However, they rebounded to garner more commercial opportunities and greater acclaim at the turn of the twenty-first century.

Independent inventors have shaped American history, with important influences on technology, culture, politics, and the economy. A deeper appreciation for their contributions and failings offers important insights into our current era, with implications for scholars, inventors, and anyone who benefits from innovation. Reflecting on their history helps us understand why individual inventors are still relevant in the twenty-first century.

*   *   *

Individual inventors matter if we want a more complete and nuanced history of American innovation. For decades, influential historians claimed that the emergence of corporate R&D labs abruptly ended the "Heroic Age of American Invention."[2] However, historians did not introduce this argument: it was originally constructed by corporate PR departments and pro-research lobbying groups. Industrial researchers generated scores of popular lectures, print advertisements, and national radio broadcasts that celebrated the R&D labs and maligned the independents, who could not counter their rhetoric on a comparable scale. Given the imbalance in available sources, historians subsequently adopted the corporate perspective. They assumed that any remaining independents played a negligible role in the economy.

By recovering the stories of a group once considered extinct, this book extends our understanding of independent inventors beyond the heroic era and into the twenty-first century. It confirms that independent inventors persisted alongside the corporate R&D labs as an appreciable, if underappreciated, source of inventions. It reframes the period from 1890 to 2015 as an era when both independent inventors and corporate R&D labs contributed substantially to technological innovation.

Indeed, if we pause to reflect on the technologies we use every day, the contributions of independent inventors are ubiquitous. When you dress for work in the morning, take a moment to appreciate how Lori Greiner organized your jewelry box and Marion O'Brien Donovan optimized the hanger space in your closet. In the kitchen, think of Joseph Friedman when you sip your breakfast smoothie from a flexible straw, or Earl Tupper when you seal your lunch in a Tupperware container. As you head out the door, stop to recall Robert Gore as you pull on your Gore-Tex coat and gloves. On the way to work, remember Frank J. Sprague when you take an elevator down to the platform and then ride a multicar subway train. If you drive, pause to appreciate Robert Kearns's intermittent windshield wipers and Jerome

Lemelson's collision avoidance system. Likewise, contemplate Charles Adler, Jr., when you stop for a flashing safety signal at a railroad crossing. At the office, you might acknowledge Chester Carlson when you make a photocopy, or Jacob Rabinow, whose automated sorting machines routed the mail. At your desk, consider Bill Gates, Paul Allen, Steve Jobs, and Steve Wozniak when you use Microsoft software to type a memo on your Apple computer. Back at home, remember Aaron Krause and Joy Mangano when you wash the dishes with your Scrub Daddy sponge and mop the floor with a self-wringing Miracle Mop. In your den, pause to appreciate Samuel Ruben and Philo T. Farnsworth when the Duracell batteries in your remote control turn on the television. Finally, think of Mark Zuckerberg when you browse a few Facebook posts before heading to bed. Most people would be unfamiliar with these independent inventors. However, it is impossible to deny the impact of their ingenuity.

*   *   *

Individual inventors also matter if we want an accurate model for understanding—and promoting—successful innovation. In *The Sources of Invention* (1958), economists John Jewkes, David Sawers, and Richard Stillerman observed that "eclecticism" is the key to providing "a social framework conducive to innovation." The only danger, they reported, would be "in plumping for one method to the exclusion of others."[3] Therefore, our best prospects for successful innovation will occur when we leverage the unique strengths of inventors and firms and encourage them to work together.

Independent inventors and industrial researchers bring different but complementary skill sets to technological development. Independent inventors tend to excel at *invention*—generating creative ideas, sketching, building prototypes, and earning patents. Independent inventors are unconstrained by corporate prerogatives, so they are free to pursue revolutionary, even radical, ideas. However, individual inventors have certain limitations. They often struggle to access the tools and financing required to realize their inventions, so they usually cede large-scale, capital-intensive projects to corporate and government labs. Instead, independent inventors typically focus on lower-cost, human-scale products and digital technologies such as toys, windshield wipers, spacesuit gloves, software, websites, and mobile apps.

In contrast, corporations tend to excel at *innovation*—the distribution of new technologies to a wide network of users. Established firms have the financial, legal, and manufacturing resources to bring new inventions to market, but they can sometimes be cautious. To protect their existing product lines, R&D labs often make incremental improvements to established technologies. Sometimes they fail to appreciate disruptive in-house inventions, as when Kodak invented—then shelved—its digital camera. This conservatism can leave high-tech firms vulnerable to upstart competitors with breakthrough technologies. When asked about the biggest threat to Microsoft, Bill Gates said, "I worry about some guy in a garage inventing something, a new technology I have never thought about." Gates would know; in the 1980s, Microsoft, Apple, and other inventor-led start-ups eroded IBM's dominant position in the computer industry.[4]

To account for their relative strengths and weaknesses, independent inventors have developed a symbiotic—if sometimes adversarial—relationship with their institutional patrons. Individual inventors can turn to venture capitalists to finance an entrepreneurial start-up, but they just as often depend on strategic partnerships with corporations, government agencies, and the military. These organizations, in turn, rely on inventors' creative, out-of-the-box ideas to gain an advantage on the battlefield and in the marketplace. When corporate and government labs acquire the inventions of independent inventors to complement the inventions they produce in-house, they increase the likelihood that one of these parallel paths will lead to a new product or improve an existing process.

Admittedly, this symbiotic relationship often leads to confrontation. Inventors and firms frequently sue one another for infringement. Corporations acquire and suppress inventors' patents to neutralize competitive threats. Federal agencies such as NASA and DARPA invite criticism when they award development contracts to blue-chip firms over entrepreneurial start-ups and vice versa. The USPTO, the Department of Justice, and the federal courts adjudicate an endless stream of patent and antitrust disputes. In the end—whether the relationship is constructive or adversarial—high-tech firms and government agencies cannot avoid engaging with independent inventors.

Technological development is a difficult and uncertain endeavor. To improve our prospects, we should encourage the combined and collaborative efforts of individual inventors, corporations, and government agencies. This open, eclectic approach is the surest path to successful innovation, and it helps explain why independent inventors have enjoyed newfound success in the twenty-first century. Federal agencies and corporations still operate their own R&D laboratories, but they have also revived the crowd-sourcing initiatives and inventor-firm alliances that independents originally forged during the early twentieth century. This trend toward open innovation bodes well for the continued relevance of independent inventors and suggests a longer historical continuity. The closed, proprietary approach to innovation that R&D firms and government agencies adopted during their postwar golden age now appears to be an anomaly.[5]

Despite recent gains, independent inventors continue to confront several persistent challenges. Like their predecessors, today's independents still struggle to sustain their fragile professional communities. Inventors are still victimized by predatory scams. Independents remain frustrated by a patent system that still favors the big corporations. Critics still use colorful rhetoric to denigrate and delegitimize certain independents as "patent trolls." Even the most successful independents still must face the risk, uncertainty, and financial precariousness of working as an individual inventor. Today, as in years past, it requires incredible tenacity to survive as an independent inventor in an era of corporate R&D.

*   *   *

Finally, individual inventors matter because they reflect certain characteristics underlying America's national identity. After the United States broke free from the British Empire, the founders built protections for inventors into Article I, Section 8, of the US Constitution. From the nation's inception, Americans combined a thirst for new technology with a belief in individual liberty and a healthy skepticism of powerful institutions. The United States emerged as a plucky, upstart nation that overcame long odds to win its freedom. Not surprisingly, Americans retain an affection for maverick inventors who persevere in the face of serious challenges. This explains why the *New York Times* ranked Thomas Edison as the greatest living American in 1922, why the nation collectively mourned Steve Jobs's death in 2011, and why millions of Americans tune in each week to watch inventor-entrepreneurs on ABC's *Shark Tank*.

Independent inventors also reflect America's diversity and democratic ideals. They include men and women of every stripe. They work in every geographic region and in every conceivable industry. They include teenagers and retirees, high school graduates and highly trained PhDs, the descendants of enslaved people and first-generation immigrants. They invent simple toys and sophisticated semiconductors. They are brash billionaires and quirky nerds. There are important caveats, of course. Women, African Americans, and minorities remain woefully underrepresented among the ranks of independents, industrial researchers, and innovators in general.[6] Yet anyone can become an independent inventor.

Despite their diverse circumstances and backgrounds, individual inventors share a common set of characteristics and ideals. They embody quintessentially American traits such as individual initiative, self-reliance, Yankee ingenuity, creativity, and persistence. If, as historian Thomas Hughes has suggested, technological innovation is the "most characteristic activity" of the American people, then independent inventors personify many of the nation's core values.[7]

Patent Commissioner Conway Coe understood this perfectly. During 1939's TNEC patent hearings—in the midst of the Great Depression—he argued that American inventors reflected "a creative faculty in our people" and demonstrated a spirit of "patience, resoluteness, sacrifice—suffering, too, if need be—in the pursuit of an ideal."[8] Coe failed to mention that independents can also be eccentric, naive, obsessive, and occasionally detached from reality. Moreover, they have consistently struggled to garner respect, sustain their professional organizations, and achieve their political goals. Yet, in 2019, Coe's successor, USPTO director Andrei Iancu, could still rightfully claim that inventors exemplify "the fundamental traits of the American character: grit, imagination, perseverance, and the willingness to take risks." In inventors, Iancu suggested, "I see America."[9]

*   *   *

"Despite the great mass organizations which compete with him," *Business Week* observed in 1929, "the individual, genius type of inventor still holds the field as the most important single source of new ideas. . . . He has a hard fight, but holds his own pretty well against the regimented scientists of industry."[10] What was true then is still true today. Independent inventors have endured many challenges over the years, but they have never disappeared. The individual genius—in the garret, in the garage, and in the dorm room—has long been, and will always remain, an important source of new technologies.

Notes

Chapter 1

1. "Says Garret Genius Has Disappeared," *New York Times*, December 11, 1927, 6. The speech was reprinted as Maurice Holland, "Research, Science, and Invention," in *A Century of Industrial Progress*, ed. Frederic William Wile (New York: Doubleday, Doran, 1928).

2. "Asserts Inventors Have Good Chance," *New York Times*, December 18, 1927, E1.

3. "The Garret Inventor," *New York Times*, December 22, 1927, 22.

4. L. Sprague de Camp, *The Heroic Age of American Invention* (Garden City, NY: Doubleday, 1961), profiles a pantheon of thirty-two nineteenth-century inventors. The heroic age ended after World War I, de Camp argued, when invention "came to be practiced by organized groups instead of by lone individuals" (p. 257).

5. Thomas P. Hughes, *American Genesis: A Century of Invention and Technological Enthusiasm, 1870–1970*, 2nd ed. (Chicago: University of Chicago Press, 2004), 138–139. See also David F. Noble; *America by Design: Science, Technology, and the Rise of Corporate Capitalism* (New York: Knopf, 1977), especially chapter 6, "The Corporation as Inventor," and chapter 7, "Science for Industry." He suggests that corporate R&D labs "completely overwhelmed the independent inventor" (p. 97).

6. I acknowledge the methodological difficulties associated with counting patents as a proxy for inventive activity. Not all inventions are patented, and many patents—such as "vanity" or "defensive" patents—are never developed into commercial products. A simple patent count gives equal weight to significant inventions (e.g., nylon) and relatively insignificant ones (e.g., a mousetrap). Nevertheless, I am sympathetic to the view of economic historian Jacob Schmookler, who remarked that "we have a choice of using patent data cautiously and learning what we can from them, or not using them and learning nothing about what they alone can teach us." See Jacob Schmookler, *Invention and Economic Growth* (Cambridge, MA: Harvard University Press, 1966), 56. See also Zvi Griliches, "Patent Statistics as Economic

Indicators: A Survey," *Journal of Economic Literature* 28, no. 4 (December 1990): 1661–1707; Adam Jaffe and Manuel Trajtenberg, eds., *Patents, Citations, and Innovations: A Window on the Knowledge Economy* (Cambridge, MA: MIT Press, 2002).

7. Individual inventors' patents are either unassigned or assigned to an individual on the date of issue. Corporate patents are those assigned to a firm when issued, signifying invention by an employee. See "Patent Applications Filed and Patents Issued, by Type of Patent and Patentee: 1790–2000," in *Historical Statistics of the United States: Earliest Times to the Present*, millennial ed., ed. Susan Carter et al. (New York: Cambridge University Press, 2006), 3:425–429, table Cg27–37.

8. For example, David A. Hounshell suggests that the "age of the heroic inventor, if that age had ever existed, was finished" by the end of World War I. See David A. Hounshell, "The Evolution of Industrial Research in the United States," in *Engines of Innovation: U.S. Industrial Research at the End of an Era*, ed. Richard S. Rosenbloom and William J. Spencer (Boston: Harvard Business School Press, 1996), 32. On individual corporate labs, see George Wise, *Willis R. Whitney, General Electric, and the Origins of U.S. Industrial Research* (New York: Columbia University Press, 1985); Leonard Reich, *The Making of American Industrial Research: Science and Business at GE and Bell, 1876–1926* (Cambridge: Cambridge University Press, 1985); Jon Gertner, *The Idea Factory: Bell Labs and the Great Age of American Innovation* (New York: Penguin, 2012); David A. Hounshell and John K. Smith, Jr., *Science and Corporate Strategy: Du Pont R & D, 1902–1980* (New York: Cambridge University Press, 1988); Reese Jenkins, *Images and Enterprise: Technology and the American Photographic Industry, 1839 to 1925* (Baltimore: Johns Hopkins University Press, 1975); Margaret B. W. Graham and Alec T. Shuldiner, *Corning and the Craft of Innovation* (New York: Oxford University Press, 2001); Margaret B. W. Graham and Bettye Pruitt, *R&D for Industry: A Century of Technical Innovation at Alcoa* (Cambridge: Cambridge University Press, 1990); Margaret B. W. Graham, *The Business of Research: RCA and the VideoDisc* (Cambridge: Cambridge University Press, 1986).

9. John Jewkes, David Sawers, and Richard Stillerman, *The Sources of Invention* (London: Macmillan, 1958), quotation at 185.

10. Jacob Schmookler, "Inventors Past and Present," *Review of Economics and Statistics* 39, no. 3 (August 1957): 321–333; Naomi R. Lamoreaux and Kenneth L. Sokoloff, "Inventors, Firms, and the Market for Technology in the Late Nineteenth and Early Twentieth Century," in *Learning by Doing in Markets, Firms, and Countries*, ed. Naomi R. Lamoreaux, Daniel M. G. Raff, and Peter Temin (Chicago: University of Chicago Press, 1999); Tom Nicholas, "The Role of Independent Invention in U.S. Technological Development, 1880–1930," *Journal of Economic History* 70, no. 1 (March 2010): 57–82.

11. Beyond the hundreds of individual biographies, popular collections include Edmund Fuller, *Tinkers and Genius: The Story of the Yankee Inventors* (New York:

Hastings House, 1955); de Camp, *The Heroic Age of American Invention*; Joseph and Frances Gies, *The Ingenious Yankees* (New York: Crowell, 1976).

12. Individual biographies of twentieth-century inventors include David Owen, *Copies in Seconds: How a Lone Inventor and an Unknown Company Created the Biggest Communication Breakthrough since Gutenberg: Chester Carlson and the Birth of the Xerox Machine* (New York: Simon and Schuster, 2004); Donald G. Godfrey, *Philo T. Farnsworth: The Father of Television* (Salt Lake City: University of Utah Press, 2001); on Farnsworth, Evan I. Schwartz, *The Last Lone Inventor: A Tale of Genius, Deceit, and the Birth of Television* (New York: Perennial, 2002); William D. Middleton and William D. Middleton III, *Frank Julian Sprague: Electrical Inventor and Engineer* (Bloomington: Indiana University Press, 2009); Frederick Dalzell, *Engineering Invention: Frank J. Sprague and the U.S. Electrical Industry* (Cambridge, MA: MIT Press, 2010); Mike Adams, *Lee de Forest: King of Radio, Television, and Film* (New York: Copernicus, 2011); Donald Godfrey, *C. Francis Jenkins, Pioneer of Film and Television* (Urbana: University of Illinois Press, 2014); Joris Mercelis, *Beyond Bakelite: Leo Baekeland and the Business of Science and Invention* (Cambridge, MA: MIT Press, 2020). For a collection of interviews, see Kenneth A. Brown, *Inventors at Work: Interviews with 16 Notable American Inventors* (Redmond, WA: Tempus Books, 1988).

13. On inventors' cognitive and workbench processes, see Eugene S. Ferguson, "The Mind's Eye: Nonverbal Thought in Technology," *Science*, n.s., 197, no. 4306 (August 26, 1977): 827–836; Thomas P. Hughes, "How Did the Heroic Inventors Do It?," *American Heritage of Invention and Technology* 1, no. 2 (Fall 1985): 18–25; Robert Friedel and Paul Israel, *Edison's Electric Light: Biography of an Invention* (New Brunswick, NJ: Rutgers University Press, 1986); Michael E. Gorman and W. Bernard Carlson, "Interpreting Invention as a Cognitive Process: Alexander Graham Bell, Thomas Edison, and the Telephone, 1876–1878," *Science, Technology, and Human Values* 15, no. 2 (Spring 1990): 131–164; Robert J. Weber and David N. Perkins, eds., *Inventive Minds: Creativity in Technology* (New York: Oxford University Press, 1992).

14. US Bureau of the Census, *Sixteenth Census of the United States: 1940, Population* (Washington, DC: Government Printing Office, 1943), vol. 4, Comparative Occupation Statistics for the United States, 1870 to 1940, table 8: "Total Gainful Workers 10 Years Old and Over, By Occupation, for the United States, 1870–1930," 111.

15. Carter et al., "Patent Applications Filed and Patents Issued."

16. Barkev Sanders, "The Number of Patentees in the United States," *IDEA* 9, no. 2 (Summer 1965): 205–221.

17. The Smithsonian's National Museum of American History has accessioned several inventors' collections over the past thirty years, including those of Charles Adler, Jr. (available 1990), Wadsworth W. Mount (1991), Leo H. Baekeland (1994), Earl Tupper (1996), Everett Bickley (1999), Marion O'Brien Donovan (2000), Joseph Friedman (2001), and Charles Eisler (2006).

18. In 1913, patent examiner Henry E. Baker documented 800 total patents earned by African American inventors since the 1830s; this represented only 0.08 percent of the approximately one million cumulative patents issued through the end of that year. See Henry E. Baker, *The Colored Inventor: A Record of Fifty Years* (New York: Crisis Publishing, 1913). Similarly, a 1923 study reported that women inventors earned approximately 1.4 percent of all US patents granted between 1905 and 1921. See US Department of Labor, *Women's Contributions in the Field of Invention*, Bulletin of the Women's Bureau no. 28 (Washington, DC: Government Printing Office, 1923), table II, 13.

19. On women inventors, see Martha Moore Trescott, ed., *Dynamos and Virgins Revisited: Women and Technological Change in History* (Metuchen, NJ: Scarecrow Press, 1979); Anne L. Macdonald, *Feminine Ingenuity: Women and Invention in America* (New York: Ballantine Books, 1992); Autumn Stanley, *Mothers and Daughters of Invention* (Metuchen, NJ: Scarecrow Press, 1993); Ethlie Ann Vare and Greg Ptacek, *Women Inventors and Their Discoveries* (Minneapolis: Oliver Press, 1993); Denise E. Pilato, *The Retrieval of a Legacy: Nineteenth-Century American Women Inventors* (Westport, CT: Praeger, 2000).

20. On African American inventors, see Portia P. James, *The Real McCoy: African-American Invention and Innovation, 1619–1930* (Washington, DC: Smithsonian Institution Press, 1989); Rayvon Fouché, *Black Inventors in the Age of Segregation: Granville T. Woods, Lewis H. Latimer, and Shelby J. Davidson* (Baltimore: Johns Hopkins University Press, 2003); Bruce Sinclair, ed., *Technology and the African-American Experience: Needs and Opportunities for Study* (Cambridge, MA: MIT Press, 2004); Patricia Carter Sluby, *The Inventive Spirit of African Americans: Patented Ingenuity* (Westport, CT: Praeger, 2004); Carroll W. Pursell, ed., *A Hammer in Their Hands: A Documentary History of Technology and the African-American Experience* (Cambridge, MA: MIT Press, 2005).

21. For immigrant inventors' autobiographies, see Michael Pupin, *From Immigrant to Inventor* (New York: Charles Scribner's Sons, 1922); Charles Eisler, *The Million-Dollar Bend* (New York: William-Frederick Press, 1960). For scholarly biographies, see W. Bernard Carlson, *Tesla: Inventor of the Electrical Age* (Princeton, NJ: Princeton University Press, 2013); Mercelis, *Beyond Bakelite*.

22. On these exclusions, see Ruth Oldenziel, *Making Technology Masculine: Men, Women and Modern Machines in America, 1870–1945* (Amsterdam: Amsterdam University Press, 2004), and several essays in Sinclair, *Technology and the African-American Experience*. Among corporate R&D labs, there were a few exceptions to the White male American rule. German Charles Steinmetz became GE's chief engineer; see Ronald R. Kline, *Steinmetz: Engineer and Socialist* (Baltimore: Johns Hopkins University Press, 1992). Norwegian chemist Fin Sparre directed DuPont's development department; see Hounshell and Smith, *Science and Corporate Strategy*, 35–36. African American Lewis H. Latimer was Thomas Edison's assistant, then managed GE's patent pool; see Fouché, *Black Inventors in the Age of Segregation*, 82–133.

23. B. Zorina Khan, *The Democratization of Invention: Patents and Copyrights in American Economic Development, 1790–1920* (Cambridge: Cambridge University Press, 2009), 124–134.

24. Howard Anderson, "Why Big Companies Can't Invent," *Technology Review* 107, no. 4 (May 2004): 56–59.

25. Robert Friedel, "Perspiration in Perspective: Changing Perceptions of Genius and Expertise in American Invention," in *Inventive Minds: Creativity in Technology*, ed. Robert J. Weber and David N. Perkins (New York: Oxford University Press, 1992), 11.

26. The American patent system enticed a broad segment of nineteenth-century Americans to become inventors; see Khan, *The Democratization of Invention*. Not all inventors were entrepreneurs; on this distinction, see Joseph A. Schumpeter, *Theory of Economic Development*, trans. Redvers Opie (Cambridge, MA: Harvard University Press, 1949), 88–89. On the market for railroad patents, see Steven W. Usselman, *Regulating Railroad Innovation: Business, Technology, and Politics in America, 1840–1920* (Cambridge: Cambridge University Press, 2002), 97–139. On the overall market for patents, see Lamoreaux and Sokoloff, "Inventors, Firms, and the Market for Technology in the Late Nineteenth and Early Twentieth Century."

27. Louis C. Hunter, "The Heroic Theory of Invention," in *Technology and Social Change in America*, ed. Edwin T. Layton, Jr. (New York: Harper and Row, 1973), 25–46; Eric S. Hintz, "'Heroes of the Laboratory and the Workshop': Invention and Technology in Books for Children, 1850–1900," in *Enterprising Youth: Social Values and Acculturation in Nineteenth-Century American Children's Literature*, ed. Monika Elbert (New York: Routledge, 2008).

28. See Carlson, *Tesla*. See also Wyn Wachhorst, *Thomas Alva Edison: An American Myth* (Cambridge, MA: MIT Press, 1981); on Edison's poll, see p. 5.

29. Carolyn C. Cooper, "Myth, Rumor, and History: The Yankee Whittling Boy as Hero and Villain," *Technology and Culture* 44, no. 1 (January 2003): 82–96.

30. Paul Israel, "Inventing Industrial Research: Thomas Edison and the Menlo Park Laboratory," *Endeavour* 26, no. 2 (June 1, 2002): 48–54; David A. Hounshell, "The Modernity of Menlo Park," in *Working at Inventing: Thomas A. Edison and the Menlo Park Experience*, ed. William Pretzer (Dearborn, MI: Henry Ford Museum and Greenfield Village, 1989).

31. On the conditions motivating the emergence of industrial research, see W. Bernard Carlson, "Innovation and the Modern Corporation: From Heroic Inventor to Industrial Science," in *Science in the Twentieth Century*, ed. John Krige and Dominique Pestre (Amsterdam: Harwood, 1997), 203–226; Hounshell, "The Evolution of Industrial Research in the United States," 13–28; David C. Mowery and Nathan Rosenberg, *Technology and the Pursuit of Economic Growth* (Cambridge: Cambridge University Press, 1989), 21–97; Reich, *The Making of American Industrial Research*,

1–41. On the formation of GE's laboratory, see Wise, *Willis R. Whitney*, 66–94; Reich, *The Making of American Industrial Research*, 42–96.

32. Carlson, "Innovation and the Modern Corporation"; Hounshell, "The Evolution of Industrial Research in the United States." On the appropriation of employees' inventions, see Steven Cherensky, "A Penny for Their Thoughts: Employee-Inventors, Pre-invention Assignment Agreements, Property, and Personhood," *California Law Review* 81, no. 2 (March 1993): 595–669; Catherine L. Fisk, *Working Knowledge: Employee Invention and the Rise of Corporate Intellectual Property, 1800–1930* (Chapel Hill: University of North Carolina Press, 2009).

33. On managerial control and vertical integration among the largest US firms, see Alfred D. Chandler, Jr., *The Visible Hand: The Managerial Revolution in American Business* (Cambridge, MA: Belknap Press, 1977). On R&D labs as a form of vertical integration, see Michael Aaron Dennis, "Accounting for Research: New Histories of Corporate Laboratories and the Social History of American Science," *Social Studies of Science* 17, no. 3 (August 1987): 479–518.

34. On Langmuir and GE, see Wise, *Willis R. Whitney*, 149–161. On Carothers and DuPont, see Hounshell and Smith, *Science and Corporate Strategy*, 229–248.

35. Richard S. Tedlow, *Keeping the Corporate Image: Public Relations and Business, 1900–1950* (Greenwich, CT: JAI Press, 1979); Roland Marchand, *Creating the Corporate Soul: The Rise of Public Relations and Corporate Imagery in American Big Business* (Berkeley: University of California Press, 1998).

36. These characterizations draw on Joseph Rossman, *The Psychology of the Inventor: A Study of the Patentee* (Washington, DC: Inventors Publishing, 1931), 27–34.

37. Reich, *The Making of American Industrial Research*, 129–217; Richard R. John, *Network Nation: Inventing American Telecommunications* (Cambridge, MA: Belknap Press, 2010), 200–406.

38. National Research Council, *Industrial Research Laboratories of the United States*, 3rd ed. (New York: R. R. Bowker, 1927).

39. Henry W. Chesbrough, *Open Innovation: The New Imperative for Creating and Profiting from Technology* (Boston: Harvard Business School Press, 2003). On the revival of earlier commercialization strategies, see David C. Mowery, "Plus ca change: Industrial R&D in the 'Third Industrial Revolution,'" *Industrial and Corporate Change* 18, no. 1 (February 2009): 1–50.

Chapter 2

1. Waldemar Kaempffert, *Invention and Society* (Chicago: American Library Association, 1930), 28, 30.

2. James B. Conant, as cited in John Jewkes, David Sawers, and Richard Stillerman, *The Sources of Invention* (London: Macmillan, 1958), 29.

3. Edmund L. Van Deusen, "The Inventor in Eclipse," *Fortune* (December 1954), 132.

4. Jacob Schmookler, "Inventors Past and Present," *Review of Economics and Statistics* 39, no. 3 (August 1957): 321–333, quotations at 321, 330.

5. Schmookler, "Inventors Past and Present," 321, emphasis in the original.

6. B. Zorina Khan, *The Democratization of Invention: Patents and Copyrights in American Economic Development, 1790–1920* (Cambridge: Cambridge University Press, 2009), 124–134.

7. Matilda Joslyn Gage, "Woman as an Inventor," *North American Review* 136, no. 318 (May 1883): 478–489, quotations at 483, 488; B. Zorina Khan, "Married Women's Property Laws and Female Commercial Activity: Evidence from United States Patent Records, 1790–1895," *Journal of Economic History* 56, no. 2 (June 1996): 356–388.

8. Charlotte Smith, "Women's Inventions," *The Woman Inventor* 1, no. 1 (April 1891): 1. For more on Smith, see chapter 4.

9. Henry E. Baker, "The Negro as an Inventor," in *Twentieth Century Negro Literature*, ed. D. W. Culp (Naperville, IL: J. L. Nichols, 1902), 399–401; Baker's biography precedes p. 399. Baker's four-volume set of patents, *Negro Patentees of the United States, 1834–1900*, is at the Moorland-Springarn Research Center, Howard University, Washington, DC.

10. Henry E. Baker, *The Colored Inventor: A Record of Fifty Years* (New York: Crisis Publishing, 1913), quotations at 6. For more on Baker, see chapter 4.

11. Charlotte Smith, "Colored Woman Inventor," *The Woman Inventor* 1, no. 1 (April 1891): 3.

12. For the paragraphs on Morgan, see Lisa D. Cook, "Overcoming Discrimination by Consumers during the Age of Segregation: The Example of Garrett Morgan," *Business History Review* 86, no. 2 (Summer 2012): 211–234; Robert L. Gale, "Morgan, Garrett A.," *American National Biography Online*, 2001, https://doi.org/10.1093/anb/9780198606697.article.1301173.

13. Anne L. Macdonald, *Feminine Ingenuity: Women and Invention in America* (New York: Ballantine Books, 1992), 295–297, 301, 308–309; Autumn Stanley, *Mothers and Daughters of Invention: Notes for a Revised History of Technology* (Metuchen, NJ: Scarecrow Press, 1993), 611–619; Cook, "Overcoming Discrimination," 230–231.

14. Wyn Wachhorst, *Thomas Alva Edison: An American Myth* (Cambridge, MA: MIT Press, 1981), 3–7.

15. Frank J. Sprague, "System of Electrical Distribution," US Patent 335,045, filed September 19, 1885, issued January 26, 1886. By the late 1800s, employers were

claiming ownership of their employees' inventions. See Catherine L. Fisk, *Working Knowledge: Employee Innovation and the Rise of Corporate Intellectual Property, 1800–1930* (Chapel Hill: University of North Carolina Press, 2009).

16. Harold C. Passer, *Frank Julian Sprague, Father of Electric Traction, 1857–1934* (Cambridge, MA: Harvard University Press, 1952); Frederick Dalzell, *Engineering Invention: Frank J. Sprague and the U.S. Electrical Industry, 1880–1900* (Cambridge, MA: MIT Press, 2009); William D. Middleton and William D. Middleton III, *Frank Julian Sprague: Electrical Inventor and Engineer* (Bloomington: Indiana University Press, 2009).

17. Matthew Josephson, *Edison: A Biography* (New York: McGraw-Hill, 1959), 238–242.

18. Frank J. Sprague to Thomas A. Edison, April 25, 1884, as quoted by Sprague in a speech accepting the AIEE's Edison Medal, May 16, 1911, in Box 101, Folder "Speeches & Writings, 1887–1934," Frank J. Sprague Papers, Manuscript Division, New York Public Library (hereafter Sprague Papers).

19. Edison interview in the *Philadelphia Press*, September 21, 1884, as quoted in Harriet Sprague, *Frank J. Sprague and the Edison Myth* (New York: William-Frederick Press, 1947), 9.

20. Middleton and Middleton, *Frank Julian Sprague*, 41–86; Dalzell, *Engineering Invention*, 70–105.

21. Middleton and Middleton, *Frank Julian Sprague*, 86; Dalzell, *Engineering Invention*, 101–105.

22. Frank J. Sprague to Board of Trustees, Sprague Railway & Motor Co., June 7, 1890, from Sprague Papers, as quoted in Paul Israel, *Edison: A Life of Invention* (New York: Wiley, 1998), 374–375.

23. Sprague's resignation letter to Edison General Electric, December 2, 1890, Sprague Papers, as quoted in Israel, *Edison*, 375. On the renaming order, see Sprague's handwritten emendations on the cover of Edison General Electric Company, *Edison System of Electric Railways*, 1891. Sprague's handwriting is confirmed by a typed note tucked in the front cover by Harriet Sprague, September 20, 1935, Box 107, "Personal Papers," Sprague Papers.

24. Frank J. Sprague, "Inventors of the Electric Railway: Frank J. Sprague Protests against the Share of the Glory Assigned in an Interview to Thomas Edison," letter to the editor, *New York Sun*, August 27, 1919, responding to Edward Marshall's interview with Edison, *New York Sun*, August 10, 1919. Both are in Box 115, "Personal Clippings, 1915–1924," Sprague Papers.

25. James C. Young, "The Magic Edison Made for the World," *New York Times Magazine*, August 12, 1928, SM3; Frank J. Sprague, "Electric Railway Not Creation of One Man," *New York Times*, September 23, 1928, X9.

26. Promotional flyer for Harriet Sprague, *Frank J. Sprague and the Edison Myth*, ca. 1947, Box 106, "Personal Papers," Sprague Papers.

27. Israel, *Edison*, 372–378, quotations at 373, 372.

28. Middleton and Middleton, *Frank Julian Sprague*; Dalzell, *Engineering Invention*.

29. Middleton and Middleton, *Frank Julian Sprague*, 103–104; Sprague, *Frank J. Sprague and the Edison Myth*, quotation at 11.

30. W. Bernard Carlson, *Innovation as a Social Process: Elihu Thomson and the Rise of General Electric, 1870–1900* (New York: Cambridge University Press, 1991).

31. George Wise, "William Stanley's Search for Immortality," *American Heritage of Invention and Technology* 4, no. 1 (Spring/Summer 1988): 42–49, quotation at 43. Stanley was unable to keep his name attached to transformers he sold to Westinghouse and General Electric.

32. Charles Adler, Jr., "The Foolish Failure or Memoirs of Charles Adler (Jr)," Box 7, Folder 1, Charles Adler, Jr., Collection, Archives Center, National Museum of American History, Smithsonian Institution, Washington, DC (hereafter Adler Collection). His obituary is "Charles Adler, Jr., Inventor of Safety Devices, Dies at 81," *Baltimore Sun*, October 24, 1980, 1.

33. Adler, "The Foolish Failure or Memoirs of Charles Adler (Jr)," 12.

34. Railway Accessories Corporation, pamphlet for "Adler Flashing Relay," ca. 1924–1926, and "New Devices: The Adler Alternating Flashing Relay," *Railway Signal Engineer* (February 1923), both clippings in Box 2, Scrapbook #1 (1919–1925, 1926), Adler Collection. See also "Saco Adler Alternate Flashing Relay," *Railway Signaling* (September 1926), advertisement in Box 2, Scrapbook #2 (1926–), Adler Collection.

35. "Noise Traffic Signal Tried," *Baltimore News*, February 22, 1928; "Auto Horn Controls Lamps to Guide Traffic," *Popular Mechanics* (March 1928); "Devices," *Time*, May 21, 1928; "New Type of Traffic Light Is Demonstrated to City Police Heads," *Schenectady Gazette*, June 29, 1928. All clippings in Box 3, Scrapbook #1 (1927–1928), Adler Collection.

36. The Adler acquisition is described in GE's internal publications, including J. G. Regan, *The Go-Getter*, no. 24 (November 9, 1928): 1–2; J. G. Regan, *The Go-Getter*, no. 25 (no date, ca. December 1928): 1; "Honk! Honk! And You Have the Right of Way," *GE Monogram* (no date, ca. 1928–1929): 14; Charles Adler, "Speeding as Well as Safeguarding Traffic at Intersections," *GE Lighting Specialist*, no. 2 (March 1929): n.p. All clippings in Box 3A, Scrapbook #1 (1928–1929), Adler Collection.

37. Charles Adler, Jr., "Traffic Signal," US Patent 2,074,246, filed October 17, 1936, issued March 16, 1937.

38. Harold C. Passer, *The Electrical Manufacturers, 1875–1900: A Study in Competition, Entrepreneurship, Technical Change, and Economic Growth* (New York: Arno Press, 1972 [c1953]), 329–334.

39. Licensing agreement, Adler Safety Control Company and the General Electric Company, September 1, 1939; licensing agreement, Adler Safety Control Company and the Westinghouse Electric and Manufacturing Company, December 15, 1939. Both are in Box 11, Folder 20, Adler Collection.

40. Carl Schoettler, "He Likes to Look Down on Trains, Traffic Lights," *Baltimore Sun*, December 13, 1977; Seymour Kopf, "Man about Town," *Baltimore News American*, May 25, 1972; quotations from William Thompson, "That First Automatic Traffic Signal Went Up 59 Years Ago," *Baltimore Evening Sun*, February 25, 1987. All clippings are in Box 4, Folder 2, Adler Collection.

41. Gaisman was well known for inventing a safety razor and merging his interests with Gillette. See "Gaisman, Who Will Head Gillette, Is an Incorrigible Inventor," *Business Week*, November 26, 1930, 20–21, and his obituary, "H. J. Gaisman, 104, Inventor, Is Dead," *New York Times*, August 7, 1974, 38.

42. "A Fortune for a Simple Invention," *Scientific American* 111, no. 6 (August 8, 1914): 101; Henry J. Gaisman, "How I Came to Invent the Autographic Kodak," *Scientific American* 111, no. 6 (August 8, 1914): 101, 106.

43. "The Autographic Kodak," Eastman Kodak Company ad, *Scientific American* 111, no. 10 (September 5, 1914): 194.

44. Richard Tedlow, *Keeping the Corporate Image: Public Relations and Business, 1900–1950* (Greenwich, CT: JAI Press, 1979); Roland Marchand, *Creating the Corporate Soul: The Rise of Public Relations and Corporate Imagery in American Big Business* (Berkeley: University of California Press, 1998); Pamela Laird, *Advertising Progress: American Business and the Rise of Consumer Marketing* (Baltimore: Johns Hopkins University Press, 2001).

45. C. E. Kenneth Mees, *The Organization of Industrial Scientific Research* (New York: McGraw-Hill, 1920), 35, 46; Marchand, *Creating the Corporate Soul*, 194.

46. Tedlow, *Keeping the Corporate Image*, 139; Marchand, *Creating the Corporate Soul*, 89, 141–145, 320.

47. Hugh G. J. Aitken, *The Continuous Wave: Technology and American Radio, 1900–1932* (Princeton, NJ: Princeton University Press, 1985); Steve J. Wurtzler, *Electric Sounds: Technological Change and the Rise of Corporate Mass Media* (New York: Columbia University Press, 2007), 101–107, quotation at 102.

48. For the "pliers and screwdriver" quotation, see the obituary, "Charles F. Kettering," *New York Times*, November 27, 1958, 28. On Kettering as a self-styled "professional amateur," see Thomas A. Boyd, *Professional Amateur: The Biography of Charles*

Franklin Kettering (New York: Dutton, 1957), 4. For the radio talk, see Charles F. Kettering, "Research Is a State of Mind," in *Short Stories of Science and Invention: A Collection of Radio Talks* (Detroit: General Motors Corporation, 1945), 10–11.

49. Roger B. White, "'An Exposition of Our Own': Corporate Identity, Consumer Advertising and Atlantic City's National Exhibits, 1898–1968," *Journal of Design History* 26, no. 1 (February 2013): 47–64; Bryant Simon, *Boardwalk of Dreams: Atlantic City and the Fate of Urban America* (New York: Oxford University Press, 2004), 107–108. On DuPont's pavilion, see "DuPont Will Move Out Its A.C. Display," *The Billboard*, May 14, 1955, 58.

50. Marchand, *Creating the Corporate Soul*, 249–311; General Motors, "Parade of Progress," 1936 brochure, GM Heritage Center, https://www.gmheritagecenter.com/gm-heritage-archive/Events/GM_Parade_of_Progress.html.

51. Senator Gerald P. Nye's (R-ND) profiteering investigations were inspired by an exposé, H. C. Engelbrecht and F. C. Hanighen, *Merchants of Death: A Study of the International Armament Industry* (New York: Dodd, Mead, 1934).

52. Marchand, *Creating the Corporate Soul*, 218–223, quotation at 221.

53. "It Started One Saturday Night," DuPont advertisement, Box 43, Folder 14, DuPont Advertising Department Records (Accession 1803), Hagley Museum and Library, Wilmington, DE (hereafter Accession no. 1803, DuPont Ad. Dept. Records).

54. "Winning Nature's Secrets," an AT&T ad in *Scientific American* 132, no. 6 (June 1925): 402; "The Greatest Chemical Industry in America," DuPont advertisement in *Scientific American* 120, no. 14 (April 5, 1919): 356.

55. Untitled DuPont advertisement from *Forbes*, December 9, 1922, 257, Box 43, Folder 3, Accession no. 1803, DuPont Ad. Dept. Records.

56. "Headquarters for Ideas," General Motors ad in *Scientific American* 131, no .7 (July 1924): 65.

57. Marchand, *Creating the Corporate Soul*, 141–145.

58. This section draws on Eric S. Hintz, "'Selling the Research Idea': The Pre-history and Genesis of the Industrial Research Institute, 1916–1945," *Research-Technology Management* 56, no. 6 (November–December 2013): 46–50.

59. Rexmond C. Cochrane, *The National Academy of Sciences: The First Hundred Years, 1863–1963* (Washington, DC: National Academy of Sciences, 1978), 200–241; Thomas P. Hughes, *American Genesis: A Century of Invention and Technological Enthusiasm, 1870–1970*, 2nd ed. (Chicago: University of Chicago Press, 2004), 121–126. See also chapter 6, this volume.

60. National Research Council, "Minutes of the Council," June 1916, as quoted in Cochrane, *National Academy of Sciences*, 209; for the NRC's committee structure, see

209–216. See also David F. Noble, *America by Design: Science, Technology, and the Rise of Corporate Capitalism* (New York: Knopf, 1977), 154.

61. George Ellery Hale, "The Purpose of the National Research Council," *Bulletin of the National Research Council* 1, no. 1 (October 1919): 1–7, quotation at 4.

62. "Tentative Draft of Committee on Relations of the Engineering Division to the Division of Industrial Research," 1919, Folder "Beginning of Division, 1919," Research Extension Files, Central Policy Files, National Academy of Sciences Archives, Washington, DC (hereafter CPF-NAS). See also Cochrane, *National Academy of Sciences*, 249–250, 289.

63. "Minutes of Meeting of NRC Interim Committee," January 2, 1924, Folder "General Correspondence, 1924," in Division of Engineering and Industrial Research Files (hereafter E&IR Files), CPF-NAS; Dugald C. Jackson, "Division of Engineering and Industrial Research," in *A History of the National Research Council, 1919–1933*, ed. National Research Council (Washington, DC: National Research Council, 1933), 17–20; Cochrane, *National Academy of Sciences*, 290. On Holland and Spraragen's hiring, see NRC secretary Vernon Kellogg to Maurice Holland, June 12, 1924; Kellogg to William Spraragen, June 11, 1924. Both are in Folder "1924, General Correspondence," E&IR Files, CPF-NAS.

64. W. H. Waggoner, "Maurice Holland, 89; Ex-engineer Founded Institute for Research," *New York Times*, March 27, 1981, D17; Benoît Godin, "The Linear Model of Innovation: Maurice Holland and the Research Cycle," *Social Science Information* 50, no. 3–4 (September 2011): 569–581.

65. Maurice Holland, "Plan for the Promotion of Research by the Division of Engineering and Industrial Research of [the] National Research Council, 7 April 1925," Folder "Plan for Promotion of Research, 1925," E&IR Files, CPF-NAS, quotations at 1–3.

66. William Spraragen, "Annual Report for the Year Ending April 1, 1925," Appendix A to Meeting Minutes, April 24, 1925, Folder "Meetings 1925–1926," E&IR Files, CPF-NAS; Maurice Holland, "The Demonstration and Dramatization of Research," a speech appended to William Spraragen, Minutes of the Meeting of the Division of Engineering and Industrial Research, November 18, 1927, Folder "Meetings, 1927–1928," E&IR Files, CPF-NAS.

67. William Spraragen, Bi-monthly Report for the NRC Executive Board, April 15, 1927, Folder "Bi-monthly Reports, 1924–1927," E&IR Files, CPF-NAS; "Science Aids Prosperity: Dr. Howe Tells Engineers Future Welfare Depends on Research," *New York Times*, March 24, 1927, 39.

68. "Says Garret Genius Has Disappeared," *New York Times*, December 11, 1927, 6. The speech was reprinted as Maurice Holland, "Research, Science, and Invention," in *A Century of Industrial Progress*, ed. Frederic William Wile (New York: Doubleday, Doran, 1928). For the ensuing debate, see chapter 1, this volume.

69. On the Washington, DC, radio talks, see an unsigned memorandum [probably W. E. Tisdale], "Science Radio Talks," May 28, 1924; Albert L. Barrows, "Report of Committee on Radio Talks, 1925–1926." Both documents are in Folder "Committee on Radio Talks, 1924–1926," in Executive Board Files, CPF-NAS. On the world's fair syndicated radio talks, see "Bimonthly Report of the Permanent Secretary to the Executive Board of the NRC," December 9, 1930; "Program: Science Advisory Board Committee Radio Talks." Both documents are in Folder "Science Advisory Committee to Trustees of Chicago World's Fair: Radio Talks, 1930–1931," in Executive Board Files, CPF-NAS. On science popularization via radio, see Marcel LaFollette, *Science on the Air: Popularizers and Personalities on Radio and Early Television* (Chicago: University of Chicago Press, 2008).

70. The pamphlet is in Folder "Brochures, 1925–1931." Spraragen's boilerplate cover letter, January 10, 1930, is in Folder "General Correspondence, 1930–1931." For context surrounding the pamphlet, see several editions of William Spraragen, "Bi-monthly Report for the NRC Executive Board," including February 4, 1928, April 9, 1928, June 4, 1928, September 18, 1928, September 24, 1929, November 30, 1929, and January 31, 1930, Folder "Bi-monthly Reports, 1928–1931." All these documents are in E&IR Files, CPF-NAS.

71. Maurice Holland and Henry F. Pringle, *Industrial Explorers* (New York: Harper and Brothers, 1929), 1–4.

72. William Spraragen, Executive Committee Minutes, September 17, 1930, Folder "Executive Committee Meetings, 1924–1931," E&IR Files, CPF-NAS; Malcolm Ross, ed., *Profitable Practice in Industrial Research: Tested Principles of Research Laboratory Organization, Administration, and Operation* (New York: Harper and Brothers, 1932).

73. William Spraragen, "Bi-monthly Report for the NRC Executive Board," September 15, 1930, Folder "Bi-monthly Reports, 1928–1931." For a roster and itinerary, see "Tour of Laboratories, October 7 to 15, 1930," Folder "Tour of Research Laboratories, 1930." The unattributed, undated newspaper headline "Visit To Research Laboratories Like a Trip to Magic Land" is cited in the NRC's pamphlet *Executives and Bankers Look at Research*, 1931, Folder "Tour of Research Laboratories, 1930." All documents are in E&IR Files, CPF-NAS.

74. "Research Results to Follow Tour," *New York Times*, October 19, 1930, 48. On the 1931 tour, see William Spraragen, "Bi-monthly Report for the NRC Executive Board," reports dated October 1, 1931, and December 1, 1931, Folder "Bi-monthly Reports, 1928–1931." On the 1935 tour, see Maurice Holland, "Notes on Activities," October 8, 1935, and December 7, 1935, Folder "Bi-monthly Reports to NRC Chairman, 1934–1935." All documents are in E&IR Files, CPF-NAS.

75. On the invited journalists, see "Prospects and Tentative Reservations for the 1935 Tour of Research Laboratories." On the resulting coverage, see "Tour of Industrial Executives and Bankers to Important Research Laboratories, October 20th to

25th, 1935: Newspaper Comments." Both documents are in Folder "Industrial Laboratories Tour, 1935," E&IR Files, CPF-NAS.

76. On enthusiasm for an independent association, see Maurice Holland to A. L. Barrows, October 16, 1935, Folder "Industrial Laboratories Tour, 1935." For quotations from the prospectus, see Maurice Holland, "National Research Laboratories Association: Acting as an Association for Industrial Research and Development," September 3, 1935, Folder "National Research Laboratories Assoc., Proposed, 1935." On the division's financial difficulties, see Vannevar Bush, "The Problem of the Division of Engineering and Industrial Research," September 29, 1937, Folder "Problem of the Division of Engineering and Industrial Research: Bush V, 1937." All documents are in E&IR Files, CPF-NAS.

77. Alfred D. Flinn, "Directors of Industrial Research: Brief History," June 1934. On skepticism of the NARL, see C. E. K. Mees to Randolph T. Major, August 23, 1937. Both documents are in Directors of Industrial Research Minutes, 1923–42 (bound volume), Box 1, Directors of Industrial Research (U.S.) Records, 1929–1982, Accession no. 1851, Hagley Museum and Library, Wilmington, DE.

78. Bush quotation in "Proceedings: Organization Meeting of the Industrial Research Institute," February 25, 1938, Folder "Industrial Research Institute: Meetings: Organization Meeting, 1938"; "Industrial Research Institute," undated brochure (ca. 1939), Folder "Industrial Research Institute, General Correspondence, 1939." Both are in E&IR Files, CPF-NAS.

79. "Constitution and Bylaws," April 5, 1941, Folder "Industrial Research Institute, Constitution, 1941." On Holland's reassignment, see NRC executive secretary Albert L. Barrows to C. G. Worthington, IRI secretary, January 7, 1942; Worthington to Barrows, January 17, 1942. Both are in Folder "Industrial Research Institute, General Correspondence, 1942." On IRI's independence, see "Dissolution of the Formal Relations between the National Research Council and the Industrial Research Institute," July 1, 1945, Folder "Industrial Research Institute, General Correspondence, 1945." These documents are in E&IR Files, CPF-NAS. On membership growth, see Industrial Research Institute, "Members of the Institute and Official Representatives and Alternates, July 1, 1945," Folder "Industrial Research Institute, 1945," Institutions-Associations-Individuals Files, CFP-NAS. The IRI has been rebranded as the Innovation Research Interchange, http://www.iriweb.org/about.

80. Frank B. Jewett, "Our Function in the Stimulation of Research in Industry," Exhibit A, attached to Minutes of the Meeting of the Division of Engineering and Industrial Research, November 18, 1927, Folder "Meetings, 1927–1928," E&IR Files, CPF-NAS. On the growth of R&D labs, see National Research Council, *Industrial Research Laboratories of the United States*, 1st–7th eds. (New York: R. R. Bowker, 1920–1940).

81. Robert Friedel, "Perspiration in Perspective: Changing Perceptions of Genius and Expertise in American Invention," in *Inventive Minds: Creativity in Technology,*

ed. Robert J. Weber and David N. Perkins (New York: Oxford University Press, 1992), 11, 22.

82. Sociologist Thomas Gieryn suggests that "boundary work" occurs "when two or more rival epistemic authorities square off for jurisdictional control over a contested ontological domain." In this case, independent inventors and industrial scientists clashed over which group possessed the more credible mode of invention. See Thomas Gieryn, *Cultural Boundaries of Science: Credibility on the Line* (Chicago: University of Chicago Press, 1999), 16. For a similar argument, see Andrew Abbott, *The System of Professions: An Essay on the Division of Expert Labor* (Chicago: University of Chicago Press, 1988).

83. James Angell, "The Development of Research in the United States," an address before the Association of Land-Grant Colleges, Chicago, November 13, 1919, published in *Reprint and Circular Series of the National Research Council* 6 (November 1919): 1–19, quotations at 9–10.

84. Charles F. Kettering, "Head Lamp of Industry," in *Prophet of Progress: Selections from the Speeches of Charles F. Kettering*, ed. T. A. Boyd (New York: Dutton, 1961), 85.

85. Tesla, quoted in Orrin E. Dunlap, Jr., "An Inventor's Seasoned Ideas," *New York Times*, April 8, 1934, X9.

86. "Inventor Adler Forgets to Eat When He Works," *Baltimore News-Post*, July 21, 1952, clipping in Box 5B, Folder 1: "Scrapbook 1952–1954," Adler Collection. Edison biographer Paul Israel affirms Adler's characterization; although assistant John Kreusi built the working model, Edison developed all the major concepts behind the phonograph. See Israel, *Edison*, 142–145.

87. Chester Carlson, interviewed by Ken and Jeandale Magner, February 24, 1966, on their radio program *Off the Wing Tip Radio*, distributed by Armed Forces Radio, transcript in Box 79 (Microfilm Reel 65), Folder "Biographical Information," The Papers of Chester Carlson, Manuscript Division, New York Public Library (hereafter Carlson Papers).

88. Chester Carlson, speech to the Foundation for Research on the Nature of Man, Durham, NC, September 1, 1966, Box 80 (Microfilm Reel 66), Folder "Speeches R–X," Carlson Papers.

89. Samuel Ruben, "Inventor of the Year Award Address," Patent, Trademark, and Copyright Research Institute, George Washington University, Washington, DC, April 14, 1966, published in *Patent, Trademark, and Copyright Journal of Research and Education* 10, no. 1 (1966): 117–121, quotations at 120–121.

90. William H. Whyte, *The Organization Man* (Philadelphia: University of Pennsylvania Press, 2002 [1956]), chapter 16, "The Fight against Genius," quotations at 213.

91. Joseph Nocera, foreword to Whyte, *The Organization Man*, vii–viii.

92. Angell, "The Development of Research in the United States," 9.

93. Konstantin K. Paluev, "How Collective Genius Contributes to Industrial Progress," *General Electric Review* 44, no. 5 (May 1941), as reprinted in Philip Alger, *The Human Side of Engineering: Tales of General Electric over 80 Years* (Schenectady, NY: Mohawk Development Service, 1972), 117–131, quotations at 121.

94. Hudson Maxim, "The Inventor and the Meaning of Aeronautical Achievements in War," address to the Aldine Association, New York City, April 28, 1908, 3–6, Box 9, Folder "Speeches, 1907–1908 (1.)," Hudson Maxim Papers, Manuscript Division, New York Public Library.

95. Hudson Maxim, "The Making of an Inventor," typescript for *The Youth's Companion*, October 28, 1911, Box 2, Folder 18: "Writings: Inventions & Progress, 1907–1919," Accession no. 2147, Hudson Maxim Papers, Manuscript Division, Hagley Museum and Library, Wilmington, DE.

96. Iain McCallum, *Blood Brothers: Hiram and Hudson Maxim, Pioneers of Modern Warfare* (London: Chatham, 1999).

97. Mees, *The Organization of Industrial Scientific Research*, 90–91, emphasis added.

98. As described in Whyte, *The Organization Man*, 214.

99. Joseph Rossman, *The Psychology of the Inventor: A Study of the Patentee* (Washington, DC: Inventors Publishing, 1931), quotations at 103–104 (Parker), 106–107 (Goldenstein), 86 (Rossman).

100. W. Bernard Carlson, "Toward a Non-linear History of R&D: Examples from American Industry, 1870–1970," in *The Creative Enterprise: Managing Innovative Organizations and People*, ed. Tony Davila, Marc J. Epstein, and Robert Shelton (Westport, CT: Praeger Perspectives, 2007), 43–76, especially 55–56.

101. Quoted in Henk van den Belt and Arie Rip, "The Nelson-Winter-Dosi Model and Synthetic Dye Chemistry," in *The Social Construction of Technological Systems*, ed. Wiebe E. Bijker, Thomas P. Hughes, and Trevor Pinch (Cambridge, MA: MIT Press, 1987), 135–158, quotation at 155.

102. Hoover's radio address, October 21, 1929, printed as "Text of President Hoover's Address at the Dearborn Fete," *New York Times*, October 22, 1929, 2. Hoover, an engineer, was an ardent supporter of industrial research; see Herbert Hoover, "The Vital Need for Greater Financial Support of Pure Scientific Research," *Reprint and Circular Series of the National Research Council* 65 (December 1925).

103. Hoover, "Text of President Hoover's Address at the Dearborn Fete."

104. Edison was featured in several GE tungsten lamp ads in 1911–1912, including "Edison's Dream Come True," *Scientific American* 105, no. 12 (September 16,

1911): 242; "The Great Inventor's Prophetic Vision," *Scientific American* 105, no. 21 (November 18, 1911): 442; "The City of Edison-Mazda-Light," *Scientific American* 106, no. 1 (January 6, 1912): 4. On development of the tungsten lamp, see George Wise, "A New Role for Professional Scientists in Industry: Industrial Research at General Electric, 1900–1916," *Technology and Culture* 21, no. 3 (July 1980): 408–429.

105. Thomas P. Hughes, "Edison's Method," in *Technology at the Turning Point*, ed. William B. Pickett (San Francisco: San Francisco Press, 1977); Frank B. Jewett, "Edison's Contributions to Science and Industry," *Science*, n.s., 75 (January 15, 1932): 65–68, quotation at 66; Karl T. Compton, "Edison's Laboratory in War Time," *Science*, n.s., 75 (January 15, 1932): 70–71, quotation at 71.

106. Waldemar Kaempffert, "Titan of the Heroic Age of Invention," *New York Times Magazine*, October 25, 1931, 22.

107. From the *Cedar Rapids Gazette*, March 12, 1966, as quoted in Samuel Ruben, *Necessity's Children: Memoirs of an Independent Inventor* (Portland, OR: Breitenbush Books, 1990), xii, emphasis in original.

Chapter 3

1. Eric S. Hintz, "'Heroes of the Laboratory and the Workshop': Invention and Technology in Books for Children, 1850–1900," in *Enterprising Youth: Social Values and Acculturation in Nineteenth-Century American Children's Literature*, ed. Monika Elbert (New York: Routledge, 2008), 197–211.

2. Munn & Co., *Patents: A Book of Facts Every Inventor Should Know* (New York: Munn & Co., 1927), 9; "Prices and Wages—Producer and Consumer Price Indexes, and Weekly Manufacturing Earnings: 1919–1939," in *Historical Statistics of the United States: Earliest Times to the Present*, millennial ed., ed. Susan Carter et al. (New York: Cambridge University Press, 2006), 3:141, table Cb71–76.

3. On inventor-entrepreneurs, see Thomas P. Hughes, "Inventors: The Problems They Choose, the Ideas They Have, and the Inventions They Make," in *Technological Innovation: A Critical Review of Current Knowledge*, ed. Patrick Kelly and Melvin Kranzberg (San Francisco: San Francisco Press, 1978), 166–182; Thomas P. Hughes, *American Genesis: A Century of Invention and Technological Enthusiasm, 1870–1970*, 2nd ed. (Chicago: University of Chicago Press, 2004), 13–95. On corporate R&D labs, see W. Bernard Carlson, "Innovation and the Modern Corporation: From Heroic Inventor to Industrial Science," in *Science in the Twentieth Century*, ed. John Krige and Dominique Pestre (Amsterdam: Harwood, 1997), 203–226; David A. Hounshell, "The Evolution of Industrial Research in the United States," in *Engines of Innovation: U.S. Industrial Research at the End of an Era*, ed. Richard S. Rosenbloom and William J. Spencer (Boston: Harvard Business School Press, 1996), 13–85.

4. Naomi R. Lamoreaux and Kenneth L. Sokoloff, "Inventors, Firms, and the Market for Technology in the Late Nineteenth and Early Twentieth Century," in *Learning by Doing in Markets, Firms, and Countries*, ed. Naomi R. Lamoreaux, Daniel M. G. Raff, and Peter Temin (Chicago: University of Chicago Press, 1999); Naomi R. Lamoreaux, Kenneth L. Sokoloff, and Dhanoos Sutthiphisal, "Patent Alchemy: The Market for Technology in US History," *Business History Review* 87, no. 1 (Spring 2013): 3–38.

5. Martha Davidson, "A Flexible Mind," *American Heritage of Invention and Technology* 21, no. 3 (Winter 2006): 55–56. On Friedman's inventions, see several boxes and folders in Series 2: "Invention & Patent Materials, 1915–1967," Joseph B. Friedman Papers, Archives Center, National Museum of American History, Smithsonian Institution, Washington, DC (hereafter Friedman Papers).

6. On Friedman's overlapping careers, see Davidson, "A Flexible Mind," 56. See also his multiple business cards in Box 6, Folder 3, Friedman Papers.

7. Davidson, "A Flexible Mind," 55; Joseph Friedman, "Drinking Tube," US Patent 2,094,268, filed November 7, 1936, issued September 28, 1937.

8. Derek Thompson, "The Amazing History and the Strange Invention of the Bendy Straw," *The Atlantic*, November 22, 2011, https://www.theatlantic.com/business/archive/2011/11/the-amazing-history-and-the-strange-invention-of-the-bendy-straw/248923/; Stone Straw Limited, "History," accessed October 15, 2020, http://stonestraw.com/history/.

9. Irwin Geiger to Joseph B. Friedman, May 3, 1938, Box 3, Folder 15, Friedman Papers. On the role of intermediaries, especially attorneys, in buying and selling patents, see Naomi R. Lamoreaux and Kenneth L. Sokoloff, "Intermediaries in the U.S. Market for Technology, 1870–1920," in *Finance, Intermediaries, and Economic Development*, ed. Stanley L. Engerman, Phillip T. Hoffman, Jean-Laurent Rosenthal, and Kenneth L. Sokoloff (New York: Cambridge University Press, 2003).

10. See the following correspondence in Friedman Papers: Box 4, Folder 10 (pencilite); Box 3, Folder 9 (ice cream dispensing device); Box 3, Folder 12 (portable screen); Box 3, Folder 6 (safety razor handle); Box 3, Folder 18 (gas range safety device); Box 3, Folder 2 (fountain pen).

11. O. W. Dieffenbach to J. B. Friedman, June 14, 1938, Box 3, Folder 15, Friedman Papers.

12. In Box 3, Folder 15, Friedman Papers, see Irwin Geiger to Joseph B. Friedman, May 10, 1938; Geiger to Friedman, May 26, 1938; Geiger to Friedman, June 1, 1938.

13. Joseph Friedman to Harry Zavin, November 3, 1938, Box 8, Folder 18, Friedman Papers.

14. Michael Masterson defines a reluctant entrepreneur as someone who has "the initiative to start his own business" but is initially unwilling "to quit his current job

and lose the income." See Michael Masterson, *The Reluctant Entrepreneur: Turning Dreams into Profits* (Hoboken, NJ: Wiley, 2012), 12.

15. See several letters between Friedman, Light, and Zavin, especially Joseph Friedman to David Light, December 6, 1938, Box 8, Folder 18, Friedman Papers, and their "Agreement," December 22, 1938, Box 12, Folder 14, Friedman Papers.

16. On the expended funds, see Joseph Friedman to David Light and Harry Zavin, February 22, 1940, Box 8, Folder 20, Friedman Papers. For quotations on real estate and manufacturing, see Joseph Friedman to Regina and David Light, July 16, 1941, Box 8, Folder 21, Friedman Papers. On wartime restrictions, see Joseph Friedman to Betty Friedman, July 21, 1945, Box 9, Folder 3, Friedman Papers.

17. Joseph Friedman to Betty Friedman, July 21, 1945, Box 9, Folder 3, Friedman Papers. See also Sampson Miller, attorney for Joseph Friedman to David Fain, attorney for Zavin and Light, November 24, 1948, Box 12, Folder 18, Friedman Papers.

18. For the improved straw patent, see Joseph B. Friedman, "Flexible Drinking Straw," US Patent 2,550,797, filed June 5, 1948, issued May 1, 1951. For the machinery patent, see Joseph B. Friedman, "Apparatus and Method for Forming Corrugations in Tubing," US Patent 2,631,645, filed September 20, 1948, issued March 17, 1953.

19. For daily production figures in September and October 1947, see Joseph Friedman's ledger, Box 10, Folder 1, Friedman Papers. On the first Flex-Straw sale (October 15, 1947), see Friedman's affidavit for Flex-Straw's patent infringement dispute with the Save-T Pacific Baking Company, n.d. (circa 1958–1960), Box 13, Folder 12, Friedman Papers. On the advantages for hospitals, see "Flexible Drinking Tube Aids Both Patients and Staff," *Hospital Merchandiser*, December 1953, in Box 13, Folder 13, Friedman Papers.

20. Joseph Friedman to Betty Friedman, August 23, 1949, Box 9, Folder 10, Friedman Papers.

21. In Box 10, Folder 7, Friedman Papers, see Flex-Straw Company, "Statement of Operations," December 31, 1953, reflecting net profits of $109,813.11; Flexible Straw Corporation, "Dividend Distribution," 1953.

22. On the role of Friedman's wife, Marjorie, see Joseph Friedman to Betty Friedman, July 18, 1948, Box 9, Folder 7, Friedman Papers. On Friedman's father, see "Dad" [Jacob David Friedman] to Joseph Friedman, February 15, 1952, Box 8, Folder 9, Friedman Papers. On daughter Judith, see Joseph Friedman to Rose Zavin, February 4, 1954, Box 8, Folder 22, Friedman Papers. On "touchiness" with Betty, see Betty Friedman to Joseph Friedman, January 24, 1949, Box 9, Folder 8, Friedman Papers. See also Andrea Colli, *The History of Family Business, 1850–2000* (Cambridge: Cambridge University Press, 2003).

23. On Betty's role, see Gay Harper, "Female Executive Says Understanding Boss Is Big Asset to Secretary," newspaper clipping, circa 1960s, in Box 13, Folder 13, Friedman Papers.

24. Quotation from Betty Friedman to Joseph Friedman, December 11, 1950, Box 9, Folder 13, Friedman Papers. See also Betty Friedman to Joseph Friedman, February 16, 1951, Box 9, Folder 14, Friedman Papers; Joseph Friedman to Betty Friedman, November 8, 1953, Box 9, Folder 15, Friedman Papers.

25. Davidson, "A Flexible Mind," 56. See a list of Friedman's patents and expiration dates, and two letters—both from David Gerber to Betty Friedman, May 2, 1969— which describe the terms of sale, all in Box 13, Folder 7, Friedman Papers.

26. W. Bernard Carlson, "At Arm's Length or Close to the Vest? A Historical Look at How Companies Locate Technological Innovation" (unpublished paper presented at the International Economic History Association Congress, Helsinki, August 2006).

27. Lamoreaux and Sokoloff, "Inventors, Firms, and the Market for Technology"; Lamoreaux, Sokoloff, and Sutthiphisal, "Patent Alchemy"; Tom Nicholas, "The Role of Independent Invention in U.S. Technological Development, 1880–1930," *Journal of Economic History* 70, no. 1 (March 2010): 57–82. On half of inventors as sellers, see B. Zorina Khan and Kenneth L. Sokoloff, "Institutions and Technological Innovation during Early Economic Growth: Evidence from the Great Inventors of the United States, 1790–1930" (NBER Working Paper no. 10966, December 2004), 21–22. On the 20–25 percent acquisition figure, see Tom Nicholas, "Spatial Diversity in Invention: Evidence from the Early R&D Labs," *Journal of Economic Geography* 9, no. 1 (January 2009): 1–31.

28. Joseph B. Friedman, "Fountain Pen," US Patent 1,412,930, filed February 15, 1921, issued April 18, 1922; Joseph B. Friedman to Mr. A. French, Eisenstadt Mfg. Co., December 12, 1925, Box 3, Folder 3, Friedman Papers; quotations from Friedman to W. A. Sheaffer, W. A. Sheaffer Pen Co., December 15, 1925, Box 3, Folder 3, Friedman Papers.

29. Quotations from Joseph B. Friedman to Mr. W. Frank Wallace, April 3, 1933, Box 3, Folder 3, Friedman Papers. See also J. Wallace (Frank's brother) to Friedman, April 21, 1933, Box 3, Folder 3, Friedman Papers.

30. Firms developed "capabilities that enabled them to learn about and assess externally generated inventions." Individuals likewise "scanned patent lists in search of developments in their fields." See Lamoreaux and Sokoloff, "Inventors, Firms, and the Market for Technology," quotations at 20, 23.

31. See annotated clipping from *Thomas' Register*, December 1933, an ad for the Sheaffer Visulated Pen, date unknown, in Box 3, Folder 3, and additional pen ads in Oversized Box 16, Folder 2, all in Friedman Papers.

32. See, for example, Joseph B. Friedman to The Wahl Company, October 19, 1934, and the other letters in Box 3, Folder 3, Friedman Papers.

33. On Wahl, in Folder 3, Box 3, Friedman Papers, see C. J. Frechette to Joseph B. Friedman, November 5, 1934; Frechette to Friedman, November 21, 1934; Friedman

to Frechette, November 30, 1934. On Sheaffer, see in same location L. J. Frantz to Friedman, November 28, 1934, and Friedman to Frantz, December 10, 1934.

34. In Box 3, Folder 3, Friedman Papers, see Joseph B. Friedman to C. J. Frechette (Wahl Co.), November 30, 1934; Friedman to L. J. Frantz (Sheaffer Pen Co.), December 10, 1934.

35. In Box 3, Folder 3, Friedman Papers, see telegrams from Marjorie Friedman to Joseph B. Friedman, December 24, 1934; Joseph Friedman to Marjorie Friedman, December 24, 1934.

36. See assignment contract between Friedman and W. A. Sheaffer Pen Co., and a note from President W. A. Sheaffer, both dated December 18, 1934. On Sheaffer's intent to determine Friedman's "date of conception and reduction to practice back as far as possible" to establish "the strength of the patent," see L. J. Frantz to Joseph B. Friedman, April 13, 1936, and Frantz to Friedman, May 4, 1936. All documents are in Box 3, Folder 2, Friedman Papers.

37. Friedman-Sheaffer assignment contract and note from President W. A. Sheaffer, December 18, 1934, Box 3, Folder 2, Friedman Papers.

38. See Box 1, Folder 1 ("Misc. Biographical Data") and Folder 3 ("Personal Write-Ups, 1942–1968"), Wadsworth W. Mount Papers, 1920–1984, Archives Center, National Museum of American History, Smithsonian Institution, Washington, DC (hereafter Mount Papers).

39. In Box 6, Folder 7, Mount Papers, see W. K. Olson to Wadsworth W. Mount, July 9, 1958; Mount to Olson, July 13, 1958; Olson to Mount, July 24, 1958; W. A. Sheaffer Pen Company, "If You Have an Idea You Wish to Submit to the W.A. Sheaffer Pen Company," n.d. (circa 1958).

40. Wadsworth W. Mount to W. K. Olson, July 28, 1958, Box 6, Folder 7, Mount Papers.

41. In Box 6, Folder 7, Mount Papers, see W. K. Olson to Wadsworth W. Mount, August 13, 1958; Mount to Olson, September 7, 1958; Olson to Mount, January 6, 1960.

42. Henry Chesbrough, *Open Innovation: The New Imperative for Creating and Profiting from Technology* (Boston: Harvard Business School Press, 2006), 30.

43. Lee Vinsel, *Moving Violations: Automobiles, Experts, and Regulations in the United States* (Baltimore: Johns Hopkins University Press, 2019). On "consumer-inventors," see Kathleen Franz, *Tinkering: Consumers Reinvent the Early Automobile* (Baltimore: Johns Hopkins University Press, 2005), especially 74–102. On the volume of letters to Ford, see Reynold W. Wik, *Henry Ford and Grassroots America: A Fascinating Account of the Model-T Era* (Ann Arbor: University of Michigan Press, 1973), 59.

44. "Hearing Inventors: General Motors Has a New Devices Section to Look Over All Inventions Submitted by Outsiders. Few Prove Helpful," *Business Week*, June 4, 1949, 24.

45. In Box 2, Folder 4, Mount Papers, see Wadsworth W. Mount to M. B. Malden, February 28, 1955, with executed "Suggestion Submission Agreement"; J. R. Lemon to Wadsworth W. Mount, March 14, 1955.

46. In Box 2, Folder 4, Mount Papers, see Wadsworth W. Mount to J. R. Lemon, April 5, 1955; Form 432-2, "Policy of the Chrysler Corporation in Regard to New Ideas," revised December 1952; George Melloan, "Public's Ideas for Auto Inventions Hit Makers at a Record Rate," *Wall Street Journal*, April 12, 1956, 1.

47. Carlson, "At Arm's Length or Close to the Vest?"

48. David C. Mowery, "The Boundaries of the U.S. Firm in R&D," in *Coordination and Information: Historical Perspectives on the Organization of Enterprise*, ed. Naomi R. Lamoreaux and Daniel M. G. Raff (Chicago: University of Chicago Press, 1995), especially 149–152.

49. On the contract R&D model, see David C. Mowery, "The Relationship between Intrafirm and Contractual Forms of Industrial Research in American Manufacturing, 1900–1940," *Explorations in Economic History* 20, no. 4 (October 1983): 351–374; Stephen R. Barley and Gideon Kunda, *Gurus, Hired Guns, and Warm Bodies: Itinerant Experts in a Knowledge Economy* (Princeton, NJ: Princeton University Press, 2004); Philip Mirowski and Robert Van Horn, "The Contract Research Organization and the Commercialization of Scientific Research," *Social Studies of Science* 35, no. 4 (August 2005): 503–548.

50. Mount to Olson, July 13, 1958, Box 6, Folder 7, Mount Papers. For Mount's clients, see in Mount Papers the contracts in Box 2, Folder 19, "Contracts and Agreements (1 of 2), 1942–1976," and Box 3, Folder 1, "Contracts and Agreements (2 of 2), 1942–1976."

51. Wadsworth W. Mount, "Ordnance," US Patent 2,399,136, filed February 17, 1943, issued April 23, 1946.

52. Mount eventually submitted his antiaircraft idea to the NIC, but the council deemed the idea of insufficient "military value . . . to warrant further investigation." See J. C. Green to Wadsworth W. Mount, August 23, 1943, Box 3, Folder 4, Mount Papers.

53. Becker's name is on the letterhead of Mount's job offer letter from George H. McCaffrey, Merchants' Association of New York (later renamed the Commerce and Industry Association of New York), March 24, 1938, Box 1, Folder 2, Mount Papers. On Intertype, see MacD. Sinclair, ed., *Intertype: Its Function, Care, Operation, and Adjustment* (Brooklyn, NY: Intertype Corporation, 1929).

54. For the agreement, see Neal Dow Becker to Wadsworth W. Mount, January 6, 1942, Box 2, Folder 19, Mount Papers.

55. Becker to Mount, January 6, 1942, Box 2, Folder 19, Mount Papers. On contracts for employee-inventors, see Catherine L. Fisk, *Working Knowledge: Employee*

Innovation and the Rise of Corporate Intellectual Property, 1800–1930 (Chapel Hill: University of North Carolina Press, 2009), 75–172.

56. Becker to Mount, January 6, 1942, Box 2, Folder 19, Mount Papers.

57. In Box 2, Folder 19, Mount Papers, see Becker to Mount, January 6, 1942; Becker to Mount, February 20, 1942.

58. In Box 1, Folder 4, Mount Papers, see Mount's status report letters to the Summit, New Jersey, draft board dated August 4, 1942, December 14, 1942, April 11, 1943, and May 18, 1943.

59. In Box 1, Folder 4, Mount Papers, see Mount to Draft Board, August 20, 1943; A. T. Mann to Draft Board, June 3, 1943.

60. See Mount's Selective Service draft card with "2B" classification (deferred for wartime production work), September 13, 1944, Box 1, Folder 4, Mount Papers. On the orders and award, see Mount, résumé, 1968, Box 1, Folder 2, Mount Papers. For the patents, see Wadsworth W. Mount, "Single Cable Advancing Mechanism and Method," US Patent 2,490,378, filed June 29, 1944, issued December 6, 1949; Wadsworth W. Mount, "Projectile," US Patent 2,522,685, filed May 8, 1945, issued September 19, 1950.

61. Mount-Intertype, "Cancellation Agreement," February 15, 1955, Box 2, Folder 19, Mount Papers.

62. In Box 1, Folder 2, Mount Papers, see Wadsworth W. Mount to F. W. Roebling III, June 5, 1952; quotations from Mount to Alan W. Hastings, Rockefeller Brothers Incorporated, May 6, 1948; Wadsworth W. Mount, résumé, 1968.

63. Wadsworth W. Mount, sample "Agreement," September 7, 1958, Box 1, Folder 2, Mount Papers.

64. For the retainer agreement, see G. Fred Ashworth, Merriman Brothers, to Wadsworth W. Mount, August 8, 1950, Box 2, Folder 19, Mount Papers. For the sublicensing agreement, see Ashworth to Neal Dow Becker, Intertype Corporation, December 17, 1951, Box 2, Folder 19, Mount Papers. For the agreement's dissolution, see Jack E. Williams to Mount, November 17, 1969, Box 2, Folder 19, Mount Papers.

65. Raymond Concrete Pile Company contract, August 16, 1956, Box 3, Folder 1, Mount Papers. For the adopted and reassigned ideas, see in Box 5, Folder 18, Mount Papers: Lindsey J. Phares to Wadsworth W. Mount, March 9, 1962; Mount to Phares, April 4, 1962; Phares to Mount, April 9, 1962.

66. This section draws on Eric S. Hintz, "The Post-heroic Generation: American Independent Inventors, 1900–1950," *Enterprise and Society* 12, no. 4 (December 2011): 732–748, especially 741–742.

67. This section draws on Eric S. Hintz, "Portable Power: Inventor Samuel Ruben and the Birth of Duracell," *Technology and Culture* 50, no. 1 (January 2009): 24–57.

68. On Ruben, see his autobiography, Samuel Ruben, *Necessity's Children: Memoirs of an Independent Inventor* (Portland, OR: Breitenbush Books, 1990); Alfred Steinberg, "Sam Ruben: Born to Invent," *Reader's Digest* 90, no. 541 (1967): 155–160; Henry B. Linford, "Samuel Ruben—Acheson Medalist," *Journal of the Electrochemical Society* 118, no.1 (1971): 11C–13C.

69. For Mallory's corporate history, see Philip R. Mallory, *Recollections: Fifty Years with the Company* (Indianapolis: P. R. Mallory Company, 1966); Edmund E. Taylor, *Bits of Mallory History: 70 Years, 1916–1986* (Indianapolis: P. R. Mallory Company, 1986).

70. Samuel Ruben, US Patent 1,751,359, "Asymmetric Electric Couple," filed August 20, 1925, issued March 18, 1930. On Ruben and Mallory's first deal, see Mallory, *Recollections*, 138; Ruben, *Necessity's Children*, 37–39. On Ruben as an inventor-consultant, see Steinberg, "Sam Ruben: Born to Invent," 159.

71. See Ruben's correspondence (1937–1984) with dozens of firms in Box 9, Folders 1–3, in the Papers of Samuel Ruben, Othmer Library of Chemical History, Science History Institute, Philadelphia, PA (hereafter Ruben Papers). On Mallory and his other clients, see Ruben, *Necessity's Children*, 37–39, quotations at 63 and 39.

72. On Ware and Hooven, see Taylor, *Bits of Mallory History*, 8; Mallory, *Recollections*, 99, 165–166.

73. Mallory noted "non-productive" collaborations with inventors John Lewis Andrews, Madison Cawein, Duncan Cox, Giles S. Moore, W. J. Polydoroff, Caton Bradley, Kurt Schimkus, Charles W. Sidney, Walter J. Six, Harry Waters, James L. Yarian, Harold Brown, and Reginald Dean. See Mallory, *Recollections*, 166–167.

74. See chapter 6 and Hintz, "Portable Power."

75. Mallory, *Recollections*, 160, 164; quotations from P. R. Mallory & Co., Inc., "1962 Annual Report," n.p.

76. P. R. Mallory & Co., Inc., "Annual Reports," 1951 to 1974 inclusive; Mallory, *Recollections*, 161; Ruben, *Necessity's Children*, 110.

77. Mallory, *Recollections*, 142. Ruben also noted that the proximity of Mallory's Tarrytown lab "allowed us to have regular meetings" to develop new battery applications. See Ruben, *Necessity's Children*, 122.

78. Emanuel Piore, "The Independent Inventor in the Contemporary World," October 16, 1963, as quoted in Ruben, *Necessity's Children*, 141–142.

79. In 1983, at age 83, Ruben contracted with Duracell to investigate possible replacements for his alkaline batteries. See D. R. Riter to R. DiPalma, June 30, 1983, Box 9, Folder 3, Ruben Papers.

80. The "long-term relationship" represents a distinctive intermediate "coordination mechanism" in which two parties "voluntarily choose to continue dealing with

each other for significant periods of time." See Naomi R. Lamoreaux, Daniel M. G. Raff, and Peter Temin, "Beyond Markets and Hierarchies: Toward a New Synthesis of American Business History," *American Historical Review* 108, no. 2 (April 2003): 404–433, especially 405–407.

81. George Wise, "Inventors and Corporations in the Maturing Electrical Industry, 1890–1940," in *Inventive Minds: Creativity in Technology*, ed. Robert J. Weber and David N. Perkins (New York: Oxford University Press, 1992), 291–310, quotation at 302. In a notable exception, African American Lewis H. Latimer managed GE's patent pool after working with Thomas Edison. See Rayvon Fouché, *Black Inventors in the Age of Segregation: Granville T. Woods, Lewis H. Latimer, and Shelby J. Davidson* (Baltimore: Johns Hopkins University Press, 2003), 82–133.

82. David F. Noble, *America by Design: Science, Technology, and the Rise of Corporate Capitalism* (New York: Knopf, 1977), 97.

83. C. E. Kenneth Mees, *The Organization of Industrial Scientific Research* (New York: McGraw-Hill, 1920), 103.

84. On Brush and Tesla, see De Forest's diary entry, February 12, 1899, in Volume 13: November 1897–June 1899. Regarding his work at Western Electric, see entries from August 12, 1899, an undated entry circa September 1899, and October 15, 1899, in Volume 14: June 1899–November 1899, all in Box 3, Papers of Lee De Forest, Manuscript Division, Library of Congress, Washington, DC.

85. Lee De Forest, *The Father of Radio: The Autobiography of Lee De Forest* (Chicago: Wilcox and Follett, 1950), 99–122, quotations at 110, 112, 117.

86. Leonard S. Reich, *The Making of American Industrial Research: Science and Business at GE and Bell, 1876–1926* (Cambridge: Cambridge University Press, 1985), 148, 155–164, 218–238. Although AT&T researcher George Campbell had invented something similar, the firm paid Pupin $455,000 between 1900 and 1917 for his loading coil patents. See James E. Brittain, "The Introduction of the Loading Coil: George A. Campbell and Michael I. Pupin," *Technology and Culture* 11, no. 1 (January 1970): 36–57.

87. Everett H. Bickley, "The Bickley Bean Sorters," an unpublished autobiography, compiled November 18, 1998, by daughter Audrey Bickley Beyer; "1910 Graduate Successful Inventor," *Carnegie Alumnus*, May 1933, 5–6, both in Box 1, Folder 2, Everett H. Bickley Collection, Archives Center, National Museum of American History, Smithsonian Institution, Washington, DC (hereafter Bickley Collection). On the motograph, see Everett H. Bickley, "Controller for Electric Signs," US Patent 1,050,203, filed July 15, 1911, issued January 14, 1913.

88. Bickley, "The Bickley Bean Sorters," quotation at 27; Everett H. Bickley, "Sorting Machine," US Patent 1,921,863, filed June 25, 1929, issued August 8, 1933.

89. Bickley, "The Bickley Bean Sorters," 54–55.

90. On "Tupper's Greenhouses, 1930–1934," see Box 3, Folder 1, and on "Tupper Tree Doctors," see Box 3, Folders 2 and 3, both in the Earl S. Tupper Papers, Archives Center, National Museum of American History, Smithsonian Institution, Washington, DC (hereafter Tupper Papers).

91. For descriptions of sixty-seven ideas, see Tupper's "Notes on Inventions, 1930s," Box 3, Folder 4, Tupper Papers. On Tupper's attempts to commercialize the convertible rumble seat, see Franz, *Tinkering*, 103–129.

92. On Tupper's position at the Doyle Works, see Alison J. Clarke, *Tupperware: The Promise of Plastic in 1950s America* (Washington, DC: Smithsonian Institution Press, 1999), 26–35. The quotation is from Tupper's diary, July 30, 1937, Box 4, Folder 1, Tupper Papers.

93. Tupper's diary, August 2, 1937, Box 4, Folder 1, Tupper Papers.

94. Tupper's stationery, described in his diary, December 10, 1937, Box 4, Folder 1, Tupper Papers.

95. Clark, *Tupperware*, 1–35.

96. Entrepreneurs typically acquire skills by first working as employees. See Pino G. Audia and Christopher I. Rider, "Entrepreneurs as Organizational Products Revisited," in *The Psychology of Entrepreneurship*, ed. J. Robert Baum, Michael Frese, and Robert A. Baron (New York: Psychology Press, 2012), 113–130.

97. George Wise, "William Stanley's Search for Immortality," *American Heritage of Invention and Technology* 4, no. 1 (Spring/Summer 1988): 42–49, quotation at 43.

Chapter 4

1. G. W. Wishard, letter to the editor, June 5, 1907, published as "The Need of an Inventors' Aid Institution," *Scientific American* 97, no. 1 (July 6, 1907): 9.

2. "Traps for the Unwary Inventor," *Scientific American* 87, no. 18 (November 1, 1902): 294.

3. The exception is Peter Whalley, "The Social Practice of Independent Inventing," *Science, Technology, and Human Values* 16, no. 2 (Spring 1991): 208–232. There is a vast body of literature on the professionalization of American science and engineering. For historical overviews, see Daniel J. Kevles, "American Science," in *The Professions in American History*, ed. Nathan O. Hatch (Notre Dame, IN: University of Notre Dame Press, 1988), 107–125; Bruce Sinclair, "Episodes in the History of the American Engineering Profession," in Hatch, *The Professions in American History*, 127–144.

4. G. E. of Paterson, New Jersey, letter to the editor, "How Shall the Inventor Introduce His Invention?," *Popular Mechanics* 6, no. 10 (October 1904): 1068; Hugo Gernsback, "Patent Advice," *Electrical Experimenter* 8, no. 2 (June 1920): 211;

Frederick Benjamin, "Hints for Inventors," *Popular Mechanics* 6, no. 6 (June 1904): 666; advertisement for the "Popular Mechanics Patent Bureau," *Popular Mechanics* 7, no. 2 (December 1905): 1279.

5. I am grateful to Paul Israel for sharing his personal research file on these organizations.

6. This section draws on Eric S. Hintz, "Failed Inventor Initiatives, from the Franklin Institute to Quirky," in *Does America Need More Innovators?*, ed. Matthew Wisnioski, Eric S. Hintz, and Marie Stettler Kleine (Cambridge, MA: MIT Press, 2019), 165–189.

7. Bruce Sinclair, *Philadelphia's Philosopher Mechanics: A History of the Franklin Institute, 1824–1865* (Baltimore: Johns Hopkins University Press, 1974); A. Michal McMahon, "'Bright Science' and the Mechanic Arts: The Franklin Institute and Science in Industrial America, 1824–1876," *Pennsylvania History* 47 (1980): 351–368; Kershaw Burbank, "Noble Ambitions: The Founding of the Franklin Institute," *Pennsylvania Heritage* 18, no. 3 (Summer 1992): 32–37.

8. For the 1895 dues, see "Article IV: Payments," in *Charter and Bylaws of the Franklin Institute*, an appendix to William Wahl, *The Franklin Institute of the State of Pennsylvania for the Promotion of the Mechanic Arts: A Sketch of Its Organization and History* (Philadelphia: Franklin Institute, 1895), 67. A sum of $8 in 1895 is equivalent to $247 in 2018. For all present values of past dollar amounts, I use a purchasing power comparator based on historical changes to the consumer price index, which is current through 2018. See Samuel H. Williamson, "Seven Ways to Compute the Relative Value of a U.S. Dollar Amount, 1774 to Present," MeasuringWorth, 2019, https://www.measuringworth.com/calculators/uscompare/.

9. R. W. Lesley, "Report of the Committee on Election and Resignation of Members," January 14, 1925, in *The Franklin Institute Year Book 1924–1925*, 88–89, The Franklin Institute Archives, Philadelphia, PA (hereafter Franklin Institute Archives).

10. Wahl, *The Franklin Institute*, 19–22, 34–39, 90–91; Sydney Wright, *The Story of the Franklin Institute* (Lancaster, PA: Lancaster Press, 1938), 16–17.

11. Wahl, *The Franklin Institute*, 23–28; Evelyn S. Paniagua, "American Inventors' Debt to the Institute," *Journal of the Franklin Institute* 247, no. 1 (January 1949): 1–6; Joseph S. Hepburn, "The Library of the Franklin Institute," *Journal of the Franklin Institute* 269, no. 3 (March 1960): 221–228.

12. A. Michal McMahon and Stephanie A. Morris, *Technology in Industrial America: The Committee on Science and the Arts of the Franklin Institute, 1824–1900* (Wilmington, DE: Scholarly Resources, 1977), especially xiii–xxxix. In 1891, the CSA began charging a $5 submission fee to cover the cost of patent searches and return postage. See CSA chairman S. Lloyd Wiegand, "Annual Report of the Committee on Science and the Arts (for 1891)," January 4, 1892, reprinted in *Journal of the Franklin Institute* 133, no. 2 (February 1892): 155–157.

13. Wahl, *The Franklin Institute*, 29; William A. McGarry, "First Aid for Inventors: The Franklin Institute and the Committee through which It Assists Meritorious Patents," *Scientific American* 123, no. 20 (November 13, 1920): 502.

14. Wahl, *The Franklin Institute*, 40–52; Wright, *The Story of the Franklin Institute*, 21–28, 49–51; Thomas Coulson, *A Short History of the Franklin Institute* (Lancaster, PA: Lancaster Press, 1957), 6–7; Petra Moser and Tom Nicholas, "Prizes, Publicity, and Patents: Non-monetary Awards as a Mechanism to Encourage Innovation," *Journal of Industrial Economics* 61, no. 3 (September 2013): 763–788.

15. For exhibitor and attendance figures, see Wahl, *The Franklin Institute*, 40–52. For the inventors at the 1884 exhibition, see *Official Catalogue of the International Electrical Exhibition*, 1884, Franklin Institute Archives.

16. W. Bernard Carlson, *Innovation as a Social Process: Elihu Thomson and the Rise of General Electric, 1870–1900* (Cambridge: Cambridge University Press, 1991); David B. Sicilia, "Thomson, Elihu," *American National Biography Online*, February 2000, https://doi.org/10.1093/anb/9780198606697.article.1001641.

17. The address, dated September 17, 1924, was published as Elihu Thomson, "One Hundred Years of the Franklin Institute," *Science* 60, no. 1555 (October 17, 1924): 343–351, quotations at 348, 344. See also Carlson, *Innovation as a Social Process*, 16–145.

18. On Banes, see McMahon and Morris, *Technology in Industrial America*, xxvii–xxx. The "patented only" policy changed sometime between 1909 and 1920. For example, compare "Regulations of the Committee on Science and the Arts of the Franklin Institute," adopted December 15, 1909, in *Journal of the Franklin Institute* 170, no. 2 (August 1910): 142–150, and McGarry, "First Aid for Inventors," 502.

19. *Regulations for the Government of the Committee on Science and the Arts*, 1893, 18–30, and *Awards of the Franklin Institute*, 1958, both in Franklin Institute Archives; "Regulations of the Committee on Science and the Arts of the Franklin Institute [1909]," Article III, Section 4.

20. CSA Case no. 2,141, Elihu Thomson, "Constant-Current Arc Light Transformer," 1901, and CSA Case no. 2,577, Elihu Thomson, "Industrial Applications of Electricity," 1912, both in Franklin Institute Archives.

21. "Every year [the Institute's] accounts present a deficiency. This cannot go on forever." See "Annual Report of Board of Managers for the Year 1887," *Journal of the Franklin Institute* 125, no. 2 (February 1888): 169–170. On the 1884 and 1885 deficits, see Jane Mork Gibson, "The International Electrical Exhibition of 1884: A Landmark for the Electrical Engineer," *IEEE Transactions of Education* E-20, no. 3 (August 1980): 169–176 at 173; "Proceedings of the Annual Meeting, held Wednesday, January 20, 1886," *Journal of the Franklin Institute* 121, no. 2 (February 1886): 156–159.

22. Wahl, *The Franklin Institute*, 51–52.

23. Michelle Tucker, "A Partnership for Public Science Education: Reinventing the Franklin Institute, 1925–1934," *Penn History Review* 8, no. 1 (Spring 2000): 9–40, quotation at 10; Wright, *The Story of the Franklin Institute*, 54, 69, 104.

24. Samuel L. Vauclain, "Personal Experiences in the Franklin Institute," November 23, 1933, reprinted in "The Historic Franklin Institute," *Journal of the Franklin Institute* 217, no. 1 (January 1934): 131–133.

25. National Institute of Inventors, "1919–1920 Year Book," 6, in Box 49, Folder 15, Wilbur and Orville Wright Papers, Manuscript Division, Library of Congress, Washington, DC (hereafter NII-LOC).

26. On Howard, see "Obituary Notes," *New York Times*, June 11, 1922, 25; "Dead Realty Man Declared a Felon," *New York Times*, October 18, 1922, 21; quotation from National Institute of Inventors, "Constitution and Bylaws of the National Institute of Inventors," 1917, 2, NII-LOC.

27. The $25 and $5 fees in 1917 are equivalent to $490 and $98 in 2018 dollars. For quotations regarding fees and services, see NII, "Constitution and Bylaws," 1917, 2–3, 10–16, NII-LOC. On the membership numbers, see NII, "1919–1920 Year Book," 23, NII-LOC. For the officers and honorary members, see George Veritzan to Benno Wieler, April 15, 1918, NII-LOC.

28. Thomas Howard to Orville Wright, April 23, 1918, with quotations from an enclosed flyer, "Somewhere in America," 1918, NII-LOC. For the $100 solicitation, see J. L. Stenquist to NII members, May 18, 1918, NII-LOC. The NII also published a fifty-page pamphlet, Ernest Gilmore Gardner, *An Outline of Plans for the Establishment and Operation of the National Laboratory for Invention and Research* (New York: National Institute of Inventors, 1918).

29. Orville Wright to Thomas Howard, October 12, 1917, NII-LOC; Omar M. Highley to Orville Wright, May 6, 1918, NII-LOC; Wright to Highley, May 7, 1918, NII-LOC.

30. Charles A. Munn to Orville Wright, May 28, 1918, NII-LOC. Thomas Edison had in fact proposed a federally funded Naval Research Laboratory in 1915; see chapter 6.

31. Samuel Biddison to Orville Wright, June 2, 1918, with pamphlet describing Biddison's "Automatic Gas Generator and Burner," NII-LOC.

32. Orville Wright to Thomas Howard, July 13, 1918, NII-LOC; Wright to P. M. Beard, July 13, 1918, NII-LOC; Wright to Charles A. Munn, July 13, 1918, NII-LOC; Munn to Wright, July 16, 1918, NII-LOC.

33. Nathan Noile to Orville Wright, November 21, 1918, NII-LOC; Wright to Noile, November 29, 1918, NII-LOC; statements by Thomas Howard in NII, "1919–1920 Year Book," 5, 19–20, NII-LOC.

34. John Johnston to F. C. Blake, February 5, 1919, in Folder "National Institute of Inventors, 1919–1920," Central Policy Files, National Research Council, National Academy of Sciences Archives, Washington, DC (hereafter NII-NAS).

35. Proudfoot's Commercial Agency, two reports dated October 6, 1917, and January 5, 1920, with quotations from the latter, both in NII-NAS.

36. "Obituary Notes"; "Dead Realty Man Declared a Felon"; postal inspectors R. P. Allen and F. D. Doyle to Orville Wright, January 14, 1925, NII-LOC.

37. Quotations from Edward W. Byrn, "The Progress of Invention during the Last Fifty Years," *Scientific American* 75, no. 4 (July 25, 1896): 82–83; R. H. Thurston, "The Care of the Patent Office," *Scientific American Supplement*, no. 842 (February 20, 1892): 13453–13454. For additional context, see "Congress and the Patent Office," *Scientific American* 51, no. 4 (July 26, 1884): 48; "Delays in the Patent Office," *Scientific American* 51, no. 18 (November 1, 1884): 282; "Shameful Treatment of Inventors," *Scientific American* 51, no. 24 (December 13, 1884): 396.

38. "Protects Inventors from Patent Thieves," *Popular Mechanics* 2, no. 14 (October 4, 1902): 620.

39. Executive Committee of the Patent Centennial Celebration, *Proceedings and Addresses: Celebration of the Beginning of the Second Century of the American Patent System at Washington City, D.C., April 8, 9, 10, 1891* (Washington, DC: Gedney and Roberts, 1892), 3–5.

40. J. E. Watkins, circular letter, February 16, 1891, reprinted in *Proceedings and Addresses: Celebration of the Beginning of the Second Century of the American Patent System*, 5–7.

41. *Proceedings and Addresses: Celebration of the Beginning of the Second Century of the American Patent System*, 21–38, 488–497; "The Patent Centennial Celebration—April 8, 9, and 10," *Scientific American* 64, no. 16 (April 18, 1891): 243–245.

42. *Proceedings and Addresses: Celebration of the Beginning of the Second Century of the American Patent System*, 37–38.

43. The $5 annual dues in 1891 are equivalent to $142 in 2018. See "Constitution of the American Association of Inventors and Manufacturers," 1891, in Series III, Box 1, Folder "American Association of Inventors and Manufacturers—Constitution," Elihu Thomson Papers, American Philosophical Society, Philadelphia, PA (hereafter Thomson Papers). On new members, see George C. Maynard, AAIM circular, July 8, 1892, in Series I, Box 1, Folder "American Association of Inventors and Manufacturers," Thomson Papers.

44. George C. Maynard, AAIM circular, April 5, 1892; quotations from J. E. Watkins, AAIM circular, January 1892, both in Series I, Box 1, Folder "American Association of Inventors and Manufacturers," Thomson Papers. See also "The American Association of Inventors and Manufacturers," *Science* 19, no. 469 (January 29, 1892): 57–58.

45. Report from the AAIM Committee on Legislation, July 8, 1892, Series I, Box 1, Folder "American Association of Inventors and Manufacturers," Thomson Papers,

emphasis in the original; "Congress on Patents, Trade Marks, and Inventions," *Scientific American* 69, no. 16 (October 14, 1893): 242.

46. Steuart to Thomson, February 15, 1899, Series I, Box 1, Folder "American Association of Inventors and Manufacturers," Thomson Papers.

47. In Series I, Box 1, Folder: "American Association of Inventors and Manufacturers," Thomson Papers, see Thomson to Steuart, January 11, 1900; Steuart to Thomson, January 19, 1900; Steuart to Thomson, March 11, 1902; Thomson to Steuart, March 20, 1902.

48. For the officers, see AAIM letterhead, Steuart to Thomson, March 11, 1902, Series I, Box 1, Folder "American Association of Inventors and Manufacturers," Thomson Papers. The AAIM's final press coverage reported on the 1898 annual meeting; see "American Association of Inventors and Manufacturers," *Scientific American* 78, no. 5 (January 29, 1898): 67.

49. The "influence" quotation is from "Notes Relevant to the Inventors Guild," *Science*, n.s., 32, no. 835 (December 30, 1910): 948. All other quotations and details are in "Inventors to Back Patent Reforms," *New York Times*, December 13, 1910, 11.

50. The recommendations were presented to the Inventors' Guild in November 1910 and subsequently reprinted in three parts. See F. L. O. Wadsworth, "Preliminary Report to the Inventors' Guild—I," *Scientific American Supplement*, no. 1940 (March 8, 1913): 146–147, quotation at 146; F. L. O. Wadsworth, "Preliminary Report to the Inventors' Guild—II," *Scientific American Supplement*, no. 1941 (March 15, 1913): 166; F. L. O. Wadsworth, "Preliminary Report to the Inventors' Guild—III," *Scientific American Supplement*, no. 1942 (March 22, 1913): 178–179.

51. "The Patent Elephant," *New York Times*, February 27, 1912, 8. On defensive patenting, see Leonard Reich, *The Making of American Industrial Research: Science and Business at GE and Bell, 1876–1926* (Cambridge: Cambridge University Press, 1985), 5–6.

52. Bauer & Cie v. O'Donnell, 229 U.S. 1 (1913); Sydney Henry et al. v. A. B. Dick Company, 224 U.S. 1 (1912); "The Rotary Mimeograph Case," *Scientific American* 106, no. 12 (March 23, 1912): 262.

53. Quotation in "Patents," *New York Times*, March 12, 1912, 12; "Topics of the Times: Inventors Have a Grievance," *New York Times*, February 16, 1912, 8.

54. The bill is summarized in Patent Commissioner Edward Moore to William A. Oldfield, April 10, 1912, reprinted in House Committee on Patents, *Oldfield Revision and Codification of the Patent Statutes*, H.R. 23417, 62nd Cong., 2nd Sess., Hearing no. 1, April 17, 1912, 3–13, quotation at 3. For a summary, see "Would Put Patents in the Trust Class," *New York Times*, July 19, 1912, 5.

55. Joris Mercelis, *Beyond Bakelite: Leo Baekeland and the Business of Science and Invention* (Cambridge, MA: MIT Press, 2020).

56. Baekeland testimony, April 26, 1912, House Committee on Patents, *Oldfield Revision and Codification*, Hearing no. 4, 28–29.

57. On the nineteen-year rule, see Moore to Oldfield, April 10, 1912, in House Committee on Patents, *Oldfield Revision and Codification*, 8. For the testimony, see Leonard, April 25, 1912, in House Committee on Patents, *Oldfield Revision and Codification*, Hearing no. 3, 18–20; Baekeland, April 26, 1912, in House Committee on Patents, *Oldfield Revision and Codification*, Hearing no. 4, 39. The USPTO converted to a first-to-file system in 2011; see chapter 7.

58. Thomas Edison to William A. Oldfield, May 17, 1912, reprinted in House Committee on Patents, *Oldfield Revision and Codification*, Hearing no. 23, May 21, 1912, 34; Leonard's testimony, April 25, 1912, in House Committee on Patents, *Oldfield Revision and Codification*, Hearing no. 3, 21–22.

59. See Leonard, April 25, 1912, in House Committee on Patents, *Oldfield Revision and Codification*, Hearing no. 3, 25; Wadsworth, May 18, 1912, in House Committee on Patents, *Oldfield Revision and Codification*, Hearing no. 21, 22–23; Baekeland, April 26, 1912, in House Committee on Patents, *Oldfield Revision and Codification*, Hearing no. 4, 42.

60. William A. Oldfield, in House Committee on Patents, *Revision of Patent Laws*, 62nd Cong., 2nd Sess., August 8, 1912, House Report no. 1161, accompanying H.R. 23417; Otto R. Barnett, "The Oldfield Bill (H. R. 23417)," *Yale Law Review* 22, no. 5 (March 1913): 383–397, quotations at 384–385, 397.

61. Leo H. Baekeland, AIChE presidential address, December 1912, reprinted as Leo H. Baekeland, "Protection of Intellectual Property in Relation to Chemical Industry," *Journal of Industrial and Engineering Chemistry* 5, no. 1 (January 1913): 51–57.

62. President's Commission on Economy and Efficiency, *Report of the Investigation of the United States Patent Office*, Doc. 1110, House of Representatives, 62nd Cong., 3rd Sess., December 9, 1912; Leo H. Baekeland, "Comments on the Report of the Investigation of the United States Patent Office," *Journal of Industrial and Engineering Chemistry* 5, no. 5 (May 1913): 417–421; "The Inventors' Guild and the Oldfield Bill," *Scientific American* 109, no. 7 (August 16, 1913): 125. On the demise of the Oldfield bill, see David F. Noble, *America by Design: Science, Technology, and the Rise of Corporate Capitalism* (New York: Knopf, 1977), 106.

63. Matilda Joslyn Gage, "Woman as an Inventor," *North American Review* 136, no. 318 (May 1883): 478–489, quotations at 478, 489.

64. Quotations from Ida M. Tarbell, "Women as Inventors," *The Chautauquan* 7, no. 6 (March 1887): 355–357. On married property laws, see Gage, "Woman as an Inventor," 488; B. Zorina Khan, "Married Women's Property Laws and Female Commercial Activity: Evidence from United States Patent Records, 1790–1895," *Journal of Economic History* 56, no. 2 (June 1996): 356–388.

65. Autumn Stanley, *Raising More Hell and Fewer Dahlias: The Public Life of Charlotte Smith, 1840–1917* (Bethlehem, PA: Lehigh University Press, 2009); Autumn Stanley, "The Champion of Women Inventors," *American Heritage of Invention and Technology* 8, no. 1 (Summer 1992): 22–26; Anne L. Macdonald, *Feminine Ingenuity: Women and Invention in America* (New York: Ballantine Books, 1992), 141–145.

66. Charlotte Smith, "Why I Became Interested in Woman Inventors," *The Woman Inventor* 1, no. 1 (April 1891): 2.

67. Smith, "Why I Became Interested in Woman Inventors," 2. The Patent Office had already circulated a few partial lists, including a roster of women patentees invited to be exhibitors at Philadelphia's 1876 Centennial Exhibition. See Macdonald, *Feminine Ingenuity*, 118–121; Deborah J. Warner, "Women Inventors at the Centennial," in *Dynamos and Virgins Revisited: Women and Technological Change in History*, ed. Martha Moore Trescott (Metuchen, NJ: Scarecrow Press, 1979), 102–119.

68. United States Patent Office, *Women Inventors to Whom Patents Have Been Granted by the United States Government, 1790 to July 1, 1888* (Washington, DC: Government Printing Office, 1888). On the list's methodology and inaccuracies, see Macdonald, *Feminine Ingenuity*, 141; Autumn Stanley, *Mothers and Daughters of Invention: Notes for a Revised History of Technology* (Metuchen, NJ: Scarecrow Press, 1993), 747–758.

69. Charlotte Smith, "Greeting," *The Woman Inventor* 1, no. 1 (April 1891): 1. Life imitated art. During the centennial's final session, Smith secured a resolution inviting women inventors to join the newly formed American Association of Inventors and Manufacturers. See Macdonald, *Feminine Ingenuity*, 171.

70. Charlotte Smith, "Recent Patents to Women," *The Woman Inventor* 1, no. 1 (April 1891): 2; Charlotte Smith, "History of Patents Granted to Women in the United States from 1809 to 1891," *The Woman Inventor* 1, no. 1 (April 1891): 3.

71. Charlotte Smith, "Greeting," 1; Charlotte Smith, "Memorial and Petition to the Members of the Centennial Celebration," *The Woman Inventor* 1, no. 1 (April 1891): 2; Charlotte Smith, "A Patent Office Building," *The Woman Inventor* 1, no. 1 (April 1891): 3.

72. The $5 annual dues in 1891 are equivalent to $142 in 2018. I was unable to locate a copy of the second issue of *The Woman Inventor* 1, no. 2 (June 1891), so I have relied on descriptions in Stanley, *Raising More Hell and Fewer Dahlias*, 147–148; Stanley, "The Champion of Women Inventors," 26; Macdonald, *Feminine Ingenuity*, 173–176.

73. Stanley, *Raising More Hell and Fewer Dahlias*, 148; Stanley, "The Champion of Women Inventors," 26; Macdonald, *Feminine Ingenuity*, 177–187, 211.

74. John F. Marszalek, *A Black Congressman in the Age of Jim Crow: South Carolina's George Washington Murray* (Gainesville: University Press of Florida, 2006); George

W. Murray, *Congressional Record*, House of Representatives, 53rd Cong., 2nd Sess., August 10, 1894, 8382–8383. Murray's eight patents include "Cotton Chopper," US Patent 520,888, and "Fertilizer Distributer," both filed September 15, 1893, and both issued June 5, 1894.

75. Rayvon Fouché, *Black Inventors in the Age of Segregation: Granville T. Woods, Lewis H. Latimer, and Shelby J. Davidson* (Baltimore: Johns Hopkins University Press, 2003); Patricia Carter Sluby, *The Inventive Spirit of African Americans: Patented Ingenuity* (Westport, CT: Praeger, 2004).

76. David E. Wharton, *A Struggle Worthy of Note: The Engineering and Technological Education of Black Americans* (Westport, CT: Greenwood Press, 1992); Lisa D. Cook, "Inventing Social Capital: Evidence from African American Inventors, 1843–1930," *Explorations in Economic History* 48, no. 4 (December 2011): 507–518; Nina Lerman, "New South, New North: Region, Ideology, and Access in Industrial Education," in *Technology and the African-American Experience: Needs and Opportunities for Study*, ed. Bruce Sinclair (Cambridge, MA: MIT Press, 2006), 77–105; Bruce Sinclair, "Integrating the Histories of Race and Technology," in Sinclair, *Technology and the African-American Experience*, 1–17, quotation at 2.

77. Henry E. Baker, "The Negro as an Inventor," in *Twentieth Century Negro Literature*, ed. D. W. Culp (Naperville, IL: J. L. Nichols, 1902), 399–401.

78. On Baker and his career, see Culp, *Twentieth Century Negro Literature*, on a plate (n.p.) inserted between 398–399; "Promotions in the Patent Office," *Scientific American* 45, no. 13 (September 24, 1881), 193; Bernard C. Nalty, *Strength for the Fight: A History of Black Americans in the Military* (New York: Free Press, 1986), 82; Frederick S. Harrod, "Jim Crow in the Navy, 1798–1941," *United States Naval Institute Proceedings* 105, no. 9 (September 1979): 46–53.

79. Baker, "The Negro as an Inventor," 401–402; W. E. B. Du Bois, "The American Negro at Paris," *American Monthly Review of Reviews* 22, no. 5 (November 1900): 575–577. Henry E. Baker, *Negro Patentees of the United States, 1834–1900*, 4 vols., is housed at the Moorland-Springarn Research Center, Howard University, Washington, DC.

80. Baker, "The Negro as an Inventor," 403–413, quotation at 405; Henry E. Baker, *The Colored Inventor: A Record of Fifty Years* (New York: Crisis Publishing, 1913).

81. Lauranett Lee, "Giles B. Jackson (1853–1924)," *Encyclopedia Virginia*, March 23, 2014, http://www.encyclopediavirginia.org/Jackson_Giles_B_1853-1924; Donna Tyler Hollie, "Grand Fountain of the United Order of True Reformers," *Encyclopedia Virginia*, October 27, 2015, http://www.encyclopediavirginia.org/Grand_Fountain_of_the_United_Order_of_True_Reformers.

82. Brian de Ruiter, "Jamestown Ter-centennial Exposition of 1907," *Encyclopedia Virginia*, October 27, 2015, http://www.encyclopediavirginia.org/Jamestown_Ter-Centennial_Exposition_of_1907; Robert Taylor, "The Jamestown Tercentennial

Exposition of 1907," *Virginia Magazine of History and Biography* 65, no. 2 (April 1957): 169–208. On the NDEC, see Giles B. Jackson and D. Webster Davis, *The Industrial History of the Negro Race of the United States* (Richmond: Virginia Press, 1908), 138–161; Lee, "Giles B. Jackson (1853–1924)"; Lucy Brown Franklin, "The Negro Exhibition of the Jamestown Ter-Centennial Exposition of 1907," *Negro History Bulletin* 38, no. 5 (June–July 1975): 408–414.

83. Quotations in Negro Development and Exposition Company of the U.S.A., *An Address to the American Negro* (Richmond, VA: NDEC, 1907), 5, 3; Jackson and Davis, *The Industrial History of the Negro Race*, 145–161.

84. Jackson and Davis, *The Industrial History of the Negro Race*, 186–349. For specific exhibitors, see Taylor (p. 209), Carr (p. 311), and Crawford (p. 332).

85. Sluby, *The Inventive Spirit of African Americans*, 89–91; Franklin, "The Negro Exhibition," 413; *Times* quotation in de Ruiter, "James Ter-centennial Exposition of 1907."

86. Mabel O. Wilson, *Negro Building: Black Americans in the World of Fairs and Museums* (Berkeley: University of California Press, 2012); "Along the Color Line," *The Crisis* 10, no. 3 (July 1915): 115.

87. Lee, "Giles B. Jackson (1853–1924)"; Sluby, *The Inventive Spirit of African Americans*, xxxiii, 78, 88.

88. "Better Mousetrap Is on Display Here," *New York Times*, June 8, 1937, 27.

89. "Inventors Prepare to Show 500 of Their Brain Children," *Los Angeles Times*, December 2, 1940, 2; "Gadgeteers Gather," *Time*, January 21, 1935, 62.

90. Jack Hitt, *Bunch of Amateurs: Inside America's Hidden World of Inventors, Tinkerers, and Job Creators* (New York: Broadway Books, 2013); Robert MacDougall, "Convention of Cranks: Why the Nineteenth Century's Golden Age of Pseudoscience May Be a Precursor of Our Own," *Scope* 2 (Spring 2011): 10–21.

91. "Happy Harmony," *Time*, June 19, 1939, 32; Naylor, quoted in "Urges Legislation to Aid Inventors," *Daily Boston Globe*, September 3, 1936, 5. All other quotations are from "Inventors Prepare to Show 500 of Their Brain Children."

92. The $5 in 1939 is equivalent to $90 in 2018. See "Dale Carnegie's Advice on Winning Success," *Wilmington Sunday Morning Star*, October 15, 1939, 36; "Beauty Enhanced by New Inventions," *New York Times*, June 9, 1937, 27; "Patent Reform Law Urged by Inventors," *New York Times*, June 10, 1937, 25; "Prizes Given 21 Inventors," *New York Times*, June 11, 1937, 16. For the $700,000 figure, see "Stratosphere Dirigible Design, Robot Caddy Shown by Inventors," *Berkeley Daily Gazette*, November 21, 1933, 11. According to WorldCat, *The Inventor* was published from 1934 to 1940.

93. For the "crackpot" quotation, see "Gadgets to Do Everything but Yawn for 1966 Citizen," *Daily Boston Globe*, August 3, 1936, 1. For "half the world's inventors," see

"Inventors Classed as 'Nuts,'" *Los Angeles Times*, April 5, 1934, 5. For "screwy" and "glory," see "Inventors Prepare to Show 500 of Their Brain Children."

94. Doris Lockerman, "David's Little, but He Offers Some Big Ideas," *Chicago Daily Tribune*, March 27, 1937, 5; "Eight-Year Old Inventor Polishes Convention Speech," *Los Angeles Times*, March 28, 1937, 9; "Invents Potato Peeler, Trick Window Washer," *Washington Post*, May 16, 1937, 1; "Scientist Says Drouth Could Have Been Averted," *Reading Eagle*, November 26, 1934, 2.

95. "Bright Ideas of Inventors Bloom Here," *Los Angeles Times*, April 2, 1933, A1; William S. Barton, "Lady Edisons," *Los Angeles Times*, December 29, 1940, F11.

96. "Inventors Are Unusual Group," *San Jose Evening News*, August 18, 1934, 3.

97. "Inventors Here Will Pick Queen," *Los Angeles Times*, March 4, 1933, A5; "Beauty Enhanced by New Inventions"; "Weekly Digest of Science and Invention," *Washington Post*, September 19, 1937, F5; "Prizes Given 21 Inventors."

98. On Burns's 7 percent estimate, see "Bright Ideas of Inventors Bloom Here." For the 42 percent figure, see Barton, "Lady Edisons." Burns's sources are unclear; however, a 1923 study suggested that women inventors earned approximately 1.4 percent of all US patents granted between 1905 and 1921 inclusive. See US Department of Labor, *Women's Contributions in the Field of Invention*, Bulletin of the Women's Bureau no. 28 (Washington, DC: Government Printing Office, 1923), table II, 13.

99. On the New Deal, see "Inventors' Congress Sees 'Push Button' Era in 15 Years," *Washington Post*, August 15, 1934, 26. On the proposed new laws, see "Inventors Ask Patents Speed," *Washington Post*, August 27, 1933, M3; "Inventors Hit 'Rackets,'" *New York Times*, August 19, 1934, N3. On the fee moratorium, see "Inventors Ask Relief on Fees," *Los Angeles Times*, April 6, 1934, 5. On the inventors' loan fund, see "Urges Legislation to Aid Inventors" and "Editorial of the Day: More Than Just Gadgets," *Grand Rapids (Mich.) Press*, reprinted in *Chicago Daily Tribune*, June 20, 1936, 12.

100. Frederic J. Haskin, "Questions of Readers Answered," *Hartford Courant*, October 8, 1939, A18; "Gadgeteers Here and Anything Goes," *New York Times*, July 10, 1941, 21.

101. "Inventors Open Exhibit of Gadgets at Biltmore," *Los Angeles Times*, September 13, 1941, 2; "Inventor Here Predicts Busy Postwar Era," *Los Angeles Times*, September 25, 1943, 3.

102. Quotation from "Inventor Here Predicts Busy Postwar Era." I have been unable to locate an obituary; however, without a source, Wikipedia lists Burns's death on December 4, 1951, accessed August 20, 2018, https://en.wikipedia.org/wiki/Albert_Garrette_Burns.

103. R. A. Summers to Orville Wright, July 24, 1918, NII-LOC.

104. Rudi Volti, *An Introduction to the Sociology of Work and Occupations*, 2nd ed. (London: Sage Publications, 2012), 153–172.

105. Amitai Etzioni, ed., *The Semi-professions and Their Organization: Teachers, Nurses, Social Workers* (New York: New Press, 1969); Stephen R. Barley and Julian E. Orr, eds., *Between Craft and Science: Technical Work in U. S. Settings* (Ithaca, NY: ILR Press, 1997); Nathan L. Ensmenger, "The 'Question of Professionalism' in the Computer Fields," *IEEE Annals of the History of Computing* 23, no. 4 (October–December 2001): 56–73.

Chapter 5

1. Franklin D. Roosevelt, "Message from the President of the United States Transmitting Recommendations Relative to the Strengthening and Enforcement of Antitrust Laws," April 29, 1938, in *Investigation of Concentration of Economic Power: Hearings before the Temporary National Economic Committee*, 31 vols. (Washington, DC: Government Printing Office, 1939–1941), pt.1, exhibit no. 1, 185–191, quotations at 185, 189. Hereafter, I will refer to the published testimony and exhibits as *TNEC*.

2. *Final Report and Recommendations of the Temporary National Economic Committee* (Washington, DC: Government Printing Office, 1941), 696, 729.

3. This chapter is an expanded version of Eric S. Hintz, "The 'Monopoly' Hearings, Their Critics, and the Limits of Patent Reform in the New Deal," in *Capital Gains: Business and Politics in Twentieth-Century America*, ed. Richard John and Kim Phillips-Fein (Philadelphia: University of Pennsylvania Press, 2017), 61–79.

4. Scholars have devoted considerable attention to the TNEC and the intensification of federal antitrust enforcement in the late 1930s. For example, see David Lynch, *Concentration of Economic Power* (New York: Columbia University Press, 1946); Ellis W. Hawley, *The New Deal and the Problem of Monopoly* (Princeton, NJ: Princeton University Press, 1966); Tony Freyer, *Regulating Big Business: Antitrust in Great Britain and America, 1880–1990* (Cambridge: Cambridge University Press, 1992); Wyatt Wells, *Antitrust and the Formation of the Postwar World* (New York: Columbia University Press, 2002); Tony Freyer, *Antitrust and Global Capitalism, 1930–2004* (Cambridge: Cambridge University Press, 2006). However, few scholars have focused specifically on the TNEC's patent hearings. Two exceptions are Larry Owens, "Patents, the 'Frontiers' of American Invention, and the Monopoly Committee of 1939: Anatomy of a Discourse," *Technology and Culture* 32, no. 4 (October 1991): 1076–1093; Richard John, "Patents and Free Enterprise: The TNEC Reconsidered" (paper presented at the Business History Conference, Portland, Oregon, April 2016).

5. B. Zorina Khan, *The Democratization of Invention: Patents and Copyrights in American Economic Development, 1790–1920* (Cambridge: Cambridge University Press, 2005), 1–19.

6. A cartel is a group of otherwise independent firms that collude to limit competition and fix prices in a given industry. Horizontal integration occurs when two firms in the same industrial sector merge to attain operational efficiencies and greater market share. Vertical integration occurs when a firm acquires new capabilities at different stages of the product life cycle. For example, a manufacturing firm might acquire its raw materials supplier or the retail chain that sells its products.

7. David M. Hart, "Antitrust and Technological Innovation in the US: Ideas, Institutions, Decisions, and Impacts, 1890–2000," *Research Policy* 30, no. 6 (June 2001): 923–936; David F. Noble, *America by Design: Science, Technology, and the Rise of Corporate Capitalism* (New York: Knopf, 1977), 84–109; Catherine L. Fisk, *Working Knowledge: Employee Innovation and the Rise of Corporate Intellectual Property, 1800–1930* (Chapel Hill: University of North Carolina Press, 2009).

8. Christopher Beauchamp, *Invented by Law: Alexander Graham Bell and the Patent That Changed America* (Cambridge, MA: Harvard University Press, 2015); Arthur Bright, *The Electric-Lamp Industry: Technological Change and Economic Development from 1800 to 1947* (New York: MacMillan, 1949), 101–104, 151; Hugh G. J. Aitken, *The Continuous Wave: Technology and American Radio, 1900–1932* (Princeton, NJ: Princeton University Press, 1985), 355–431; L. W. Moffett, as quoted in Noble, *America by Design*, 85.

9. Charles Eisler, *The Million-Dollar Bend: The Autobiography of the Benefactor of the Radio Tube and Lamp Industry* (New York: William-Frederick Press, 1960), especially 101–107. On the incandescent lamp cartel, see Leonard S. Reich, "Lighting the Path to Profit: GE's Control of the Electric Lamp Industry, 1882–1941," *Business History Review* 66, no. 2 (Summer 1992): 305–334.

10. Eisler, *The Million-Dollar Bend*, 115–119, quotation at 118–119.

11. Eisler, *The Million-Dollar Bend*, 119–126, 255–263; Bright, *The Electric Lamp Industry*, 253.

12. Eisler, *The Million-Dollar Bend*, 122–132, quotation at 123; Bright, *The Electric Lamp Industry*, 252–253, 276–277. The crucial article was Charles Eisler, "Tungsten Lamp Manufacture," *Machinery* 23, no. 3 (December 1916): 21–29. On GE's buyout offer, see Charles Eisler to William Harris, Box 41, Folder 16, Eisler Engineering Company Records, Archives Center, National Museum of American History, Smithsonian Institution, Washington, DC.

13. Donald G. Godfrey, *Philo T. Farnsworth: The Father of Television* (Salt Lake City: University of Utah Press, 2001), 1–36; Evan I. Schwartz, *The Last Lone Inventor: A Tale of Genius, Deceit, and the Birth of Television* (New York: HarperCollins, 2002), 1–141; Earle Ennis, "S. F. Man's Invention to Revolutionize Television," *San Francisco Chronicle*, September 3, 1928, sec. 2, 11.

14. Kenneth Bilby, *The General: David Sarnoff and the Rise of the Communications Industry* (New York: Harper and Row, 1986); Tom Lewis, *Empire of the Air: The Men*

Who Made Radio (New York: HarperCollins, 1991), 151–166, 223–225. On "package licensing," see Margaret B. W. Graham, "When the Corporation Almost Displaced the Entrepreneur: Rethinking the Political Economy of Research and Development," *Enterprise and Society* 18, no. 2 (June 2017): 245–281, especially 262–268; Schwartz, *The Last Lone Inventor*, 159.

15. Schwartz, *The Last Lone Inventor*, 54–55, 143–157; Godfrey, *Philo T. Farnsworth*, 43–46.

16. Farnsworth's wife, Pem Farnsworth, believed that RCA gave Philco an ultimatum, but this interpretation may be somewhat overblown; see Godfrey, *Philo T. Farnsworth*, 47–58.

17. Godfrey, *Philo T. Farnsworth*, 43–58, 71–76; Schwartz, *The Last Lone Inventor*, 154–157, 179–188, 196–209, 233–235.

18. Reich, "Lighting the Path to Profit"; Aitken, *The Continuous Wave*, 497–509.

19. Hawley, *The New Deal and the Problem of Monopoly*, 7–146, Hawley quotation at 14.

20. Hawley, *The New Deal and the Problem of Monopoly*, 383–419; Lynch, *Concentration of Economic Power*, 8–25.

21. Franklin D. Roosevelt, "Message to Congress on Stimulating Recovery," April 14, 1938, in *The American Presidency Project*, ed. John T. Woolley and Gerhard Peters, accessed August 7, 2019, https://www.presidency.ucsb.edu/node/209601; Roosevelt, "Message from the President of the United States Transmitting Recommendations Relative to the Strengthening and Enforcement of Antitrust Laws."

22. Laurence Stern, "Outline of a New Economic Order," *Magazine of Wall Street* 62 (September 24, 1938): 670–672, as quoted in Lynch, *Concentration of Economic Power*, 54.

23. Lynch, *Concentration of Economic Power*, 35–50; Schwartz, *The Last Lone Inventor*, 251; Malcolm Rutherford, *The Institutionalist Movement in American Economics, 1918–1947: Science and Social Control* (Cambridge: Cambridge University Press, 2011).

24. For a summary of the early hearings, see Joseph C. O'Mahoney, *Preliminary Report of Temporary National Economic Committee*, 76th Cong., 1st Sess., 1939, S. Doc. 95, 5–22.

25. On Arnold, see Corwin D. Edwards, "Thurman Arnold and the Antitrust Laws," *Political Science Quarterly* 58, no. 3 (September 1943): 338–355; Gene M. Gressley, "Thurman Arnold, Antitrust, and the New Deal," *Business History Review* 38, no. 2 (Summer 1964): 214–231; Wilson Miscamble, "Thurman Arnold Goes to Washington: A Look at Antitrust Policy in the Later New Deal," *Business History Review* 56, no. 1 (Spring 1982): 1–15; Alan Brinkley, "The Antimonopoly Ideal and the Liberal State: The Case of Thurman Arnold," *Journal of American History* 80, no. 2 (September 1993): 557–579. For Arnold's quotations, see *TNEC*, pt. 2, 256, 665.

26. "Patent troll" is a pejorative term, coined in the early 1990s, to describe an inventor who does not manufacture his patent but instead threatens infringement lawsuits to extract retroactive royalties and damages from wealthy corporations who were unwittingly using the patent. Alternative terms include "patent shark," "patent pirate," or "non-practicing entity." See Robert P. Merges, "The Trouble with Trolls: Innovation, Rent-Seeking, and Patent Law Reform," *Berkeley Technology Law Journal* 24, no. 4 (Fall 2009): 1583–1614.

27. For testimony on the Selden affair, see *TNEC*, pt. 2, 267–272, 287, 361–363. See also William Greenleaf, *Monopoly on Wheels: Henry Ford and the Selden Automobile Patent* (Detroit: Wayne State University Press, 2011).

28. For Reeves's testimony, see *TNEC*, pt. 2, 285–303. For the AMA's member firms, see exhibit no. 92, "Members of the Automobile Manufacturers Association Corrected to November 1938," *TNEC*, pt. 2, 682–683; exhibit no. 101, "Growth of Membership, Automobile Manufacturers Association," *TNEC*, pt. 2, 691.

29. For testimony by Edsel Ford and patent counsel Joseph Farley, see *TNEC*, pt. 2, 256–285. For Tibbetts's testimony, see *TNEC*, pt. 2, 304–328.

30. Farley's testimony in *TNEC*, pt. 2, 262–263. For a summary of the automobile hearings, see George E. Folk, *Patents and Industrial Progress: A Summary, Analysis, and Evaluation of the Record on Patents of the Temporary National Economic Committee*, 2nd ed. (New York: Harper and Brothers, 1942), 28.

31. *TNEC*, pt. 2, 351.

32. Kettering quotations in *TNEC*, pt. 2, 342, 356; Stuart W. Leslie, *Boss Kettering* (New York: Columbia University Press, 1983).

33. This evidence had been uncovered via subpoenas in Arnold's pending anti-trust prosecution against the "Glass Trust." After appealing to the Supreme Court, Hartford-Empire was eventually convicted of antitrust violations. See "Hartford-Empire v. United States: Integration of the Anti-trust and Patent Laws," *Columbia Law Review* 45, no. 4 (July 1945): 601–625.

34. See *TNEC*, pt. 2, 377–667, as well as exhibit nos. 112–162, 303, and 431 in the appendix, 737–834. For the percentages, see p. 383 and "Major Inter-company Relations in the Glass Container Industry," exhibit no. 113, *TNEC*, pt. 2, 762. For helpful summaries, see Folk, *Patents and Industrial Progress*, 29–61; Lynch, *Concentration of Economic Power*, 162–164, 227–231, 273–279. On the Corning-GE partnership, see Margaret B. W. Graham and Alec T. Shuldiner, *Corning and the Craft of Innovation* (Oxford: Oxford University Press, 2001), 12–13, 90–92, 281–286.

35. "Hartford-Empire Company—Annual Receipts from Royalties and License Fees," exhibit no. 115, *TNEC*, pt. 2, 764.

36. Hartford-Empire delayed the issuance of the Steimer "gob-feed" patent for about twenty years. See *TNEC*, pt. 2, 438–441; *TNEC*, pt. 3, exhibit no. 199, 1134; Lynch, *The Concentration of Economic Power*, 230.

37. *TNEC*, pt. 2, 624–637. After its 1945 antitrust conviction, Hartford-Empire was forced to drop its infringement lawsuits and pay $1.2 million in damages to settle Obear-Nester's counterclaims. The cases are described in "Hartford-Empire v. United States" and Obear-Nester Glass Company v. Commissioner of Internal Revenue, 20 T. C. 1102 (1953).

38. *TNEC*, pt. 2, 400–426; "Memorandum as to Hartford-Fairmont and Hartford-Empire History and Policy," March 26, 1928, exhibit no. 124, *TNEC*, pt. 2, 768–771.

39. "Memorandum on Policy of Hartford-Empire Company," February 18, 1930, exhibit no. 125, *TNEC*, pt. 2, 771–780. For the testimony, see *TNEC*, pt. 2, 386–396, 449–452; Lynch, *Concentration of Economic Power*, 162–163.

40. *TNEC*, pt. 2, 455–456.

41. *TNEC*, pt. 3, 838–867, 1043–1044, quotation at 859; see also exhibit nos. 179–205, pp. 1123–1137.

42. *TNEC*, pt. 3, 853–857, 860–862; exhibit no. 193, p. 1130; exhibit nos. 198–202, pp. 1133–1136.

43. *TNEC*, pt. 3, 862.

44. *TNEC*, pt. 3, 838–843, quotations at 842. For the sixteen "classic patents," see exhibit nos. 163–178, pp. 1107–1122.

45. *TNEC*, pt. 3, 863.

46. On Coe's reluctance to fundamentally alter the patent laws, see "Map Wide Inquiry on Patent System," *New York Times*, August 7, 1938, 18.

47. For Arnold's proposals, see *Final Report and Recommendations*, 36–37.

48. *TNEC*, pt. 3, 837.

49. *TNEC*, pt. 3, 938–945.

50. *TNEC*, pt. 3, 947.

51. Godfrey, *Philo T. Farnsworth*, 109–121; Schwartz, *The Last Lone Inventor*, 254–255.

52. *TNEC*, pt. 3, 81–1007, quotations at 1001, 1005.

53. *TNEC*, pt. 3, 1004; Schwartz, *The Last Lone Inventor*, 254–257.

54. *TNEC*, pt. 3, 912, 916, 975; *TNEC*, pt. 2, 341–343, 356, 359.

55. *TNEC*, pt. 3, 871–872.

56. *TNEC*, pt. 3, 920–921.

57. *TNEC*, pt. 3, 899–900.

58. On the "pioneering" rhetoric of the corporate witnesses, see Owens, "Patents, the 'Frontiers' of American Invention, and the Monopoly Committee of 1939." On Bush, Holland, and the NRC's promotion of industrial research, see chapter 2. Bush invoked the same "frontier" discourse in his call for postwar federal research funding; see Vannevar Bush, *Science, the Endless Frontier: A Report to the President* (Washington, DC: Government Printing Office, 1945).

59. *TNEC*, pt. 3, quotations at 988, 1062, 1073.

60. *TNEC*, pt. 3, 971–972.

61. On this point, see Thomas P. Hughes, *American Genesis: A Century of Invention and Technological Enthusiasm, 1870–1970*, 2nd ed. (Chicago: University of Chicago Press, 2004), 138–139.

62. On the attorneys' opposition, see "Map Wide Inquiry on Patent System." On the independents' support for Coe's reforms, see the testimony of Farnsworth, John Graham, and Maurice Graham, *TNEC*, pt. 3, 981–1006, 938–947, and 1070–1077, respectively. Corporate support for Coe's proposals is reflected in Vannevar Bush's testimony (pp. 870–911) and a report he and Jewett had coauthored for the Science Advisory Board, "Report of the Committee on the Relation of Patent System to the Stimulation of New Industries," April 1, 1935, exhibit no. 206, 1139–1148.

63. For testimony and exhibits opposing compulsory licenses, see in *TNEC*, pt. 3, Bush (p. 890), Farnsworth (pp. 995–1000), Carlton (pp. 1053–1055), and the Science Advisory Board report (pp. 1143–1145).

64. Conway Coe to Joseph O'Mahoney, March 7, 1941, reprinted in *TNEC*, pt. 31-A, 18475–18478.

65. *TNEC*, pt. 3, 1034.

66. Thurman Arnold to Joseph O'Mahoney, March 29, 1941, reprinted in *TNEC*, pt. 31-A, 18483–18489, quotation at 18486.

67. *TNEC*, pt. 3, 977, 1006.

68. Schwartz, *The Last Lone Inventor*, 259–269; Godfrey, *Philo T. Farnsworth*, 109–128. RCA leveraged its own products—radio and television—to persuade the public that "the site of invention" had shifted decisively "from the workbench of the individual inventor to the well-capitalized research and development laboratory." See Steve J. Wurtzler, *Electric Sounds: Technological Change and the Rise of Corporate Mass Media* (New York: Columbia University Press, 2007), 101–107, quotation at 102.

69. *Final Report and Recommendations*, 36.

70. For NAM member firms, see minutes of the NAM Patent Committee, November 16, 1938, and December 7, 1939, in the chronological "NAM Committee Minutes" folders from 1938–1939, Box 150, National Association of Manufacturers Records, Hagley Museum and Library, Wilmington, DE (hereafter NAM Records). Scholarly histories include Albion G. Taylor, *Labor Policies of the National Association of Manufacturers* (Urbana: University of Illinois, 1928); Albert K. Steigerwalt, *The National Association of Manufacturers, 1895–1914: A Study in Business Leadership* (Ann Arbor: University of Michigan, 1964); Alfred S. Cleveland, "Some Political Aspects of Organized Industry" (PhD diss., Harvard University, 1947); John N. Stalker, "The National Association of Manufacturers: A Study in Ideology" (PhD diss., University of Wisconsin, 1951).

71. John Scoville and Noel Sargent, *Fact and Fancy in the TNEC Monographs* (New York: National Association of Manufacturers, 1942).

72. I have found no evidence that the NAM's lobbying influenced Coe's TNEC testimony or recommendations; rather, the Coe-NAM relationship apparently developed after the hearings. On the town hall meetings, see "Minutes of the Modern Pioneers Executive Committee," July 18, 1939, Box 150, Folder "NAM Committee Minutes 1939, July–October," NAM Records. On NAM's Patent Office tour, see "Report of the Meeting of the Patent Procedure Subcommittee with Conway P. Coe, United States Commissioner of Patents," Box 150, Folder "NAM Committee Minutes 1939, April–June," NAM Records.

73. For the NAM's positions, see "Minutes of Patent Legislation Subcommittee of the Patents and Trademarks Committee," March 16, 1939, Box 150, Folder "NAM Committee Minutes 1939, January–April," NAM Records. See also Robert Lund's address before NAM's 44th Congress of American Industry, December 7, 1939, reprinted as Robert L. Lund, *Patents and Free Enterprise* (New York: NAM, 1940).

74. Representatives of NAM member firms testifying before the TNEC patent hearings included George Baekeland (Bakelite Corporation), William Coolidge (General Electric), F. Goodwin Smith (Hartford-Empire), Milton Tibbetts (Packard Motor Car Co.), and Clarence C. Carlton (Motor Wheel Corporation). On coaching the witnesses, see minutes of the "Special Research Subcommittee Meeting to Consider the Patent Hearings," October 31, 1938, Box 150, Folder "NAM Committee Minutes 1938, July–November," NAM Records.

75. See Folk, *Patents and Industrial Progress*. Pamphlets include *How to Analyze the Patent Situation in Your Company* (New York: NAM, 1939); *Limitations of the Right to License Patents* (New York: NAM, 1939?); *You and Patents* (New York: NAM, 1939?); *Inventive America* (New York: NAM, 1941); *Patents and Invention* (New York: NAM, 1942); George E. Folk, *A Review of Proposals for Revision of the United States Patent System* (New York: NAM, 1946).

76. "Minutes of the Meeting of N.A.M. Committee on Patents and Trademarks," April 26, 1938, Box 150, Folder "NAM Committee Minutes 1939, January–April," NAM Records.

77. I have been unable to locate the NICB-AEC report in the NAM Records. Its principal findings are summarized in NAM, "Telling America the Truth about Our Patent System," March 1940, Box 195, Folder "Modern Pioneers Folder #2," NAM Records. The quotation is from the AEC's Fairfield E. Raymond, "Minutes of Meeting of NAM Committee on Patents and Research," April 12, 1940, Box 151, Folder "NAM Committee Minutes 1940, April," NAM Records.

78. Richard Tedlow, "The National Association of Manufacturers and Public Relations during the New Deal," *Business History Review* 50, no. 1 (Spring 1976): 25–45; NAM, "Telling America the Truth about Our Patent System."

79. "Minutes of Meeting of the National Modern Pioneers Executive Committee," July 18, 1939, Box 150, Folder "NAM Committee Minutes 1939, July–October," NAM Records. See also "Entries for NAM Awards to Inventors Close December 1," press release, November 28, 1939, Box 195, Folder "Modern Pioneers Folder #1," NAM Records. Finally, see "Nomination for Modern Pioneer on the American Frontier of Industry," an entry blank, 1939, Box 195, Folder "Modern Pioneers Folder #2," NAM Records.

80. Quotations from the press release "Entries for NAM Awards to Inventors Close December 1." Historian Frederick Jackson Turner believed that the struggle to overcome the American frontier inspired the nation's exceptional and democratic qualities; see Frederick Jackson Turner, *The Frontier in American History* (New York: Henry Holt, 1920).

81. Quotation from the press release "Entries for NAM Awards to Inventors Close December 1." For the award winners, see the banquet program, "National Modern Pioneers Banquet," February 27, 1940, Box 195, Folder "Modern Pioneers Folder #2," NAM Records.

82. See NAM, "Telling America the Truth about Our Patent System"; "Summary of the 13 Regional Modern Pioneers Banquets," 1940, Box 195, Folder "Modern Pioneers Folder #1," NAM Records.

83. Robert L. Lund, "Patents Make Jobs," NBC radio address, February 26, 1940, printed in *Our Modern Pioneers and the American Patent System: A Series of Addresses Given at the National Modern Pioneers Banquet,* ed. National Association of Manufacturers (New York: National Association of Manufacturers, 1940), 23–29, quotations at 26, 29.

84. See the program "National Modern Pioneers Banquet" and "Seating List: National Modern Pioneers Banquet," both February 27, 1940, Box 195, Folder "Modern Pioneers Folder #2," NAM Records. For the speeches, see National Association of Manufacturers, ed., *Our Modern Pioneers and the American Patent System.*

85. The publicity figures and Lund's quotations are in NAM, "Telling America the Truth about Our Patent System." See also "Summary of the 13 Regional Modern Pioneers Banquets."

86. *Final Report and Recommendations*, 36–37, 357–359; Lynch, *Concentration of Economic Power*, 359, 376.

87. On the earlier reform efforts by the AAIM and the Inventors' Guild, see chapter 4. On the eventual implementation of the Court of Appeals for the Federal Circuit (1982) and the twenty-year rule for patent terms (1995), see chapter 7.

88. Alan Brinkley, *The End of Reform: New Deal Liberalism in Recession and War* (New York: Knopf, 1995), 106–136.

89. David M. Hart, *Forged Consensus: Science, Technology, and Economic Policy in the United States, 1921–1953* (Princeton, NJ: Princeton University Press, 1998), 90–96; Freyer, *Antitrust and Global Capitalism*, 32–59; Wells, *Antitrust and the Formation of the Postwar World*, 27–136.

90. Lawrence Langner to Franklin D. Roosevelt, May 14, 1940, reprinted in Senate Subcommittee, *National Inventors and Engineers Commission*, Hearings on S. 2078, 77th Cong., 2nd Sess., January 19, 1942, 3–4.

Chapter 6

1. Merritt Roe Smith, ed., *Military Enterprise and Technological Change: Perspectives on the American Experience* (Cambridge, MA: MIT Press, 1985).

2. Thomas P. Hughes, *American Genesis: A Century of Invention and Technological Enthusiasm, 1870–1970*, 2nd ed. (Chicago: University of Chicago Press, 2004), 138–139; Daniel E. Worthington, "Inventive Genius or Scientific Research?," *Swords and Ploughshares* 6, nos. 1–2 (Fall–Winter 1991): 11–14.

3. James Phinney Baxter, *Scientists against Time: The Story of the Office of Scientific Research and Development during World War II* (Boston: Little, Brown, 1946); Daniel J. Kevles, *The Physicists: The History of a Scientific Community in Modern America*, 2nd ed. (Cambridge, MA: Harvard University Press, 1995 [1977]), 287–348; Richard Rhodes, *The Making of the Atomic Bomb* (New York: Simon and Schuster, 1986); John A. Miller, *Men and Volts at War: The Story of General Electric in World War II* (New York: McGraw-Hill, 1947); Pap Ndiaye, *Nylon and Bombs: DuPont and the March of Modern America*, trans. Elborg Forster (Baltimore: Johns Hopkins University Press, 2007).

4. Jon Cavicchi, "The Mystery of the Classified and Missing National Inventors Council (NIC) Files," Pierce Law IP Mall, circa 2003, archived May 22, 2011, https://web.archive.org/web/20110522081003/http://ipmall.info/hosted_resources/nic.asp.

5. "Boycott Germany, Edison Advises," *New York Times*, May 22, 1915, 3; Thomas A. Bailey and Paul B. Ryan, *The Lusitania Disaster: An Episode in Modern Warfare and Diplomacy* (New York: Free Press, 1975).

6. On technoscientific warfare, see William H. McNeill, *The Pursuit of Power: Technology, Armed Forces, and Society since A.D. 1000* (Chicago: University of Chicago Press, 1982), 262–345; Edison quoted in "Boycott Germany, Edison Advises," 3.

7. Edison, quoted in an interview by Edward Marshall, "Edison's Plan for Preparedness," *New York Times Magazine*, May 30, 1915, 6–7.

8. Josephus Daniels to Thomas A. Edison, July 7, 1915, reprinted in Lloyd N. Scott, *Naval Consulting Board of the United States* (Washington, DC: Government Printing Office, 1920), 286–288; "Edison Will Head Navy Test Board," *New York Times*, July 13, 1915, 1.

9. Josephus Daniels, *The Wilson Era: Years of Peace, 1910–1917* (Chapel Hill: University of North Carolina Press, 1946), 490–491. On Hutchison's role, see Hutchison to Daniels, September 12, 1935, in Hutchison Correspondence, Thomas A. Edison Papers digital edition, Doc. X042G2EU, as quoted in Thomas E. Jeffrey, "'Commodore' Edison Joins the Navy: Thomas Alva Edison and the Naval Consulting Board," *Journal of Military History* 80, no. 2 (April 2016): 411–445, quotations at 418–419.

10. The NCB reflected a "not-so-subtle interweaving of public and private interests." See David K. van Keuren, "Science, Progressivism, and Military Preparedness: The Case of the Naval Research Laboratory, 1915–1923," *Technology and Culture* 33, no. 4 (October 1992): 710–736, quotation at 720.

11. Sprague to Daniels, July 19, 1915, as cited in David K. Allison, *New Eye for the Navy: The Origin of Radar at the Naval Research Laboratory*, NRL Report no. 8466 (Washington, DC: Naval Research Laboratory, 1981), 22, fn21.

12. Miller Reese Hutchison to Josephus Daniels, November 6, 1915, as quoted in Kevles, *The Physicists*, 108–109, emphasis in original; Hughes, *American Genesis*, 121–122, quotation at 121.

13. Scott, *Naval Consulting Board*, 11–13; "Navy Test Board Success Assured," *New York Times*, July 14, 1915, 2; "Daniels Names Naval Advisers," *New York Times*, September 13, 1915, 1; "Developing Our Navy," *New York Times*, September 13, 1915, 8.

14. On the appropriation, see van Keuren, "Science, Progressivism, and Military Preparedness," 727–728; William D. Middleton and William D. Middleton III, *Frank Julian Sprague: Electrical Inventor and Engineer* (Bloomington: Indiana University Press, 2009), 174.

15. Hudson Maxim, *Defenseless America* (New York: Hearst's International Library, 1915). See several letters, especially P. S. du Pont to Hudson Maxim, September 18, 1916, with attached bill from Hearst International Library Company, in Box 1,

Folder 8, Accession no. 2147, Hudson Maxim Papers, Hagley Museum and Library, Wilmington, DE.

16. Scott, *Naval Consulting Board*, 7–25.

17. Scott, *Naval Consulting Board*, chap. 2, "Industrial Preparedness Campaign," 26–55; "135,683 Serious, Earnest Americans Emphasize Demand for Preparedness in Parade That Marches for Twelve Hours," *New York Times*, May 14, 1916, 1.

18. Scott, *Naval Consulting Board*, 122–123.

19. For the bulletins, see Scott, *Naval Consulting Board*, 252–286. On the naive suggestions, see Middleton and Middleton, *Frank Julian Sprague*, 180.

20. Hudson Maxim, *Reminiscences and Comments, as Reported by Clifton Johnson* (New York: Doubleday, Page, 1924), 192; Middleton and Middleton, *Frank Julian Sprague*, 180–182, quotation at 182.

21. Scott, *Naval Consulting Board*, 125; C. H. Claudy, "Ideas That Will Not Work," *Scientific American* 117, no. 13 (September 29, 1917), 226, 235, quotation at 235.

22. "Inventions Section of the General Staff of the Department of War," *Science*, n.s., 47, no. 1221 (May 24, 1918): 509–510. For statistics comparing military crowdsourcing initiatives, see National Inventors Council, *Administrative History of the National Inventors Council* (Washington, DC: Department of Commerce, 1946), 10.

23. Scott, *Naval Consulting Board*, 138, quotation at 128.

24. Daniels to Edison, July 7, 1915, as reprinted in Scott, *Naval Consulting Board*, 287. On Maxim, see Scott, *Naval Consulting Board*, 197–199; Maxim, *Reminiscences and Comments*, 192.

25. Scott, *Naval Consulting Board*, 213–217. On nitrates, see George Wise, *Willis R. Whitney, General Electric, and the Origins of U.S. Industrial Research* (New York: Columbia University, 1985), 195.

26. Scott, *Naval Consulting Board*, 199–202; Middleton and Middleton, *Frank Julian Sprague*, 177–187; Frederick Dalzell, *Engineering Invention: Frank J. Sprague and the U. S. Electrical Industry* (Cambridge, MA: MIT Press, 2010), 224–225. Desmond Sprague's quotation is from an unpublished manuscript by Frank Rowsome, Jr., "The Man Who Invented Commuting," as quoted in Dalzell, *Engineering Invention*, 225.

27. Thomas P. Hughes, *Elmer Sperry: Inventor and Engineer* (Baltimore: Johns Hopkins University Press, 1971), xiii–xiv, 201–274; Hughes, *American Genesis*, 106–111, 126–127, 135–137; Scott, *Naval Consulting Board*, 193–197.

28. Hughes, *Elmer Sperry*, 173–200, 258–274; Hughes, *American Genesis*, 126–135; Stuart W. Leslie, *Boss Kettering* (New York: Columbia University Press, 1983), 70–71, 80–87.

29. On the NRC, see Rexmond Cochrane, *The National Academy of Sciences: The First Hundred Years, 1863–1963* (Washington, DC: National Academy of Sciences, 1978), 208–241; Hughes, *American Genesis*, 121–122. On the Saunders exchange, see George Ellery Hale to Evelina Hale, April 21, 1916, as cited in Kevles, *The Physicists*, 114.

30. Hughes, *American Genesis*, 122–124; Scott, *Naval Consulting Board*, 67–69; Hughes, *Elmer Sperry*, 258; Wise, *Willis R. Whitney*, 187–189; Kevles, *The Physicists*, 119–120.

31. Kevles, *The Physicists*, 119–120; Hughes, *American Genesis*, 123–124.

32. Quotations in Leo H. Baekeland, Diary 22, June 17, 1917, 156–157, Box 19, Folder 6, Leo H. Baekeland Papers, Archives Center, National Museum of American History, Smithsonian Institution, Washington, DC (hereafter Baekeland Papers); Baekeland to Marston Bogert, May 1, 1917, as quoted in Joris Mercelis, *Beyond Bakelite: Leo Baekeland and the Business of Science and Invention* (Cambridge, MA: MIT Press, 2020), 130.

33. Benjamin G. Lamme to Robert Millikan, June 13, 1917, and Willis R. Whitney to Josephus Daniels, June 16, 1917, as quoted in Kevles, *The Physicists*, 120–121.

34. Kevles, *The Physicists*, 122–126.

35. Hughes, *American Genesis*, 124; Kevles, *The Physicists*, 124–126.

36. For Edison's inventive work, see Scott, *Naval Consulting Board*, 160–192. The quotation is from the *New York World*, February 13, 1923, as cited in Matthew Josephson, *Edison: A Biography* (New York: McGraw-Hill, 1959), 454.

37. Jeffrey, "'Commodore' Edison Joins the Navy."

38. Frank J. Sprague and Lawrence Addicks, "Naval Laboratory: Report of Committee on Sites" (pp. 225–230), and Edison's "Minority Report" (pp. 230–232), both in Scott, *Naval Consulting Board*; van Keuren, "Science, Progressivism, and Military Preparedness," 728–736; Allison, *New Eye for the Navy*, 25–38; Thomas A. Edison to Josephus Daniels, March 4, 1918, as quoted in Paul Israel, *Edison: A Life of Invention* (New York: Wiley, 1998), 450.

39. van Keuren, "Science, Progressivism, and Military Preparedness," 728–736; Israel, *Edison*, 448–450.

40. On the reunions, see Thomas Robins to Leo Baekeland, November 14, 1928, and Thomas Robins to Howard E. Coffin, November 13, 1928, both in Box 17, Folder 26, Baekeland Papers. On the NRC's publicity campaign, see chapter 2.

41. Middleton and Middleton, *Frank Julian Sprague*, 173, 192–196.

42. Scott, *Naval Consulting Board*, 125, quotations by Daniels at 286 and by Scott at 222.

43. Middleton and Middleton, *Frank Julian Sprague*, 193–194; Scott, *Naval Consulting Board*, 222–223.

44. For example, see Kevles, *The Physicists*, 138; A. Hunter Dupree, *Science in the Federal Government: A History of Policies and Activities to 1940* (New York: Harper and Row, 1964 [1957]), 306–308; Israel, *Edison*, 446–452. For a more sanguine view of the Naval Consulting Board, see Hughes, *Elmer Sperry*, 243–274; Hughes, *American Genesis*, 118–137, especially 126.

45. Robert A. Divine, *The Illusion of Neutrality: Franklin D. Roosevelt and the Struggle over the Arms Embargo* (Chicago: University of Chicago Press, 1962); Lawrence Langner to Franklin D. Roosevelt, May 14, 1940, reprinted in *National Inventors and Engineers Commission*, Hearings on S. 2078 Before a Subcommittee of the Committee on Military Affairs, 77th Cong., 2nd Sess., January 19, 1942, 3–4.

46. Lawrence Langner, *The Magic Curtain: The Story of a Life in Two Fields, Theatre and Invention* (New York: Dutton, 1951); "Langner Is Dead: Theater Figure," *New York Times*, December 28, 1962, 8. For Langner's TNEC testimony, see chapter 5.

47. Lawrence Langner to Franklin D. Roosevelt, May 14, 1940, reprinted in *National Inventors and Engineers Commission*, 3–4.

48. Lawrence Langner to Conway P. Coe, May 15, 1940 (p. 4) and Coe to Langner, May 16, 1940 (p. 5), both reprinted in *National Inventors and Engineers Commission*. On the Patent Office meeting, see Langner, *Magic Curtain*, 355.

49. Vannevar Bush, *Pieces of the Action* (New York: Morrow, 1970), 32–33; Franklin D. Roosevelt to Vannevar Bush, June 15, 1940, Official File 4010, "National Defense Research Committee," Papers as President, Official File (OF), Franklin D. Roosevelt Library, Hyde Park, NY; Langner, *Magic Curtain*, 355–356.

50. Stuart W. Leslie, *Boss Kettering*; Langner, *Magic Curtain*, 51–55, 66, 342–344; United States Patent Office, *United States Patent Law Sesquicentennial Celebration* (Washington, DC: Government Printing Office, 1941). On the nomination, see Harry L. Hopkins to Charles F. Kettering, July 11, 1940, Box 2, Folder "History: Organization—Staff, 1940–1946," General Subject Files, 1940–1972, National Inventors Council, Records of the National Institute of Standards and Technology, 1830–1994, Record Group 167, National Archives at College Park, College Park, MD (hereafter NIC Subject Files).

51. On Kettering's suitability to lead the NIC, see Leslie, *Boss Kettering*, 307–308, 340–341.

52. Hopkins to Kettering, July 11, 1940, NIC Subject Files. Eight of the NIC's fifteen charter members—Kettering, Midgley, Langner, Coe, Coolidge, Davis, Jones, and Wright—had served as co-organizers of the patent law sesquicentennial; see United States Patent Office, *United States Patent Law Sesquicentennial Celebration*, vi, 34, 69.

53. For the charter members, see Robert H. Hinckley to President Franklin D. Roosevelt, August 26, 1940, Box 2, Folder "History: Organization—Staff, 1940–1946," NIC Subject Files.

54. Lawrence Langner, "Minutes of the Organization Meeting of the National Inventors Council," 2, August 6, 1940, Box 3, Minutes File, NIC Subject Files; Lawrence Langner, "Organization Plan," n.d. (circa 1940), Box 2, Folder "History: Organization—Staff, 1940–1946," NIC Subject Files, with quotations from Scott, *Naval Consulting Board*, 123–125.

55. National Inventors Council, *Administrative History*, 2–4, quotation in foreword; Lawrence Langner, "Meeting of September 24, 1940: Report of Secretary," 3, and attached appendix, "Proposed Committees of the National Inventors Council," Box 3, Minutes File, NIC Subject Files; Lawrence Langner, "Temporary Procedures for Chairmen of Technical Committees," October 3, 1940, Box 1, Folder "Committees and Procedures for Committees," NIC Subject Files.

56. "Consultants: Committees 5 and 7," n.d., Box 4, Folder "Personnel—NIC Consultants, Staff, etc Records," NIC Subject Files.

57. Langner, "Organization Plan," NIC Subject Files; National Inventors Council, *Administrative History*, 4, 10.

58. "Statistics of NIC Budgets and Appropriations," n.d. (circa 1946), Box 1, Folder "Budget and Organization," NIC Subject Files.

59. On the bulletins, see Langner, "Minutes of the Organization Meeting of the National Inventors Council," 2–7, August 6, 1940, NIC Subject Files; National Inventors Council, *Information Bulletin No. 1* (Washington, DC: Government Printing Office, 1941), quotation at 4, Box 2, Folder "Information Bulletins, 1941–1955," NIC Subject Files.

60. On the evaluation process, see National Inventors Council, *Information Bulletin No. 2: How Inventors Can Aid the National Defense* (Washington, DC: Government Printing Office, 1941), 2–3, Box 2, Folder "Information Bulletins, 1941–1955," NIC Subject Files. See also "Procedures," n.d. (circa 1940), Box 4, Folder "Policies and Procedures," NIC Subject Files. Quotation in National Inventors Council, *Administrative History*, 4.

61. Thomas R. Taylor to George W. Lewis, October 18, 1941, and the attached "List of Suggestions Outstanding to Dr. Lewis," October 17, 1941, Box 1, Lewis correspondence folder, NIC Subject Files.

62. "Procedures," NIC Subject Files. On feedback regarding suggestions 8,693 and 14,322, see L. B. Lent to Fin Sparre, April 30, 1941, Box 1, Sparre correspondence file, NIC Subject Files.

63. Quotation from Lawrence Langner, "Meeting of September 24, 1940: Report of Secretary," 3, Box 3, Minutes File, NIC Subject Files. On inventor recruitment and the military's wish lists, see NIC Meeting Minutes, August 6, 1940 (p. 7), September 24, 1940 (pp. 3–5), December 4, 1940 (pp. 5–6), all in Box 3, Minutes File, NIC Subject Files. I have tried in vain to locate the list of 2,000 inventors in the NIC Subject Files.

64. On the procurement function, see Lawrence Langner, NIC Meeting Minutes, December 4, 1940 (pp. 5–6), and his appended "Report of the Secretary," November 30, 1940 (pp. 2–3), Box 3, Minutes File, NIC Subject Files. See also Chief Engineer L. B. Lent to Thomas R. Taylor, February 11, 1941, Box 4, Folder "Policies and Procedures," NIC Subject Files.

65. "Crackpots' Haven," *Time*, August 19, 1940, 20; "Idea Mill for War," *Business Week*, April 5, 1941, 47–48.

66. On the PR campaign, see Lawrence Langner, "Report of the Secretary," November 30, 1940; NIC Meeting Minutes, December 4, 1940, Box 3, Minutes File, NIC Subject Files; Franklin D. Roosevelt, "President's Call for Full Response on Defense," reprinted in *New York Times*, December 30, 1940, 6; Stuart Chase, "Calling All Inventors," *Reader's Digest* 38, no. 225 (January 1941): 15–18.

67. Chase reported the increased submissions in a follow-up article, Stuart Chase, "Second Call for Inventions!," *Reader's Digest* 40, no. 239 (March 1942): 19–22. On Davis's PR efforts, see "National Inventors Council Publicity File Index," 1955, in folder of the same name, Box 6, NIC Subject Files.

68. On the radio talks, see the following in Box 6, Folder "Publicity 1942," NIC Subject Files: Lawrence Langner to William Coolidge, January 8, 1942; Coolidge to Langner, January 9, 1942; Davis-Langner transcript, "Adventures in Science," February 14, 1942. See also Charles Kettering, *Short Stories of Science and Invention: A Collection of Radio Talks by C. F. Kettering*, ed. Public Relations Staff, General Motors (Detroit: General Motors, 1959), 60–61, 76–77.

69. For a sample of his many speeches, see L. B. Lent, "Inventions for Defense," delivered before the American Society of Mechanical Engineers, December 3, 1941. See also multiple speeches in the 1941, 1942, and 1943 "Publicity" folders, Box 6, NIC Subject Files.

70. On the posters, see in Box 6, Folder "Publicity 1941," NIC Subject Files: Thomas R. Taylor to Malcolm E. Kerlin, November 17, 1941; W. J. Tate to Thomas R. Taylor, December 16, 1941; W. J. Tate to Watson Davis, December 31, 1941; W. J. Tate to Walter Myers, December 31, 1941.

71. A. E. Hotchner, "Wanted: War Inventions," *Los Angeles Times*, June 24, 1951, H7, H16–H17; C. A. Hedden, "Transformer for a Metal Locator," US Patent 2,129,058, filed July 11, 1936, issued September 6, 1938; "Amateur Inventors Produce New Devices to Aid U.S. at War," *Wall Street Journal*, August 25, 1943, 1.

72. Hotchner, "Wanted: War Inventions"; National Inventors Council, *Administrative History*, 16. For the quotation, see Lawrence Langner, "Secretary's Report," May 18, 1943, 5–6, Box 3, Minutes File, NIC Subject Files.

73. This section draws on Eric S. Hintz, "Portable Power: Inventor Samuel Ruben and the Birth of Duracell," *Technology and Culture* 50, no. 1 (January 2009): 24–57.

On the invitation and meeting with Taylor, see Samuel Ruben, *Necessity's Children: Memoirs of an Independent Inventor* (Portland, OR: Breitenbush Books, 1990), 84–85. On the incendiary bomb (October 20, 1941), poison-gas detector (December 10, 1941), low-temperature batteries (April 1, 1942), and several other concepts, see Ruben's "Laboratory Notebook, March 1941–September 1943," Box 3, Folder 3, Papers of Samuel Ruben, Science History Institute, Philadelphia, PA.

74. Ruben, *Necessity's Children*, 85–89; Maurice Friedman and Charles E. McCauley, "The Ruben Cell—a New Alkaline Primary Dry Cell Battery," *Transactions of the Electrochemical Society* 92 (1947): 195–215; Samuel Ruben, "Alkaline Dry Cell," US Patent 2,422,045, filed July 10, 1945, issued June 10, 1947.

75. Ruben, *Necessity's Children*, 89–94; P. R. Mallory & Co., Inc., "Annual Report, Year Ended December 31st 1944," 32–33; Philip Rogers Mallory, *Recollections: Fifty Years with the Company* (Indianapolis: P. R. Mallory Company, 1966), 101; National Inventors Council, *Administrative History*, 13.

76. On the $4.50 savings, see J. C. Green, "NIC Staff Operations," attached to NIC Meeting Minutes, November 22, 1944, Box 3, Minutes File, NIC Subject Files. On Ruben's long-term partnership with Mallory and Duracell, see chapter 3 and Hintz, "Portable Power."

77. For his unpublished autobiography, see Everett Bickley, "The Bickley Bean Sorter," compiled by Audrey Bickley Beyer, November 18, 1998, Box 1, Folder 2, Everett H. Bickley Collection, Archives Center, National Museum of American History, Smithsonian Institution, Washington, DC (hereafter Bickley Collection). See also chapter 3.

78. "Brief Biographies of the Council Members and Lists of Experts Serving with Them on the Council Committees," n.d., Box 1, Folder "Committees and Procedures for Committees," NIC Subject Files; Everett Bickley, "Everett H. Bickley a $1.00 a Year Man Made These Disclosures in Behalf of the War Effort to the National Inventors Council," Box 1, Folder 2, Bickley Collection. On the Cat's Eye, see Box 16, Folder 5, Bickley Collection. On the Talking Fish, see Box 16, Folder 14, Bickley Collection.

79. The quotations are from Everett H. Bickley to Henry A. Wallace, March 17, 1945. On the patent extension bills, see Audrey's handwritten notes, May 18, 1949, and a copy of House Subcommittee on Patents, Trademarks, and Copyrights, Judiciary Committee, *Extension of Patents in Certain Cases*, Hearings regarding H.R. 65, H.R. 124, H.R. 1107, and H.R. 1984, 80th Cong., 1st Sess., held April 9, May 9, and May 14, 1947. All documents are in Box 16, Folder 2, Bickley Collection.

80. "Hedy Lamarr Inventor," *New York Times*, October 1, 1941, 24; Hans-Joachim Braun, "Advanced Weaponry of the Stars," *American Heritage of Invention and Technology* 12, no. 4 (Spring 1997): 10–16; Richard Rhodes, *Hedy's Folly: The Life and Breakthrough Inventions of Hedy Lamarr, the Most Beautiful Woman in the World* (New

York: Doubleday, 2011); Hedy Lamarr, *Ecstasy and Me* (New York: Bartholomew House, 1966).

81. Rhodes, *Hedy's Folly*, 30–77, 133–136. For his autobiography, see George Antheil, *Bad Boy of Music* (Garden City, NY: Doubleday, 1945), quotation at 330.

82. Rhodes, *Hedy's Folly*, 147–169, 184–185; Braun, "Advanced Weaponry of the Stars." Lamarr patented under the last name of her second husband, Gene Markey; see Hedy Kiesler Markey and George Antheil, "Secret Communication System," US Patent 2,292,387, filed June 10, 1941, issued August 11, 1942.

83. Rhodes, *Hedy's Folly*, 185–214; the quotations are from George Antheil to William C. Bullitt, July 13, 1942, at 186–187. On spread spectrum, see Braun, "Advanced Weaponry of the Stars," 193.

84. For the two-year statistics, see Leonard Lent, "Chief Engineer's Report," 11, and on the luncheon generally, see Lawrence Langner, "National Inventors Council, Second Anniversary Luncheon, Rose Room, Hotel Washington," 22–23, both reports attached to NIC Meeting Minutes, September 22, 1942, Box 3, Minutes File, NIC Subject Files. See also "Flood Tide of Invention," *New York Times*, September 19, 1943, E8.

85. Alfred Toombs, "Battle of Brains," *American Magazine* 133, no. 3 (March 1942): 41, 52; Taylor quoted in Chase, "Second Call for Inventions!," 19.

86. Waldemar Kaempffert, "American Science Enlists," *New York Times Magazine*, November 2, 1941, 4, 29; NIC press release, "Torpedo Net or Screen for Ships Is Most Frequent Idea Submitted to National Inventor Council," September 22, 1942, Box 6, Folder "Publicity 1942," NIC Subject Files; Carl Dreher, "Inventions Can Win the War," *Popular Science* 141, no. 4 (October 1942): 64–69, 220, 222; National Inventors Council, *Information Bulletin No. 2*, 3–4, 18–22, Box 2, Folder "Information Bulletins, 1941–1955," NIC Subject Files.

87. C. C. Henry to Conway Coe, January 14, 1941, and William F. Roeser to Thomas R. Taylor, July 18, 1942, both in Box 5, Folder "Inventive Problems 1941–1942: Memorandum on Status of Procurement Problems," NIC Subject Files. On the use of development contracts in the NDRC and OSRD, see Larry Owens, "The Counterproductive Management of Science in the Second World War: Vannevar Bush and the Office of Scientific Research and Development," *Business History Review* 68, no. 4 (Winter 1994): 515–576.

88. E. L. Gustus to Colonel [Georges] Doriot, March 20, 1944, Box 5, Folder "Problem List 1943," NIC Subject Files.

89. On the navy's hesitance, see H. L. Phelps to L. B. Lent, June 21, 1943, Box 6, Folder "Problem List 1943," NIC Subject Files. On the army's vetting, see L. B. Lent to Walter A. Wood, Jr., May 29, 1943, and Wood to Lent, June 2, 1943, Box 6, Folder

"Problem List 1943," NIC Subject Files. The army list is "Some Problems in Which the Army Is Interested," July 1943, and the distribution list is "Technical Societies Which Received List of Special Army Problems," July 1943, both in Box 6, Folder "Problem List 1943," NIC Subject Files. On the promising response, see John C. Green to NIC Technical Staff, November 30, 1943, and L. B. Lent's form letter "To Liaison Officers with NIC," March 10, 1944, Box 5, Folder "Problem List 1943," NIC Subject Files.

90. "Inventions Wanted: Inventors Council Asks for More Ideas to Aid Armed Forces," *New York Times*, May 28, 1944, E9; V. C. Jones, "Inventions Wanted: Army and Navy Seeking Technical Help on 55 Battlefront Problems," *Wall Street Journal*, May 27, 1944, 1. For all NIC problem lists, see Box 5, Folder "Master List, 1943–1956," NIC Subject Files.

91. *National Inventors and Engineers Commission; Technological Mobilization*, 3 vols., Hearings on S. 2721, Before a Subcommittee of the Committee on Military Affairs, 77th Cong., 2nd Sess., October–December 1942.

92. For Kaempffert's testimony, see *Technological Mobilization*, 1:67–80, quotation at 68–69. For Sheridan's testimony, see *Technological Mobilization*, 1:152–172, quotations at 166, 157–158.

93. *National Inventors and Engineers Commission*, especially Johnson's remarks at 10, 17, 18, 26; W. Houston Kenyon, Jr., "Inventors' Agency Proposed," *New York Times*, April 9, 1942, 18.

94. "Statement on Behalf of the National Inventors Council with Respect to S. 2078," in *National Inventors and Engineers Commission*, 8–9; for Langner's testimony, including plans for a model-building fund, see pp. 2–20.

95. Lawrence Langner, "Information Regarding Construction of Trial Devices by the National Bureau of Standards for the National Inventors Council," n.d. (circa March–April 1942), Box 4, Folder "Model Building Fund—1942," NIC Subject Files. On the Allwood fuses, see R. W. Lewis to R. M. Osborne, November 3, 1943, Box 4, Folder "Model Building Fund—1942," NIC Subject Files. On the Isteg reinforcing bars, see Frederick Feiker to L. B. Lent, August 25, 1942, Box 4, Folder "Model Building Fund—1942," NIC Subject Files. On the total expenditures, see National Bureau of Standards, "Cooperation with the National Inventors Council," n.d. (circa 1945–1946), Box 4, Folder "Model Building Fund—1942," NIC Subject Files.

96. On deliberations regarding the peacetime council, see Lawrence Langner to Orville Wright, October 11, 1944, and for the quotation, see Charles F. Kettering to Jesse H. Jones, October 11, 1944, both in Box 49, Folder 17, Wilbur and Orville Wright Papers, Manuscript Division, Library of Congress, Washington, DC. See also Lawrence Langner, "Plan for Continuance of the National Inventors Council and Extension of the Facilities of the Council and Staff for Peace Usages," attached to NIC Meeting Minutes, May 22, 1945, Box 3, Minutes File, NIC Subject Files.

97. *To Establish an Office of Technical Services in the Department of Commerce*, Hearings on S. 1248, Before a Subcommittee of the Committee on Commerce, 79th Cong., 1st Sess., December 12–14, 1945; Fulbright's "applied science" quotation is at p. 8. For his National Science Foundation proposal, see Vannevar Bush, *Science, the Endless Frontier* (Washington, DC: Government Printing Office, 1945). For an overview, see Daniel L. Kleinman, *Politics on the Endless Frontier: Postwar Research Policy in the United States* (Durham, NC: Duke University Press, 1995).

98. See the OTS organizational chart, October 1946, in Department of Commerce, "Office of Technical Services: Salaries and Expenses," 40, and also Inventions and Engineering Division, Inventors Service, "Purpose and Activities Summary," 1–3, both in Box 1, Folder "Budget and Organization," NIC Subject Files.

99. National Inventors Council, *Administrative History*, 9–12.

100. National Inventors Council, *Administrative History*, 9–11.

101. National Inventors Council, *Administrative History*, 11.

102. "In days of war, when novel needs define themselves and the urgency of the times sharpens men's wits, the individual inventor comes into his own. For then he is looked on as the servant of the community and not as the butt of its ridicule, his wasteful mistakes are condoned because of his occasional successes, his singleness of purpose is considered a virtue and not a nuisance." See John Jewkes, David Sawers, and Richard Stillerman, *The Sources of Invention* (London: Macmillan, 1958), 255.

103. Jenn Gustetic, "Innovation for Every American," in *Does America Need More Innovators?*, ed. Matthew Wisnioski, Eric S. Hintz, and Marie Stettler Kleine (Cambridge, MA: MIT Press, 2019); David C. Mowery, "Plus ca change: Industrial R&D in the 'Third Industrial Revolution,'" *Industrial and Corporate Change* 18, no. 1 (February 2009): 1–50.

Chapter 7

1. John Kenneth Galbraith, *American Capitalism: The Concept of Countervailing Power* (New York: Houghton Mifflin, 1952), 86–87; Ian Mount, "The Return of the Lone Inventor," *Fortune Small Business* (March 2005): 18; "Out of the Dusty Labs: The Rise and Fall of Corporate R&D," *The Economist*, March 3, 2007, 74–76.

2. David A. Hounshell, "The Evolution of Industrial Research in the United States," in *Engines of Innovation: U.S. Industrial Research at the End of an Era*, ed. Richard S. Rosenbloom and William J. Spencer (Boston: Harvard Business School Press, 1996), 41–58; David C. Mowery and Nathan Rosenberg, *Paths of Innovation: Technological Change in 20th-Century America* (Cambridge: Cambridge University Press, 1998), 30–46; David C. Mowery, "Plus ca change: Industrial R&D in the 'Third Industrial Revolution,'" *Industrial and Corporate Change* 18, no. 1 (February 2009): 1–50; Sally

H. Clarke, Naomi R. Lamoreaux, and Steven W. Usselman, eds., *The Challenge of Remaining Innovative: Insights from Twentieth-Century American Business* (Stanford, CA: Stanford Business Books, 2009).

3. Barkev S. Sanders, "The Number of Patentees in the United States," *IDEA* 9, no. 2 (Summer 1965): 205–221; Organisation for Economic Co-operation and Development, "R-D Personnel by Sector of Employment and Occupation, US Researchers (Excluding Technicians and Support Personnel), Business Enterprise," May 17, 2017, https://stats.oecd.org/Index.aspx?DataSetCode=PERS_FUNC.

4. "Patent Applications Filed and Patents Issued, by Type of Patent and Patentee: 1790–2000," in *Historical Statistics of the United States: Earliest Times to the Present*, millennial ed., ed. Susan Carter et al. (New York: Cambridge University Press, 2006), 3: 425–429, table Cg27–37; US Patent and Trademark Office, "Patenting by Organizations (Utility Patents) 2015," table A1–1b: Breakout by Ownership Category, accessed September 6, 2019, https://www.uspto.gov/web/offices/ac/ido/oeip/taf/topo_15.htm#PartA1_1b; Eric Dahlin, "Are Independent Inventors a Relic of the Past? A Study of Independent Patent Inventors in the United States, 1975–2014" (unpublished manuscript, August 5, 2016), Microsoft Word file.

5. Mowery, "Plus ca change," 7–9.

6. Hounshell, "The Evolution of Industrial Research in the United States," 41–58; Peter Galison and Bruce Hevly, eds., *Big Science: The Growth of Large-Scale Research* (Stanford, CA: Stanford University Press, 1992); Larry Owens, "The Counterproductive Management of Science in the Second World War: Vannevar Bush and the Office of Scientific Research and Development," *Business History Review* 68, no. 4 (Winter 1994): 515–576; Margaret B. W. Graham, "Entrepreneurship in the United States, 1920–2000," in *The Invention of Enterprise: Entrepreneurship from Ancient Mesopotamia to Modern Times*, ed. David S. Landes, Joel Mokyr, and William J. Baumol (Princeton, NJ: Princeton University Press, 2010), 410, 422–423.

7. David M. Hart, "Antitrust and Technological Innovation in the US: Ideas, Institutions, Decisions, and Impacts, 1890–2000," *Research Policy* 30, no. 6 (June 2001): 923–936; Mowery and Rosenberg, *Paths of Innovation*, 39–40; David A. Hounshell and John K. Smith, Jr., *Science and Corporate Strategy: Du Pont R&D, 1902–1980* (New York: Cambridge University Press, 1988), 331–365; Internal Revenue Service, "U.S. Corporation Income Tax: Tax Brackets and Rates, 1909–2010," accessed September 8, 2019, https://www.irs.gov/pub/irs-soi/histabb.xls.

8. A compulsory license forces a firm to offer licenses on its proprietary, patented technologies, the very patents that helped the firm sustain its monopoly. The licensing terms and royalty provisions are determined via a negotiated consent decree or by the judge (or jury) following conviction in an antitrust trial.

9. Pat Choate, *Saving Capitalism: Keeping America Strong* (New York: Vintage, 2009), 116–117; Martin Watzinger, Thomas A. Fackler, Markus Nagler, and Monica

Schnitzer, "How Antitrust Enforcement Can Spur Innovation: Bell Labs and the 1956 Consent Decree," Centre for Economic Policy Research discussion paper no. DP11793, January 9, 2017, https://cepr.org/active/publications/discussion_papers /dp.php?dpno=11793.

10. Jon Gertner, *The Idea Factory: Bell Labs and the Great Age of American Innovation* (New York: Penguin, 2012); Watzinger et al., "How Antitrust Enforcement Can Spur Innovation," 7–8.

11. Watzinger et al., "How Antitrust Enforcement Can Spur Innovation"; Choate, *Saving Capitalism*, 114–116.

12. Michael Hiltzik, *Dealers of Lightning: Xerox PARC and the Dawn of the Computer Age* (New York: Harper Business, 1999); Douglas K. Smith and Robert C. Alexander, *Fumbling the Future: How Xerox Invented, Then Ignored, the Personal Computer* (New York: William Morrow, 1988); Henry Chesbrough, "Graceful Exits and Missed Opportunities: Xerox's Management of Its Technology Spin-off Organizations," *Business History Review* 76, no. 4 (Winter 2002): 803–837.

13. Margaret B. W. Graham, *RCA and the VideoDisc: The Business of Research* (Cambridge: Cambridge University Press, 1986); Ronald K. Fierstein, *A Triumph of Genius: Edwin Land, Polaroid, and the Kodak Patent War* (Chicago: Ankerwycke, 2015); Gertner, *The Idea Factory*, 338.

14. Richard S. Rosenbloom and William J. Spencer, "Introduction: Technology's Vanishing Wellspring," in *Engines of Innovation: U. S. Industrial Research at the End of an Era*, ed. Richard S. Rosenbloom and William J. Spencer (Boston: Harvard Business School Press, 1996); Graham, "Entrepreneurship in the United States, 1920–2000," 422–428; Mowery, "Plus ca change"; Henry W. Chesbrough, *Open Innovation: The New Imperative for Creating and Profiting from Technology* (Boston: Harvard Business School Press, 2003).

15. Robert W. Gore, *The Early Days of W. L. Gore & Associates, Inc.* (Newark, DE: W. L. Gore & Associates, 2008); Robert Sorrels, "The Culture of Innovation: An Ethnography of W. L. Gore and Associates" (EdD thesis, Northern Arizona University, 1986).

16. Lillian Hoddeson and Peter Garrett, *The Man Who Saw Tomorrow: The Life and Inventions of Stanford R. Ovshinsky* (Cambridge, MA: MIT Press, 2018).

17. Robert McG. Thomas, Jr., "Marion Donovan, 81, Solver of the Damp-Diaper Problem," *New York Times*, November 18, 1998, B15; Darice Bailer, "Inventor Sinks Teeth into Work," *Greenwich Time*, June 13, 1993, B7. On the Boater, see several documents in Box 2, Folders 1–6, Marion O'Brien Donovan Papers, Archives Center, National Museum of American History, Smithsonian Institution, Washington, DC.

18. See the memoir, Jacob Rabinow, *Inventing for Fun and Profit* (San Francisco: San Francisco Press, 1990); Kenneth A. Brown, "Jacob Rabinow," in *Inventors at Work: Interviews with 16 Notable American Inventors*, ed. Kenneth A. Brown (Redmond, WA: Microsoft Press, 1988),183–217.

19. Steven W. Usselman, "Unbundling IBM: Antitrust and the Incentives to Innovation in American Computing," in Clarke, Lamoreaux, and Usselman, *The Challenge of Remaining Innovative*, 249–279.

20. Usselman, "Unbundling IBM"; John Markoff, "Alan F. Shugart, 76, a Developer of Disk Drive Industry, Dies," *New York Times*, December 15, 2006, C13; Clayton M. Christensen, "The Rigid Disk Drive Industry: A History of Commercial and Technological Turbulence," *Business History Review* 67, no. 4 (Winter 1993): 531–588.

21. Paul Freiberger and Michael Swaine, *Fire in the Valley: The Making of the Personal Computer* (Berkeley, CA: Osborne / McGraw-Hill, 1984); Everett M. Rogers and Judith K. Larsen, *Silicon Valley Fever: Growth of High Technology Culture* (New York: Basic Books, 1984).

22. Michael Malone, *Infinite Loop: How the World's Most Insanely Great Computer Company Went Insane* (New York: Currency / Doubleday, 1999), 51–81.

23. Paul Ceruzzi, *A History of Modern Computing*, 2nd ed. (Cambridge, MA: MIT Press, 2003), 233–236, 263–280; Graham, "Entrepreneurship in the United States, 1920–2000," 417.

24. Chesbrough, *Open Innovation*; Dietmar Harhoff and Karim R. Lakhani, eds., *Revolutionizing Innovation: Users, Communities, and Open Innovation* (Cambridge, MA: MIT Press, 2016).

25. Steve Lohr, "The Corporate Lab as Ringmaster," *New York Times*, August 16, 2009, B3; "Out of the Dusty Labs"; Mowery, "Plus ca change," 13–35.

26. Larry Huston and Nabil Sakkab, "Connect and Develop: Inside Procter & Gamble's New Model for Innovation," *Harvard Business Review* 84, no. 3 (March 2006): 58–66.

27. Dan Alexander, "The Invention Machine: Cleveland Duo Churns Out Ideas Worth Billions," *Forbes* 193, no. 3 (March 3, 2014): 34–36; Marcia Pledger, "Cleveland-Based Nottingham Spirk Proves to Be a Quiet but Serious Force for Product Developments Nationwide," *Cleveland Plain Dealer*, July 20, 2015, https://www.cleveland.com/business/2015/07/cleveland-based_nottingham_spi.html.

28. Karim R. Lakhani, "InnoCentive.com," Harvard Business School Case 608–170, June 2008, revised October 2009; Cornelia Dean, "If You Have a Problem, Ask Everyone," *New York Times*, July 22, 2008, F1.

29. Eric S. Hintz, "Creative Financing: The Rise of Cash Prizes for Innovation Is a Response to Changing Business Conditions—and a Return to a Winning Strategy," *Wall Street Journal*, September 27, 2010, R8; Jacob Silverman, "The Crowdsourcing Scam: Why Do You Deceive Yourself?," *The Baffler* 26 (October 2014), https://thebaffler.com/salvos/crowdsourcing-scam.

30. For the statistics, see InnoCentive, "About Us," accessed September 11, 2019, https://www.innocentive.com/about-us/.

31. Dean, "If You Have a Problem, Ask Everyone"; James L. Kearns, LinkedIn, accessed September 11, 2019, https://www.linkedin.com/in/dbajameslkearns; InnoCentive, "InnoCentive—Top Solvers," 2017, https://innocentive.com/ar/challenge/topSolvers.

32. Green's testimony in *Inventive Contributions Awards Board*, Hearings Before Subcommittee No. 2 of the House Judiciary Committee on H.R. 7316, May 14, 1952, 82nd Cong., 2nd Sess., 6–11, quotation at 10. See also House Judiciary Committee, *Inventive Contributions Awards*, House Report no. 148, 85th Cong., 1st Sess., February 21, 1957, 7–9; and the NIC-commissioned study, James N. Mosel, Barkev S. Sanders, and Irving H. Siegel, "Incentives and Deterrents to Inventing for National Defense," *Patent, Trademark, and Copyright Journal of Research and Education* 1, no. 2 (December 1957): 185–215.

33. National Bureau of Economic Research, ed., *The Rate and Direction of Inventive Activity: Economic and Social Factors* (Princeton, NJ: Princeton University Press, 1962); David A. Hounshell, "The Medium Is the Message, or How Context Matters: The RAND Corporation Builds an Economics of Innovation, 1946–1962," in *Systems, Experts, and Computers: The Systems Approach in Management and Engineering, World War II and After*, ed. Agatha C. Hughes and Thomas P. Hughes (Cambridge, MA: MIT Press, 2000), 255–310.

34. On the innovation experts, see Matthew Wisnioski, *Every American an Innovator* (Cambridge, MA: MIT Press, forthcoming), chapter 4, "A Nation of Innovators." Humphrey's quotation, on January 21, 1966, is from "More Inventive America," *Science News Letter* 89, no. 6 (February 5, 1966): 84. The key policy statement was Department of Commerce, *Technological Innovation: Its Environment and Management* (Washington, DC: US Government Printing Office, 1967).

35. Richard M. Nixon, "Special Message to the Congress on Science and Technology," March 16, 1972, in *The American Presidency Project*, ed. Gerhard Peters and John T. Woolley, accessed September 11, 2019, https://www.presidency.ucsb.edu/node/255178.

36. Deborah Shapley, "The Presidential Prize Caper," *Science* 183, no. 4128 (March 8, 1974): 938. On the NIC's privatization, see John Cavicchi, "The Mystery of the Classified and Missing National Inventors Council (NIC) Files," IPMall, accessed September 11, 2019, http://www.ipmall.info/content/mystery-classified-and-missing-national-inventors-council-nic-files.

37. Judith Stein, *Pivotal Decade: How the United States Traded Factories for Finance in the Seventies* (New Haven, CT: Yale University Press, 2010); Barry Bluestone and Bennett Harrison, *The Deindustrialization of America: Plant Closings, Community Abandonment, and the Dismantling of Basic Industry* (New York: Basic Books, 1984).

38. MIT economist David L. Birch was the key theorist of small business job creation. See his white paper, David L. Birch, *The Job Generation Process* (Cambridge,

MA: MIT Program on Neighborhood and Regional Change, 1979); David L. Birch, "Who Creates Jobs?," *Public Interest* 65, no. 3 (Fall 1981): 3–14; David L. Birch, *Job Creation in America: How Our Smallest Companies Put the Most People to Work* (New York: Free Press, 1987).

39. Josh Lerner, "The Government as Venture Capitalist: The Long-Run Impact of the SBIR Program," *Journal of Business* 72, no. 3 (July 1999): 285–318; David B. Audretsch, "Standing on the Shoulders of Midgets: The U.S. Small Business Innovation Research Program (SBIR)," *Small Business Economics* 20, no. 2 (March 2003): 129–135.

40. Tom Nicholas, *VC: An American History* (Cambridge, MA: Harvard University Press, 2019), 173–181.

41. David C. Mowery, Richard R. Nelson, Bhaven N. Sampat, and Arvids A. Ziedonis, *Ivory Tower and Industrial Innovation: University-Industry Technology Transfer Before and After the Bayh-Dole Act* (Stanford, CA: Stanford University Press, 2004); Adam B. Jaffe and Josh Lerner, "Reinventing Public R&D: Patent Policy and the Commercialization of National Laboratory Technologies," *RAND Journal of Economics* 32, no. 1 (Spring 2001): 167–198; Janet Abbate, *Inventing the Internet* (Cambridge, MA: MIT Press, 1999), 113–146.

42. Sally Smith Hughes, *Genentech: The Beginnings of Biotech* (Chicago: University of Chicago Press, 2011).

43. Paul E. Ceruzzi, *GPS* (Cambridge, MA: MIT Press, 2018), 105–122.

44. Richard L. Brandt, *One Click: Jeff Bezos and the Rise of Amazon.com* (New York: Portfolio / Penguin, 2011); Brad Stone, *The Everything Store: Jeff Bezos and the Age of Amazon* (New York: Little, Brown, 2014). The "one-click" patent is Peri Hartman, Jeffrey P. Bezos, Shel Kaphan, and Joel Speigel, "Method and System for Placing a Purchase Order via a Communications Network," US Patent 5,960,411, filed September 12, 1997, issued September 28, 1999.

45. David A. Vise and Mark Malseed, *The Google Story: Inside the Hottest Business, Media, and Technology Success of Our Time* (New York: Delacorte Press, 2005); Lawrence Page, "Method for Node Ranking in a Linked Database," US Patent 6,285,999, filed January 9, 1998, issued September 4, 2001. On Page and Brin's NSF funding, see David Hart, "On the Origins of Google," August 17, 2004, https://www.nsf.gov/discoveries/disc_summ.jsp?cntn_id=100660.

46. Jack Hitt, "The Amateur Future of Space Travel," *New York Times Magazine*, July 1, 2007, 40–47, 66, 80; Flagsuit LLC, "About Flagsuit," accessed September 13, 2019, http://www.flagsuit.com/About.html; Peter K. Homer, "Glove Reinforcement and Method Thereof," US Patent 8,181,276 B2, filed April 30, 2008, issued May 22, 2012.

47. The original America COMPETES Act of 2007 was designed to "Create Opportunities to Meaningfully Promote Excellence in Technology, Education, and Science."

On government crowdsourcing initiatives, see Jenn Gustetic, "Innovation for Every American," in *Does America Need More Innovators?*, ed. Matthew Wisnioski, Eric S. Hintz, and Marie Stettler Kleine (Cambridge, MA: MIT Press, 2019), 112–117.

48. Pamphlet, United Inventors and Scientists of America, n.d. (circa 1950s), Box 5, Folder 12, Joseph B. Friedman Papers, Archives Center, National Museum of American History, Smithsonian Institution, Washington, DC (hereafter Friedman Papers). On Resnick, see Charles Hillinger, "Inventor, 90, in No Hurry to Retire or Marry," *Los Angeles Times*, February 11, 1974, A1; quotation in Lee Shippey, "Leeside," *Los Angeles Times*, July 26, 1945, A4.

49. UISA pamphlet, Friedman Papers; Chester Vonier, "The Inventor's Lot Is Not, Alas, a Happy One," *Los Angeles Times*, May 3, 1949, A5; Sue Banashek, "Inventors' Dreams Put on Display," *Los Angeles Times*, July 30, 1974, B1; quotation in Hillinger, "Inventor, 90, in No Hurry to Retire."

50. Peter Whalley, "The Social Practice of Independent Inventing," *Science, Technology, and Human Values* 16, no. 2 (Spring 1991): 208–232 at 211; Don Lancaster, "Inventors Associations," *The Blatant Opportunist* 18 (November–December 1992), www.tinaja.com/glib/invenorg.pdf. For Anchorage's year 2000 population, see US Census Bureau, search utility, accessed October 20, 2020, https://data.census.gov/cedsci/table?g=1600000US0203000&tid=DECENNIALSF12000.P001&hidePreview=false.

51. United Inventors Association, "History," April 17, 2009, https://web.archive.org/web/20090417111335/http://www.uiausa.org/Default.aspx?page=93.

52. On the Chicago Inventors' Council, see Whalley, "The Social Practice of Independent Inventing," 220; Inventors Network of the Capital Area, "Welcome," accessed September 16, 2019, http://www.dcinventors.org/; "Member Inventions," accessed September 16, 2019, http://www.dcinventors.org/inventions-from-our-members/.

53. Enventys Partners, "Bunch O Balloons," accessed September 16, 2019, https://enventyspartners.com/case-study/bunch-o-balloons/; Katie Kindelan, "Father of Eight Creates Genius Water Balloon Invention," *ABC News*, July 24, 2019, https://abcnews.go.com/Lifestyle/father-creates-genius-water-balloon-invention/story?id=24698486.

54. Kickstarter, "Stats," accessed September 19, 2019, https://www.kickstarter.com/help/stats. In comparison, traditional venture capitalists invested $84 billion in 2017 alone; see National Venture Capital Association, "Record Unicorn Financings Drove 2017 Total Venture Capital Investment to $84 Billion, the Largest Amount Since Dot-Com Era," January 9, 2018, https://nvca.org/pressreleases/record-unicorn-financings-drove-2017-total-venture-capital-investments-84-billion-largest-amount-since-dot-com-era/.

55. Facebook, "Company Info," statistics as of June 30, 2019, https://newsroom.fb.com/company-info/; J. Clement, "Number of Apps Available in Leading App

Stores as of 3rd Quarter 2019," Statista, October 9, 2019, https://www.statista.com /statistics/276623/number-of-apps-available-in-leading-app-stores/.

56. Ashlee Vance, "Inventors Wanted. Cool Tools Provided," *New York Times*, April 11, 2010, BU4; TechShop, "Welcome," October 9, 2016, https://web.archive.org/web /20161009152058/http://www.techshop.ws/. On the machine shop as incubator, see Paul Israel, *From Machine Shop to Industrial Laboratory: Telegraphy and the Changing Context of American Invention, 1830–1920* (Baltimore: Johns Hopkins University Press, 1992).

57. This section draws on Eric S. Hintz, "Failed Inventor Initiatives, from the Franklin Institute to Quirky," in Wisnioski, Hintz, and Kleine, *Does America Need More Innovators?*, 165–189.

58. Steve Lohr, "The Invention Mob Brought to You by Quirky," *New York Times*, February 15, 2015, BU1; Andy Jordan, "Tech Diary: 'Quirky' Inventions Get a Home Online," *Wall Street Journal*, August 24, 2011, http://blogs.wsj.com/digits/2011/08 /24/tech-diary-quirky-inventions-get-a-home-online/; Kaufman quotation in Kevin Chupfka, "Quirky Allows Anyone to Become an Inventor," *Yahoo! Finance*, September 17, 2013, http://finance.yahoo.com/blogs/breakout/quirky-allows-anyone -become-inventor-160333336.html.

59. Adrienne Burk, "Inventor of a Smart Air Conditioner Will Get a Cool Half Million," *Yahoo! Small Business*, May 2014, https://smallbusiness.yahoo.com/advisor /inventor-of-a-smart-air-conditioner-will-get-a-cool-half-million-181054486.html; Pete Pichaskie, "Columbia Man's Air Conditioner Invention Picked Up by Quirky," *Baltimore Sun*, May 1, 2014, http://www.baltimoresun.com/news/maryland/howard /columbia/ph-ho-cf-leslie-0501-20140529-story.html; Allison Bailes, "Is the Aros Smart Window Air Conditioner Worth the Hype?," *Energy Vanguard*, July 15, 2014, https://www.energyvanguard.com/blog/76125/Is-the-Aros-Smart-Window-Air -Conditioner-Worth-the-Hype.

60. Jack Anzarouth, "Quirky Appoints New President, Is Primed for Relaunch," *PRWeb*, March 16, 2017, http://www.prweb.com/releases/2017/03/prweb14155691 .htm; "June Update: What's Going On," *Quirky* blog, June 12, 2017, https://web .archive.org/web/20170719113258/http://aquirkyblog.squarespace.com/home/2017 /6/june-update-whats-going-on.

61. Stephanie Gleason, "Invention Platform Quirky Relaunches," *Wall Street Journal*, March 17, 2017, https://www.wsj.com/articles/invention-platform-quirky-relaunches -1489787519; Taron Foxworth, "What the Return of Quirky Means for Us," *Medium*, April 7, 2017, https://medium.com/stay-connected/what-the-return-of-quirky-means -for-us-42c31d6210d5; Eagledancing, comment dated April 7, 2017, on Gina Waldhorn, "A Letter from Our President, Gina Waldhorn," *Quirky* blog, March 21, 2017, https://shop.quirky.com/blogs/news/a-letter-from-our-president-gina-waldhorn.

62. Courtney Linder, "TechShop to Close All U.S. Locations Immediately, Filing Chapter 7 Bankruptcy," *Pittsburgh Post-Gazette*, November 15, 2017, http://www

.post-gazette.com/business/tech-news/2017/11/15/TechShop-to-close-all-locations
-immediately-following-Chapter-7-bankruptcy/stories/201711150170.

63. *Caveat Inventor: Invention Marketing Scams*, Hearing Before the Subcommittee on Regulation and Government Information of the Committee on Governmental Affairs, United States Senate, 103rd Cong., 2nd Sess., September 2, 1994; Edward R. Ergenzinger, Jr., "The American Inventor's Protection Act: A Legislative History," *Wake Forest Intellectual Property Law Journal* 7, no. 1 (2006–2007): 145–172.

64. Brittany Shammas, "A Miami Beach Scam Took Millions of Dollars from Thousands of Inventors Feds Say," *Miami New Times*, August 22, 2017, https://www.miaminewtimes.com/news/world-patent-marketing-scam-took-millions-of-dollars-from-inventors-9605870; Adam Goldman and Frances Robles, "Acting Attorney General Was on Board of Company Closed for Fraud," *New York Times*, November 9, 2018, A19.

65. Patent Act of 1952, Public Law No. 82–593, 1952, quotations from 35 U.S.C. § 103; "The Standard of Patentability: Judicial Interpretation of Section 103 of the Patent Act," *Columbia Law Review* 63, no. 2 (February 1963): 306–325; Paul Israel, "The Flash of Genius: Defining Invention in the Era of Corporate Research," unpublished paper, ConTexts of Invention Conference, Cleveland, OH, April 2006, revision note quoted at 14.

66. Gerardo Con Díaz, *Software Rights: How Patent Law Transformed Software Development in America* (New Haven, CT: Yale University Press, 2019).

67. Daniel J. Kevles, "Ananda Chakrabarty Wins a Patent: Biotechnology, Law, and Society, 1972–1980," *Historical Studies in the Physical and Biological Sciences* 25, no. 1 (1994): 111–135.

68. Rochelle Cooper Dreyfuss, "The Federal Circuit: A Case Study in Specialized Courts," *New York University Law Review* 64, no. 1 (April 1989): 1–77; Gerald Sobel, "The Court of Appeals for the Federal Circuit: A Fifth Anniversary Look at Its Impact on Patent Law and Litigation," *American University Law Review* 37, no. 4 (1988): 1087–1139. For the validity statistics, see William M. Landes and Richard A. Posner, *The Economic Structure of Intellectual Property Law* (Cambridge, MA: Belknap Press, 2003), 336–339.

69. John Seabrook, "The Flash of Genius," in *Flash of Genius and Other True Stories of Invention* (New York: St. Martin's, 2008), 1–31.

70. The views and opinions expressed about Jerome Lemelson's career are the author's and do not necessarily reflect the official policy or position of the Smithsonian Institution's Jerome and Dorothy Lemelson Center for the Study of Invention and Innovation.

71. Tom Wolfe, "Land of Wizards," *Popular Mechanics* 163, no. 7 (July 1986): 126–139; Kenneth A. Brown, "Jerome Lemelson," in Brown, *Inventors at Work*, 121–146,

quotation at 140; William Green, "Patently Creative: Is Jerry Lemelson a Great Inventor, or Just a Great Litigator?," *Boston Globe Magazine*, May 1, 1994, 14–15, 28–34; Lawrence D. Maloney, "Lone Wolf of the Sierras," *Design News*, March 6, 1995, 71–84.

72. Wolfe, "Land of Wizards"; Brown, "Jerome Lemelson"; Green, "Patently Creative"; Maloney, "Lone Wolf of the Sierras." For a list of Lemelson's 606 US patents, see Martha Davidson, "Jerome Lemelson," Lemelson Center for the Study of Invention and Innovation, Smithsonian Institution, accessed May 9, 2020, https://invention.si.edu/jerome-lemelson.

73. Brown, "Jerome Lemelson," 130–131; Wolfe, "Land of Wizards," 134; Maloney, "Lone Wolf of the Sierras," 80. On the IBM job offer and licensing deal, see Lemelson's attorney Gerald D. Hosier, "Oral History Interview with Gerald D. Hosier," conducted by Arthur Daemmrich and Arthur Molella, December 13–14, 2016, and October 17, 2017, 55, 110, Archives Center, National Museum of American History, Smithsonian Institution, Washington, DC (hereafter Hosier Oral History).

74. For Lemelson's testimony, see *Patent Policy*, Hearings (on S.1215) Before the Subcommittee on Science, Technology, and Space of the Committee on Commerce, Science, and Transportation, United States Senate, 96th Cong., 1st Sess., July 27, 1979, 320–324. The relevant cases are Lemelson v. Kellogg Company, 440 F.2d 986 (2d Cir. 1971); Lemelson v. TRW, Inc., 760 F.2d 1254 (Fed. Cir. 1985); Centseable Products, Inc. v. Lemelson, 591 F.2d 400 (7th Cir. 1962). See Lemelson's 1970s patents in Davidson, "Jerome Lemelson."

75. Jerome H. Lemelson, "Combination Tools," US Patent 3,227,012, filed December 24, 1954, issued January 4, 1966; Jerome H. Lemelson, "Automatic Measurement Apparatus," US Patent, 3,081,379, filed December 4, 1956, issued March 12, 1963.

76. Jerome H. Lemelson, "Scanning Apparatus and Method," US Patent 4,338,626, filed February 16, 1979, issued July 6, 1982. On Lemelson's machine vision strategy, see Hosier Oral History, 58–73, 82–92, 112–123.

77. Hosier Oral History, 70, 82–88.

78. Hosier Oral History, 83–91, 115–119; Lemelson quoted in Green, "Patently Creative," 28. Critical articles include Edmund L. Andrews, "Rich in the 90s on Ideas Hatched in the 50s," *New York Times*, November 13, 1992, A1; Otis Port, "Inspiration, Perspiration, or Manipulation?," *Business Week*, April 3, 1995, 56–57; Bernard Wysocki, Jr., "Engineer Makes a Fortune on Patent Infringement Suits," *Wall Street Journal*, April 9, 1997, 1; Nicholas Varchaver, "The Patent King," *Fortune*, May 14, 2001, 203–216.

79. Robert McG. Thomas, Jr., "Jerome H. Lemelson, an Inventor, Dies at 74," *New York Times*, October 4, 1997, 16; Hosier Oral History, 89–91, 112; Prager, quoted in Maloney, "Lone Wolf of the Sierras," 73.

80. Gilbert P. Hyatt, "Single Chip Integrated Circuit Computer Architecture," US Patent 4,942,516, filed June 17, 1988, issued July 17, 1990; Dean Takahashi, "Chip

Designer's 20-Year Quest," *Los Angeles Times*, October 21, 1990, OCD1; Dean Taka-hashi, "Gil Hyatt Interview," *VentureBeat*, August 31, 2018, https://venturebeat.com /2018/08/31/gil-hyatt-interview-why-patent-examiners-gave-controversial-patents-a -scarlet-letter/. On Detkin, see Seabrook, "Introduction," in *Flash of Genius*, xx.

81. Stephen A. Merrill, Richard C. Levin, and Mark B. Myers, eds., *A Patent System for the 21st Century* (Washington, DC: National Academies Press, 2004); Adam B. Jaffe and Josh Lerner, *Innovation and Its Discontents: How Our Broken Patent System Is Endangering Innovation and Progress, and What to Do About It* (Princeton, NJ: Princeton University Press, 2007); James Bessen and Michael J. Meurer, *Patent Failure: How Judges, Bureaucrats, and Lawyers Put Innovators at Risk* (Princeton, NJ: Princeton University Press, 2008).

82. Arthur A. Daemmrich, "Stalemate at the WTO: TRIPS, Agricultural Subsidies, and the Doha Round," Harvard Business School teaching note 9–711–043, revised May 31, 2011; Ergenzinger, "The American Inventor's Protection Act," 151–153; Robert P. Merges, Peter S. Menell, and Mark A. Lemley, *Intellectual Property in the New Technological Age*, 6th ed. (New York: Wolters Kluwer Law and Business, 2012), 397–399.

83. Eric Schmitt, "Senate Panel Approves Bill on Patent Revisions," *New York Times*, November 3, 1999, C1; Selby, quoted in John Schwartz, "Inventors Say Proposed Patent Law Will Lead to Stealing Ideas," *Washington Post*, November 4, 1999, A8.

84. Saul Hansell, "As Patents Multiply, Web Sites Find Lawsuits Are a Click Away," *New York Times*, December 11, 1999, A1; Stephen D. Messer, "Data Processing System for Integrated Tracking and Management of Commerce Related Activities on a Public Access Network," US Patent 5,991,740A, filed June 10, 1997, issued November 23, 1999.

85. Merges, Menell, and Lemley, *Intellectual Property in the New Technological Age*, 290; Schmitt, "Senate Panel Approves Bill on Patent Revisions."

86. National Economic Council and Office of Science and Technology Policy, *A Strategy for American Innovation*, September 2009; Arti Rai, Stuart Graham, and Mark Doms, *Patent Reform: Unleashing Innovation, Promoting Economic Growth & Producing High-Paying Jobs* (Washington, DC: Department of Commerce, 2010), quotation at 1; Public Law 112–29, 125 Stat. 284, *Leahy-Smith America Invents Act*, 112th Cong., 1st Sess., enacted September 12, 2011.

87. Edward Wyatt, "Fighting Backlog in Patents, Senate Approves Overhaul," *New York Times*, September 9, 2011, B4.

88. Amanda Becker, "Patent Reform Measure Ignited Fierce Lobbying Effort," *Washington Post*, March 27, 2011, https://www.washingtonpost.com/capital_business/patent -reform-measure-ignited-fierce-lobbying-effort/2011/03/25/AFzD9VkB_story.html.

89. Becker, "Patent Reform Measure Ignited Fierce Lobbying Effort"; Gaydos quoted in Wyatt, "Fighting Backlog in Patents, Senate Approves Overhaul"; Poltorak quoted in

Hayley Tsukayama, "Q&A: Small Inventors Raise Patent Reform Concerns," *Washington Post*, March 28, 2011, https://www.washingtonpost.com/blogs/post-tech/post/qanda-small-inventors-raise-patent-reform-concerns/2011/03/28/AFLJ9NpB_blog.html.

90. Pat Choate and Joan Maginnis, "Raid on Gibraltar: How the U.S. Patent System Was Rigged against Independent Inventors," *IPWatchdog* blog, September 17, 2017, http://www.ipwatchdog.com/2017/09/17/u-s-patent-system-rigged-against-independent-inventors/id=88008/, citing United States Patent and Trademark Office, *Trial Statistics: IPR, PGR, CBM, Patent Trial and Appeal Board*, May 2017, https://www.uspto.gov/sites/default/files/documents/Trial_Statistics_2017-05-31.pdf.

91. Michelle Malkin, "The Bipartisan 'Stick It to Individual Inventors' Act," *National Review*, May 22, 2015, https://www.nationalreview.com/2015/05/how-obama-radically-transformed-americas-patent-system-michelle-malkin/.

92. A Google N-Gram of the word "crackpot" shows its peak usage during the 1940s and 1950s; see https://books.google.com/ngrams, accessed September 20, 2019.

93. Kathryn Steen, "What's 'The Big Idea'?: Patents and the Ideology of Invention," unpublished paper, Society for the History of Technology meeting, October 2009, Pittsburgh, PA.

94. James Phinney Baxter, *Scientists against Time* (New York: Little, Brown, 1946); Leslie Groves, *Now It Can Be Told: The Story of the Manhattan Project* (New York: Harper, 1962); Industry on Parade Film Collection, Archives Center, National Museum of American History, Smithsonian Institution, Washington, DC; Gertner, *The Idea Factory*, 180–182.

95. Time, Inc., *Official Guide: New York World's Fair, 1964–1965* (New York: Time-Life Books, 1964); Scott G. Knowles and Stuart W. Leslie, "'Industrial Versailles': Eero Saarinen's Corporate Campuses for GM, IBM, and AT&T," *Isis* 92, no. 1 (March 2001): 1–33; William H. Whyte, *The Organization Man* (Philadelphia: University of Pennsylvania Press, 2002 [1956]).

96. Hounshell, "The Evolution of Industrial Research in the United States"; Stein, *Pivotal Decade*; Matthew Wisnioski, *Engineers for Change: Competing Visions of Technology in 1960s America* (Cambridge, MA: MIT Press, 2012).

97. "Fortune 500," *Fortune*, March 29, 2019, https://fortune.com/fortune500/2019/search/?mktval=desc. As measured by market capitalization, the top firms are (1) Microsoft, (2) Apple, (3) Amazon.com, (4) Alphabet, Google's parent company, (5) Berkshire Hathaway, and (6) Facebook.

98. Andrew I. Gavil, *The Microsoft Antitrust Cases* (Cambridge, MA: MIT Press, 2014); Siva Viadhyanathan, *Antisocial Media: How Facebook Disconnects Us and Undermines Democracy* (New York: Oxford University Press, 2018); Walter Isaacson, *Steve Jobs* (New York: Simon and Schuster, 2011), 575–577.

99. Robert X. Cringely, *Accidental Empires: How the Boys of Silicon Valley Make Their Millions, Battle Foreign Competition, and Still Can't Get a Date* (Boston: Addison-Wesley, 1992); Walter Isaacson, *The Innovators: How a Group of Hackers, Geniuses, and Geeks Created the Digital Revolution* (New York: Simon and Schuster, 2014).

100. Adams Nager, David M. Hart, Stephen Ezell, and Robert D. Atkinson, *The Demographics of Innovation in the United States*, Information Technology and Innovation Foundation, February 24, 2016, https://itif.org/publications/2016/02/24/demo graphics-innovation-united-states.

101. See Lisa D. Cook, "The Innovation Gap in Pink and Black," in Wisnioski, Hintz, and Kleine, *Does America Need More Innovators?*, 221–247; Lucinda M. Sanders and Catherine Ashcraft, "Confronting the Absence of Women in Technology Innovation," in Wisnioski, Hintz, and Kleine, *Does America Need More Innovators?*, 323–343.

102. Microsoft, #MakeWhatsNext commercial, 2016, https://www.youtube.com /watch?v=tNqSzUdYazw.

103. Karyn D. Collins, "Lori Greiner Marks a Tidy Ten Years on QVC," *Philadelphia Inquirer*, April 2, 2010, https://www.inquirer.com/philly/home/20100402_Lori_Greiner _marks_a_tidy_10_years_on_QVC.html; ABC, "About Shark Tank," accessed September 23, 2019, http://abc.go.com/shows/shark-tank/about-the-show; Jared Shelly, "From 'Junk' to *Shark Tank*'s Biggest Success Story," *Philadelphia Magazine*, June 1, 2015, https://www.phillymag.com/business/2015/06/01/scrub-daddy-shark-tanks -biggest-success-story/.

104. Internet Movie Database, accessed March 3, 2020, https://www.imdb.com/; Jobs quoted in Andy Hertzfeld, *Revolution in the Valley: The Insanely Great Story of How the Mac Was Made* (Sebastopol, CA: O'Reilly Media, 2005), 166–167.

Chapter 8

1. John Seabrook, *Flash of Genius and Other True Stories of Invention* (New York: St. Martin's, 2008), xxiv.

2. L. Sprague de Camp, *The Heroic Age of American Invention* (Garden City, NY: Doubleday, 1961), 257; David F. Noble, *America by Design: Science, Technology, and the Rise of Corporate Capitalism* (New York: Knopf, 1977), 97; Thomas P. Hughes, *American Genesis: A Century of Invention and Technological Enthusiasm, 1870–1970*, 2nd ed. (Chicago: University of Chicago Press, 2004), 138–139; David A. Hounshell, "The Evolution of Industrial Research in the United States," in *Engines of Innovation: U.S. Industrial Research at the End of an Era*, ed. Richard S. Rosenbloom and William J. Spencer (Boston: Harvard Business School Press, 1996), 32.

3. John Jewkes, David Sawers, and Richard Stillerman, *The Sources of Invention* (London: Macmillan, 1958), 246.

Notes

4. Ken Auletta recalling his 1998 interview with Gates, as cited in Ken Auletta, "Luncheon Address," *International Antitrust Law & Policy: Annual Proceedings of the Fordham Competition Law Institute*, ed. Barry E. Hawk (Huntington, NY: Juris Publications, 2011), quotation at 410. On disruptive innovation, see Clayton M. Christensen, *The Innovator's Dilemma: When New Technologies Cause Great Firms to Fail* (Boston: Harvard Business Review Press, 1997).

5. David C. Mowery, "Plus ca change: Industrial R&D in the 'Third Industrial Revolution,'" *Industrial and Corporate Change* 18, no. 1 (February 2009): 1–50.

6. Lisa D. Cook, "The Innovation Gap in Pink and Black," in *Does America Need More Innovators?*, ed. Matthew Wisnioski, Eric S. Hintz, and Marie Stettler Kleine (Cambridge, MA: MIT Press, 2019), 221–247.

7. Hughes, *American Genesis*, 2. Different nations develop distinctive inventive cultures. See Christine MacLeod, "The Paradoxes of Patenting: Invention and Its Diffusion in 18th- and 19th-Century Britain, France, and North America," *Technology and Culture* 32, no. 4 (October 1991): 885–910; Jung Lee, "Invention without Science: 'Korean Edisons' and the Changing Understanding of Technology in Colonial Korea," *Technology and Culture* 54, no. 4 (October 2013): 782–814.

8. Coe, quoted in *Investigation of Concentration of Economic Power: Hearings Before the Temporary National Economic Committee* (Washington, DC: Government Printing Office, 1939), pt. 3, 863.

9. Andrei Iancu, "Remarks by Director Iancu at the U.S. Chamber of Commerce: Driving American Innovation Policy Conference," Atlanta, GA, February 25, 2019, https://www.uspto.gov/about-us/news-updates/remarks-director-iancu-us-chamber-commerce-driving-american-innovation-policy.

10. "Lo, the Poor Inventor!," *Business Week*, December 21, 1929, 22–23.

Index

AEC. *See* American Engineering
Council (AEC) or Atomic Energy
Commission (AEC)
Aerial torpedo, 179, 183, 186, 197,
204
African American inventors, 7–8.
See also names of individual inventors
banding together for recognition and
respect, 17, 94, 117, 122–128
comprehensive list of Black
patentees, 24, 123–125, 254n18,
257n9, 284n79
concealed identities, 16, 22–26, 54
at fairs and expositions, 123–128
inventing for Black consumers, 26
inventing for social mobility, 8
myth of Black disingenuity, 122–123
National Inventors Congress
(Inventors of America) and, 131–132
patents issued to, 24, 122
racial discrimination faced by, 22–26,
84, 122–128
underrepresentation of, 7, 22–23, 54,
241, 250, 254n18
Air conditioning, 5, 165
Airplane safety beacons, 32–34
Air raid shelter, 133
Alcoa, 4, 43, 167, 210
Alexanderson, Ernst, 144
Allen, Paul, 216–217, 223, 240, 247
Alliance for American Innovation, 237
Alliances. *See* "Ally" economic strategy
Allwood, William, 202
"Ally" economic strategy, 58–59, 72–84,
90–91, 213, 217–219, 247–249.
See also Commercialization strate-
gies; Inventor-consultants
Alto (Xerox), 211
Amateur inventors, 17, 129, 135
National Inventors Congress
(Inventors of America), 128–133
National Inventors Council (NIC),
190–191, 198, *199*, 203

Naval Consulting Board (NCB),
176–177, 181, 183
Quirky, 228
Amazon.com, 223, 227, 229, 240
America COMPETES Act (2007), 310n47
America COMPETES Reauthorization
Act (2010), 224–225
America Invents Act (AIA), 237–239
American Aeronautical Society, 173
American Association for the
Advancement of Science, 172–173
American Association of Inventors and
Manufacturers (AAIM)
dues, 107, 109, 280n43
establishment and agenda of, 17,
106–110, 111, 116, 121, 132, 138,
167
American Capitalism (Galbraith), 207
American Chemical Society, 173
American Electrochemical Society, 173
American Engineering Council (AEC),
163, 166
American identity, independent
inventors as representative of, 153–
154, 164–166, 249–250. *See also* "Pio-
neering" rhetoric
American Innovators for Patent Reform,
226, 238
American Institute of Chemical
Engineers (AIChE), 115–116, 134
American Institute of Electrical
Engineers, 43, 173
American Institute of Mining Engineers,
173
American Inventors Protection Act
(AIPA), 230–231, 236–237
American Marconi, 140
American Mathematical Society, 173
American Physical Society, 134, 172–173
American Society of Aeronautic
Engineers, 173
American Society of Automotive
Engineers, 173